Recent Advances in Intelligent Control Systems

Wen Yu
Editor

Recent Advances in Intelligent Control Systems

Springer

Editor
Wen Yu, PhD
Departamento de Control Automatico
CINVESTAV-IPN
Av. IPN 2508
México D.F. 07360
México
yuw@ctrl.cinvestav.mx

ISBN 978-1-4471-5769-4 ISBN 978-1-84882-548-2 (eBook)
DOI 10.1007/978-1-978-1-84882-548-2
Springer Dordrecht Heidelberg London New York

British Library Cataloguing in Publication Data
A catalogue record for this book is available from the British Library

Softcover re-print of the Hardcover 1st edition 2009

Cover design: eStudioCalamar, Figueres/Berlin

Printed on acid-free paper

Springer is part of Springer Science+Business Media (www.springer.com)

To my daughters
Huijia and Lisa

Preface

Intelligent control has dramatically changed the face of industrial control engineering; it is a rapidly developing, complex and challenging field with great practical importance and potential. Because of the rapidly developing and interdisciplinary nature of the subject, there are only a few books including the general know-how in designing, implementing and operating intelligent control systems. The book introduces the traditional control techniques and contrasts them with intelligent control, and presents several recent methods in how to design intelligent identifiers and controllers via neural networks, fuzzy logic, and neural fuzzy systems. A variety of applications, most in the areas of robotics and mechatronics, but with others including building structure and process/production control are also illustrated as design examples. The book is suitable for advanced undergraduate students and graduate engineering students. In addition, practicing engineers will find it appropriate for self-study.

Mexico City
February 2009

Wen Yu

Contents

List of Contributors

Yeesock Kim
Texas A&M University, College Station, TX 77843-3136, USA, e-mail: controlga@tamu.edu

Stefan Hurlebaus
Texas A&M University, College Station, TX 77843-3136, USA, e-mail: shurlebaus@tamu.edu

Reza Langari
Texas A&M University, College Station, TX 77843-3123, USA, e-mail: rlangari@tamu.edu

Jianbin Qiu
Department of Manufacturing Engineering and Engineering Management, University of Science and Technology of China & City University of Hong Kong Joint Advanced Research Center, Dushu Lake Higher Education Town, Suzhou Industrial Park, 215123, P.R. China, e-mail: jbqiu@mail.ustc.edu.cn

Gang Feng
Department of Manufacturing Engineering and Engineering Management, City University of Hong Kong, 83 Tat Chee Avenue, Kowloon, Hong Kong, e-mail: megfeng@cityu.edu.hk

Jie Yang
Department of Precision Machinery and Precision Instrumentation, University of Science and Technology of China, Hefei, 230026, P.R. China, e-mail: jieyang@ustc.edu.cn

Hongli Dong
Space Control and Inertial Technology Research Center, Harbin Institute of Technology, Harbin 150001, P.R. China. H. Dong is also with the College of Electrical and Information Engineering, Daqing Petroleum Institute, Daqing 163318, P.R. China, e-mail: shiningdhl@gmail.com

Zidong Wang
Department of Information Systems and Computing, Brunel University, Uxbridge, Middlesex, UB8 3PH, United Kingdom, e-mail: zidong.wang@brunel.ac.uk

Huijun Gao
Space Control and Inertial Technology Research Center, Harbin Institute of Technology, Harbin 150001, P.R. China, e-mail: hjgao@hit.edu.cn

Wen Yu
Departamento de Control Automatico, CINVESTAV-IPN, A.P. 14-740, Av.IPN 2508, México D.F., 07360, México, e-mail: yuw@ctrl.cinvestav.mx

Xiaoou Li
Departamento de Computación, CINVESTAV-IPN, A.P. 14-740, Av.IPN 2508, México D.F., 07360, México, e-mail: lixo@cs.cinvestav.mx

Balaje T. Thumati
Department of Electrical and Computer Engineering, Missouri University of Science and Technology, Rolla, MO, USA, e-mail: bttr74@mst.edu

Jagannathan Sarangapani
Department of Electrical and Computer Engineering, Missouri University of Science and Technology, Rolla, MO, USA, e-mail: sarangap@umr.edu

Xubin Sun
Institute of Automation, Chinese Academy of Sciences, Beijing, P.R. China, e-mail: xubin.sun@ia.ac.cn

Jinliang Ding
Key Laboratory of Integrated Automation of Process Industry, Northeastern University Shenyang, P.R. China, e-mail: tychai@mail.neu.edu.cn

Tianyou Chai
Key Laboratory of Integrated Automation of Process Industry, Northeastern University Shenyang, P.R. China, e-mail: tychai@mail.neu.edu.cn

Hong Wang
The University of Manchester, Manchester, United Kingdom, e-mail: hong.wang@manchester.ac.uk

Alejandro García
Departamento de Control Automatico, CINVESTAV-IPN, A.P. 14-740, Av.IPN 2508, México D.F., 07360, México, e-mail: agarcia@ctrl.cinvestav.mx

Isaac Chairez
Bioelectronic Section, UPIBI-IPN, México, e-mail: apoznyak@ctrl.cinvestav.mx

Alexander Poznyak
Departamento de Control Automatico, CINVESTAV-IPN, A.P. 14-740, Av.IPN 2508, México D.F., 07360, México, e-mail: apoznyak@ctrl.cinvestav.mx

Jun Wang
Department of Mechanical and Automation Engineering, The Chinese University of Hong Kong, Shatin, N.T., Hong Kong, e-mail: jwang@mae.cuhk.edu.hk

Xiaolin Hu
State Key Laboratory of Intelligent Technology and Systems, Tsinghua National Laboratory for Information Science and Technology (TNList), and Department of Computer Science and Technology, Tsinghua University, Beijing 100084, P.R. China, e-mail: xiaolin.hu@gmail.com

Bo Zhang
State Key Laboratory of Intelligent Technology and Systems, Tsinghua National Laboratory for Information Science and Technology (TNList), and Department of Computer Science and Technology, Tsinghua University, Beijing 100084, P.R. China, e-mail: xiaolin.hu@gmail.com

Mou Chen
Department of Electrical & Computer Engineering, National University of Singapore, 117576, Singapore, e-mail: elechenm@nus.edu.sg

Shuzhi Sam Ge
Department of Electrical & Computer Engineering, National University of Singapore, 117576, Singapore, e-mail: samge@nus.edu.sg

Bernard Voon Ee How
Department of Electrical & Computer Engineering, National University of Singapore, 117576, Singapore, e-mail: bernardhow@nus.edu.sg

Ning Wang
Dalian Maritime University, Dalian 116026, P.R. China, e-mail: n.wang.dmu.cn@gmail.com

Meng Joo Er
Nanyang Technological University, 639798, Singapore, e-mail: emjer@ntu.edu.sg

Xianyao Meng
Dalian Maritime University, Dalian 116026, P.R. China, e-mail: n.wang.dmu.cn@gmail.com

Chin-Teng Lin
Department of Electrical and Control Engineering, National Chiao-Tung University, Hsinchu 300, Taiwan, R.O.C., e-mail: ctlin@nctu.edu.tw

Cheng-Hung Chen
Department of Electrical and Control Engineering, National Chiao-Tung University, Hsinchu 300, Taiwan, R.O.C., e-mail: chchen.ece93g@nctu.edu.tw

Cheng-Jian Lin
Department of Computer Science and Information Engineering, National Chin-Yi University of Technology, Taiping City, Taichung County 411, Taiwan, R.O.C., e-mail: cjlin@ncut.edu.tw

Anmin Zhu
The College of Information Engineering, Shenzhen University, Shenzhen 518060, P.R. China, e-mail: azhu@szu.edu.cn

Simon X. Yang
The Advanced Robotics and Intelligent System (ARIS) Laboratory, School of Engineering, University of Guelph, Guelph, ON N1G 2W1, Canada, e-mail: syang@uoguelph.ca

Aninda Bhattacharya
Remote Prognostics Lab, GE Global Research, Bangalore, India, e-mail: aninda.bhattacharya1@ge.com

Alexander G. Parlos
Department of Mechanical Engineering, Texas A&M University, TX 77843, USA, e-mail: a-parlos@tamu.edu

Kyriakos G. Vamvoudakis
Automation and Robotics Research Institute, University of Texas at Arlington, Arlington, TX, USA, e-mail: kyriakos@arri.uta.edu

Frank L. Lewis
Automation and Robotics Research Institute, University of Texas at Arlington, Arlington, TX, USA, e-mail: lewis@uta.edu

Part I
Fuzzy Control

Chapter 1
Fuzzy Control of Large Civil Structures Subjected to Natural Hazards

Yeesock Kim, Stefan Hurlebaus and Reza Langari

Abstract In this chapter, a new semiactive nonlinear fuzzy control (SNFC) system design framework is proposed through integration of a set of Lyapunov-based state feedback controllers and Kalman filters. A nonlinear multi-input multi-output (MIMO) autoregressive exogenous (ARX) Takagi-Sugeno (T-S) fuzzy model is constructed out of a set of linear dynamic models. Subsequently, multiple Lyapunov-based state feedback controllers are formulated in terms of linear matrix inequalities (LMIs) such that the resulting system is globally asymptotically stable. The resulting state feedback controllers are integrated with Kalman filters and a converting algorithm using a T-S fuzzy interpolation method to construct a semiactive output feedback controller. To demonstrate the effectiveness of the proposed design framework, the resulting scheme is applied to a three- and a twenty-story building employing nonlinear hysteretic control devices. It is demonstrated from numerical simulations that the proposed approach is effective in controlling the responses of seismically excited large civil structures equipped with magnetorheological (MR) dampers: both displacement and acceleration responses of both three- and twenty-story buildings subjected to the 1940 El-Centro earthquake disturbance are dramatically reduced when the proposed control approach is applied.

Yeesock Kim
Texas A&M University, College Station, TX 77843-3136, USA, e-mail: controlga@tamu.edu

Stefan Hurlebaus
Texas A&M University, College Station, TX 77843-3136, USA, e-mail: shurlebaus@tamu.edu

Reza Langari
Texas A&M University, College Station, TX 77843-3123, USA, e-mail: rlangari@tamu.edu

1.1 Introduction

1.1.1 Structural Vibration Control

In recent years, structural vibration control of large civil structures has received a great deal of attention [2, 32, 35]. The control strategies in the field of large civil structures might be categorized as, 1. passive damping, 2. active control, 3. semiactive control and, 4. hybrid control. Passive damping systems, including base isolation of buildings or bridges and/or the use of viscoelastic dampers, tuned mass dampers, and tuned liquid dampers, have been widely accepted by the civil engineering society as an appropriate means of mitigating damage to structures subjected to destructive natural hazards such as strong winds and earthquakes. Passive systems, while relatively inexpensive in comparison with active ones, cannot adapt well to nonlinear and/or time-varying properties of control devices, structures, and environmental loadings. For instance, passive damping systems are not effective in mitigating the vibration of structures if the magnitude and frequencies of earthquake signals exceed the capacity of passive systems. To overcome the drawbacks of the passive systems, many studies have been carried out to apply active control systems to large structures. However, the active systems have not been widely accepted in the field of large civil structures due to 1. high cost and 2. low reliability. Therefore, in recent years, semiactive control systems, including variable-orifice dampers, variable-stiffness devices, smart tuned mass dampers, variable-friction dampers, and controllable-fluid dampers, have received great attention. In particular, MR dampers have received attention due to the following: 1. low power requirement, 2. low manufacturing cost, 3. reliable operation, and 4. fast response time [33]. However, it is difficult to design an effective semiactive controller using MR dampers because of the nonlinear hysteretic behavior of the device. Fuzzy logic is potentially a useful tool in this regard since it is often used in systems where nonlinear behavior is present. With this in mind, we briefly review some recent studies in this area.

1.1.2 Fuzzy Control Approaches in Large Civil Structures

Since Zadeh's paper [49], fuzzy logic has attracted great attention in control engineering in recent years [17,26–28,46]. A number of design methodologies for fuzzy logic controllers have been successively applied to large civil structures. These include: trial-and-error-based methodologies [1,9,10,29,30,37,38]; a self-organizing approach [7]; training using linear Quadratic Gaussian (LQG) data [6]; neural networks-based learning [14, 15, 31, 42]; adaptive fuzzy [47]; genetic algorithms-based training [3–5, 21, 45]; fuzzy sliding mode control [8, 22, 43]; etc. However, only a few studies have focused on the design of SNFC for large civil structures equipped with highly nonlinear hysteretic control devices [23, 25].

This is partly due to the lack of systematic design of fuzzy controllers for large scale systems. To be sure, viable approaches exist for active nonlinear fuzzy control (ANFC) design via the so-called parallel distributed compensation (PDC) approach, which employs multiple linear controllers [17] [19]. Moreover, Tanaka and Sano (1994) have proposed a theorem on the stability analysis of an ANFC system using the Lyapunov direct method. This theorem states a sufficient condition for an ANFC system to be globally asymptotically stable by finding a common symmetric positive definite matrix such that a set of simultaneous Lyapunov inequalities are satisfied. However, there have been few, if any, systematic methods to design SNFC systems for building structures equipped with nonlinear semiactive devices [23, 25]. With this in mind, this chapter is organized as follows. Section 1.2 describes smart control devices. In Sections 1.3 and 1.4, a novel model-based semiactive nonlinear fuzzy control system design framework is presented. In Section 1.5, simulation results are described. Concluding remarks are given in Section 1.6.

1.2 Smart Control Device: MR Damper

In recent years, the idea of smart structures received attention because the performance of structural systems can be improved without either significantly increasing the structure mass or requiring high control power. They may be called intelligent structures, adaptive structures, active structures, and the related technologies adaptronics, structronics, etc. These terminologies refer to a smart structure which is an integration of actuators, sensors, control units, and signal processing units with a structural system. The materials that are usually used to make a smart structure are: piezoelectrics, shape memory alloys, electrostrictive/magnetostrictive materials, polymer gels, and MR/electrorheological fluids [18]. Semiactive devices have been applied to large civil structures. Semiactive control strategies combine favorable features of both active and passive control systems. Semiactive control systems include devices such as variable orifice dampers, variable-stiffness devices, variable friction dampers, controllable fluid dampers, shape memory alloy actuators, and piezoelectrics [18]. In particular, one of the controllable-fluid dampers, MR damper has attracted attention in recent years because it has many attractive characteristics. In general, a MR damper consists of a hydraulic cylinder, magnetic coils, and MR fluids that consist of micron-sized magnetically polarizable particles floating within oil-type fluids [33]. The MR damper is operated as a passive damper; however, when a magnetic field is applied to the MR fluids, the MR fluids are changed into a semi-active device in a few milliseconds. Its characteristics are summarized, 1. a MR damper is operated with low power sources, e.g., SD-1000 MR damper manufactured by Lord Corp. can generate a force up to 3000 N using a small battery with capacity less than 10 W, 2. it has high yield strength level, e.g., its maximum yield strength is beyond 80 kPa, 3. the performance is stable in a broad temperature range, e.g., MR fluids operates at the temperature between $-40^{o}C$ and $150^{o}C$, 4. the response time is a few milliseconds, 5. the performance is not sensitive to contamination during manufacturing the MR damper. Moreover, the operating point of the MR damper, which is a current-controlled device, can be changed by a permanent magnet. However, it is difficult to design an effective controller to operate the MR damper. The reason is that the MR damper is a highly nonlinear hysteretic control device. With this in mind, we review the use of fuzzy logic in conjunction with MR dampers.

1.3 Background Materials

1.3.1 Takagi-Sugeno Fuzzy Model

In 1985, Takagi and Sugeno [39] proposed an effective way to represent a fuzzy model of a dynamic system. A continuous T-S fuzzy model is composed of N_r fuzzy rules that can represent the nonlinear behavior of large civil structures that can be represented as

$$\begin{aligned} \mathrm{R_j} \ &: \text{if } z_{FZ}^1(t) \text{ is } p_{1,j} \text{ and } z_{FZ}^2(t) \text{ is } p_{2,j} \text{ and } \cdots \text{ and } z_{FZ}^n(t) \text{ is } p_{n,j} \\ &\quad \text{then } \dot{x}(t) = A_j x(t) + B_j u(t), \quad j = 1,2,\cdots,N_r \end{aligned} \tag{1.1}$$

where

z_{FZ}^i : i^{th} linguistic variable,
p_i : fuzzy term set of z_{FZ}^i,
$p_{i,j}$: a fuzzy term of p_i selected for j^{th} rule R_j,
$x(t)$: state vector and $x(t) = [x_1(t)...x_n(t)]^T \in R^n$,
$u(t)$: input vector and $u(t) = [u_1(t)...u_m(t)]^T \in R^m$,
$A_j \in R^{n\times n}$,
$B_j \in R^{n\times m}$.

Note that z_{FZ}^i may be the same as state variables. For any current state vector $x(t)$ and input vector $u(t)$, the T-S fuzzy model infers $\dot{x}(t)$ as the output of the fuzzy model as follows:

$$\dot{x}(t) = \frac{\Sigma_{j=1}^{N_r} w_j \left[A_j x(t) + B_j u(t)\right]}{\Sigma_{i=1}^{N_r} w_j} \tag{1.2}$$

where

$$w_j = \prod_{i=1}^{N_r} \mu_{p_{i,j}}^i (z_{FZ}^i(t)). \tag{1.3}$$

For an unforced dynamic system (i.e., $u(t) \equiv 0$), (1.2) can be written as

$$\dot{x}(t) = \frac{\sum_{j=1}^{N_r} w_j A_j x(t)}{\sum_{j=1}^{N_r} w_j}. \tag{1.4}$$

1.3.2 Sufficient Stability Condition of T-S Fuzzy Model

Tanaka and Sugeno [41] suggested here an important criterion for the stability of the T-S fuzzy model.

Theorem 1.1. *(Stability Criterion for Continuous T-S Fuzzy Model) The equilibrium of the continuous T-S fuzzy model (1.4) (namely, $x = 0$) is globally asymptotically stable if there exists a common symmetric positive definite matrix P such that*

$$A_j^T P + PA_j < 0 \text{ for all } j = 1, 2, \cdots, N_r. \tag{1.5}$$

It should be noted that (1.5) is a *sufficient condition* for stability but not necessary condition. Therefore there may exist better criterion for the stability of the T-S fuzzy model.

1.3.3 Parallel Distributed Compensation

A continuous T-S fuzzy controller which uses full state feedback is composed of N_s control rules that can be represented as

$$\begin{aligned} \mathrm{R_j} :\ & \text{if } z_{FZ}^1(t) \text{ is } p_{1,j} \text{ and } z_{FZ}^2(t) \text{ is } p_{2,j} \text{ and } \cdots \text{ and } z_{FZ}^n(t) \text{ is } p_{n,j} \\ & \text{then } u(t) = K_j x(t), \quad j = 1, 2, \cdots, N_s \end{aligned} \tag{1.6}$$

where N_s means that the number of the control rule is not necessarily equal to the number of rule N_r of the dynamic system. Note that z_{FZ}^i refers to the same as in (1.1).

For any current state vector $x(t)$, the T-S fuzzy controller infers $u(t)$ as the output of the fuzzy controller as follows:

$$u(t) = \frac{\sum_{j=1}^{N_s} w_j K_j x(t)}{\sum_{j=1}^{N_s} w_j}. \tag{1.7}$$

The T-S fuzzy controller is constructed using the premise parameters of the existing fuzzy model representing the dynamic system. In this case, we can use a proper linear control method for each pair of control rules and dynamic system rules. [44] named it PDC. It has very important advantages because it makes possible application of (1.7) to (1.2). Therefore, the closed-loop behavior of the continuous T-S fuzzy model (1.1) with the T-S fuzzy controller (1.7) using PDC can be obtained by substituting (1.7) into (1.2) as follows:

$$\dot{x}(t) = \frac{\sum_{j=1}^{N_r}\sum_{q=1}^{N_r} w_j w_q (A_j + B_j K_q) x(t)}{\sum_{j=1}^{N_r}\sum_{q=1}^{N_r} w_j w_q}. \tag{1.8}$$

The corresponding sufficient condition for the stability of (1.8) can be easily obtained and summarized in Theorem 1.2.

Theorem 1.2. *[44] The equilibrium of the closed-loop system of the dynamic systems (1.8) (namely, $x = 0$) is globally asymptotically stable if there exists a common symmetric positive definite matrix P such that*

$$(A_j + B_j K_q)^T P + P(A_j + B_j K_q) < 0 \text{ for all } j, q = 1, 2, \cdots, N_r. \tag{1.9}$$

It is suggested that P can be determined numerically by solving LMIs in (1.9). It should be noted that (1.9) has r^2 LMIs. Wang [44] rewrote (1.9) by regrouping the terms as follows:

$$\dot{x}(t) = \frac{\sum_{j=1}^{N_r} w_j w_j (A_j + B_j K_j) x(t) + 2\sum_{j<q}^{N_r} w_j w_q G_{jq} x(t)}{\sum_{j=1}^{N_r}\sum_{q=1}^{N_r} w_j w_q} \tag{1.10}$$

where

$$G_{jq} = \frac{(A_j + B_j K_q) + (A_q + B_q K_j)}{2}, j < q \le N_r. \tag{1.11}$$

The corresponding sufficient condition for stability of (1.10) is summarized in Corollary 1.1.

Corollary 1.1. *The equilibrium of the closed-loop control system (1.10) (namely, $x = 0$) is globally asymptotically stable if there exists a common symmetric positive definite matrix P such that*

$$\begin{array}{ll} (A_j + B_j K_j)^T P + P(A_j + B_j K_j) < 0, & j = 1, 2, \cdots N_r \\ G_{jq}^T P + P G_{jq} < 0, & j < q \le N_r. \end{array} \tag{1.12}$$

The number of LMIs for (1.12) is $r(r+1)/2$. Therefore, the number of LMIs to be solved is reduced by approximately $1/2$.

Remark 1.1. It is noted that the advantage of (1.10) and (1.12) over (1.9) is not only the reduction of the number of LMIs to be solved but less conservativeness of the stability criterion of (1.10) and (1.12). Therefore, the stability criterion in (1.10) and (1.12) are used throughout this chapter.

Remark 1.2. The sufficient condition for stability (1.10) and (1.12) can be used only for the purpose of checking the stability of the T-S fuzzy control system in which feedback gains K_i's, $i = 1,2,\cdots,N_r$, is predetermined by a proper linear control system design method.

1.4 LMI Formulation of Semiactive Nonlinear Fuzzy Control Systems

It should be emphasized that the stability criterions (1.10) and (1.12) are LMIs since feedback gains K_is are predetermined using the proper design method [44]. It is, however, necessary to treat K_i's and P as matrix variables when we try to find a systematic control system design method which guarantees the stability criteria (1.10) and (1.12), i.e., they should be determined simultaneously. In that case, the stability criteria (1.10) and (1.12) are not LMIs, since they contain terms like $K_i^T B_i^T P$ and $K_i^T B_i^T P B_i K_i$, i.e., they are nonlinear matrix inequalities (NMIs).

1.4.1 Stability LMIs of T-S Fuzzy Control Systems

In this section, the NMIs are converted to LMIs for continuous T-S fuzzy control systems.

Theorem 1.3. *([19] Stability LMIs for Nonlinear Dynamic Systems) The equilibrium of the continuous T-S fuzzy system (1.8) (namely, $x = 0$) is globally asymptotically stable if there exists a common symmetric positive definite matrix $Q = P^{-1} > 0$ which satisfies*

$$
\begin{aligned}
&QA_j^T + A_jQ + M_j^T B_j^T + B_jM_j < 0, j = 1,2,\cdots,N_r \\
&QA_j^T + A_jQ + QA_q^T + A_qQ + M_q^T B_j^T + B_jM_q + M_j^T B_q^T + B_qM_j < 0, \quad j < q \le N_r
\end{aligned}
\tag{1.13}
$$

where Q and $M_j = K_iQ$, $j = 1,2,\cdots,N_r$ are new matrix variables of LMIs.

Proof. Substitution of (1.12) into (1.13) yields

$$
\begin{aligned}
&(A_j + B_jK_j)^T P + P(A_j + B_jK_j) < 0, j = 1,2,\cdots,N_r \\
&(A_j + B_jK_q)^T P + (A_q + B_qK_j)^T P + P(A_j + B_jK_q) + P(A_q + B_qK_j) < 0 \\
&j < q \le N_r.
\end{aligned}
\tag{1.14}
$$

If we let $Q = P^{-1}$ and pre- and post-multiply (1.14) by $Q > 0$ then the negative definiteness of (1.14) will not be changed due to Sylvester's law of inertia [36] since Q is symmetric and positive definite. (1.14) becomes

$$\begin{aligned} &QA_j^T + A_jQ + QK_j^TB_j^T + B_jK_jQ < 0, j = 1,2,\cdots,N_r \\ &QA_j^T + A_jQ + QA_q^T + A_qQ + QK_q^TB_j^T + B_jK_qQ + QK_j^TB_q^T + B_qK_jQ < 0 \\ &j < q \le N_r. \end{aligned} \tag{1.15}$$

Now, we can obtain LMIs by letting $K_jQ = M_j$ as follows:

$$\begin{aligned} &QA_j^T + A_jQ + M_j^TB_j^T + B_jM_j < 0, j = 1,2,\cdots,N_r \\ &QA_j^T + A_jQ + QA_q^T + A_qQ + M_q^TB_j^T + B_jM_q + M_j^TB_q^T + B_qM_j < 0, j < Q \le N_r \end{aligned} \tag{1.16}$$

where Q and M_j, $j = 1,2,\cdots,N_r$ are new matrix variables of LMIs. (1.16) is certainly LMIs for matrix variables Q, M_j, and M_q. Note that these LMIs deal with the performance on transient responses of dynamic systems. □

1.4.2 Performance LMIs

In structural engineering, the performance on transient responses is an important issue; however, the stability LMI formulation does not directly deal with the issue. Therefore, in this section, the pole-assignment concept is recast by LMI formulation. The formulation of the pole-placement in terms of a LMI is motivated by [11]. A D-stable region is defined first.

Definition 1.1. Let D be a subset of the complex plane that represents the behavior of a dynamic system. If all the poles (or eigenvalues) of the dynamic system are located in the subset region D, the dynamic system (or the system matrix) is D-stable.

Definition 1.2. The closed-loop poles (or eigenvalues) of a dynamic system are located in the LMI stability region

$$D = \{s \in C \mid f_D(s) := \alpha + \beta s + \beta^T \bar{s} < 0\} \tag{1.17}$$

if and only if there exists a symmetric positive definite matrix $\alpha = [\alpha_{kl}] \in R^{m\times m}$ and a matrix $\beta = [\beta_{kl}] \in R^{m\times m}$. The characteristic function $f_D(s)$ is a $m \times m$ Hermitian matrix.

Hong and Langari [17] applied a circular LMI region D that is convex and symmetric with respect to the real axis such that it is associated with Chilali and Gahinet's theorem [11]. A circular LMI region that has a center at $(-q_c, 0)$ and a radius $r_c > 0$ is defined

$$D = \{x + jy \in C : (x + q_c)^2 + y^2 < r_c^2\}. \tag{1.18}$$

The associated characteristic function $f_D(s)$ is given by

$$f_D(s) = \begin{pmatrix} -r_c & \bar{s}+q_c \\ s+q_c & -r_c \end{pmatrix} < 0. \tag{1.19}$$

Figure 1.1 shows a schematic of the circular LMI region. This circular LMI region can be related to a LMI stability region described in terms of an $m \times m$ block matrix.

Theorem 1.4. *[11] The system dynamics of $\dot{x} = Ax$ is D-stable if and only if there exists a symmetric matrix Q such that*

$$\begin{aligned} M_D(A,X) &:= \alpha \otimes Q + \beta \otimes (AQ) + \beta^T \otimes (AQ)^T \\ &= [\alpha_{kl} Q + \beta_{kl} AQ + \beta_{kl} QA^T]_{1 \le k,l \le m} < 0, Q > 0. \end{aligned} \tag{1.20}$$

It should be noted that $M_D(A,Q)$ and $f_D(s)$ have a close relationship, i.e., replacing $(1,s,\bar{s})$ of $f_D(s)$ by (Q,AQ,QA^T) of $M_D(A,Q)$ yields

$$\begin{pmatrix} -r_c Q & +q_c Q + QA^T \\ q_c Q + AQ & -r_c Q \end{pmatrix} < 0. \tag{1.21}$$

From (1.21), a LMI for the pole-placement controller is derived.

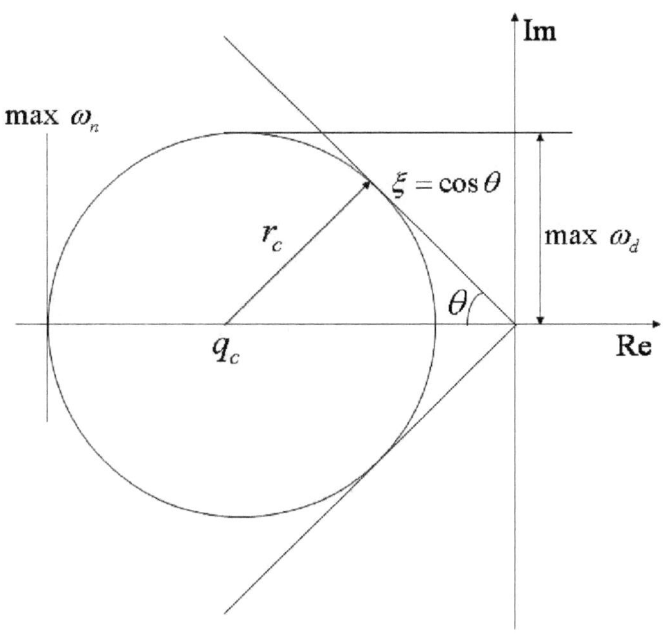

Fig. 1.1 Circular LMI region

Theorem 1.5. *[17] The continuous closed-loop T-S fuzzy control system is D-stable if and only if there exists a positive symmetric matrix Q such that*

$$\begin{pmatrix} -r_c Q & +q_c Q + Q(A_j + B_j K_q)^T \\ q_c Q + (A_j + B_j K_q)Q & -r_c Q \end{pmatrix} < 0. \tag{1.22}$$

Remark 1.3. It is noted that the inequality (1.22) is not a LMI because the matrices Q and K_j are coupled. This nonlinear matrix inequality can be transformed into a LMI by defining a new matrix variable, i.e., $M_j = K_j Q$.

Corollary 1.2. *The continuous closed-loop T-S fuzzy control system is D-stable if and only if there exists a positive symmetric matrix Q such that*

$$\begin{pmatrix} -r_c Q & +q_c Q + Q A_j^T + M_j^T B_j^T \\ q_c Q + A_j Q + B_j K_j M_j & -r_c Q \end{pmatrix} < 0. \tag{1.23}$$

This LMI (1.23) directly deals with the performance on the transient responses of the dynamic system.

In summary, LMIs (1.16) and (1.23) are solved simultaneously to obtain Q and M_j. Then, the common symmetric positive definite matrix P and state feedback control gains K_j are determined

$$\begin{aligned} &P = Q^{-1} \\ &K_j = M_j Q^{-1} = M_j P,\ j = 1, 2, \cdots, N_r. \end{aligned} \tag{1.24}$$

These state feedback control gains are integrated with state estimators to construct output feedback controllers.

1.4.3 State Estimator

From a practical point of view, it is challenging to implement the full state feedback strategies because not all the states are readily available; however, this issue can be resolved by the aid of estimation theories.

Consider the following state space equation

$$\begin{aligned} \dot{x} &= A_j x + B_j u, \\ y &= C_j x + D_j u. \end{aligned} \tag{1.25}$$

By adding and subtracting a term $L_j y$ into (1.25),

$$\dot{x} = A_j x + B_j u + L_j y - L_j y. \tag{1.26}$$

The substitution of (1.26) into (1.25) yields

$$\dot{x} = (A_j + L_jC_j)x + (B_j + L_jD_j)u - L_jy. \tag{1.27}$$

A feedback controller is given by

$$u = K_jx. \tag{1.28}$$

By substituting (1.28) into (1.27),

$$\dot{x} = (A_j + L_jC_j)x + (B_j + L_jD_j)K_jx - L_jy. \tag{1.29}$$

Then, a continuous-time state observer model of a dynamic system is derived

$$\begin{aligned} \dot{\hat{x}} &= (A_j + L_jC_j)\hat{x} + (B_j + L_jD_j)K_j\hat{x} - L_jy, \\ u &= K_j\hat{x}. \end{aligned} \tag{1.30}$$

However, it is difficult to directly control a MR damper via the output feedback controller. The reason is that only current signals can be controlled using the output feedback controller. Thus, the output feedback control gains are integrated with a converting algorithm.

1.4.4 Semiactive Converting Algorithm

Once the output feedback-based ANFC is designed, a semiactive converter and a MR damper are integrated with the output ANFC to develop a SNFC system. In general, a MR damper cannot be directly controlled by a control algorithm. The reason is that a controller generates force signals while a MR damper requires voltage or current signals to be operated. Therefore, a control component that converts a control force signal into a voltage or current signal should be integrated with the ANFC system to construct a SNFC system. Such a control component would either use an inverse MR damper model or implement a converting algorithm. Candidates for the inverse MR damper models may include Bingham, polynomial, Bouc-Wen, and modified Bouc-Wen models [33]. A good candidate for the conversion algorithm is a clipped algorithm [48]

$$v = V_a([f_{ANFC} - f_m]f_m] \tag{1.31}$$

where v is the applied voltage level, V_a is the arbitrary voltage level, H is a Heaviside step function, f_m is a measured MR damper force, and f_{ANFC} is a control force signal generated by an ANFC system

$$f_{ANFC} = \frac{\sum_{j=1}^{N_r} \prod_{i=1}^{N_r} \mu^i_{p_{i,j}}(z^i_{FZ}(t))\,[K_j\hat{x}]}{\sum_{i=1}^{N_r} \prod_{i=1}^{N_r} \mu^i_{p_{i,j}}(z^i_{FZ}(t))}. \tag{1.32}$$

By substitution of (1.32) into (1.31), the final voltage equation can be written as

$$v = V_a\left(\left[\frac{\sum_{j=1}^{N_r}\prod_{i=1}^{N_r}\mu_{p_{i,j}}^{i}(z_{FZ}^{i}(t))\,[K_j\hat{x}]}{\sum_{i=1}^{N_r}\prod_{i=1}^{N_r}\mu_{p_{i,j}}^{i}(z_{FZ}^{i}(t))} - f_m\right]f_m\right]. \tag{1.33}$$

This SNFC system is applied to a three-story building employing a single MR damper and a twenty-story building structure equipped with multiple MR dampers to demonstrate the performance of the proposed strategy.

1.5 Examples

In this section, large civil structures employing MR dampers are presented. The governing equations of motion of the large structure-MR damper systems are given by

$$M\ddot{x} + C\dot{x} + Kx = \Gamma f_{MR}(t,x,\dot{x}) - M\Lambda\ddot{w}_g \tag{1.34}$$

where

M : the system mass matrix,
C : the damping matrix,
K : the stiffness matrix,
f_{MR} : the MR damper force input matrix,
$\ddot{w}_g$: the ground acceleration record,
x: the displacement of the structure-MR damper system,
$\dot{x}$: the velocity of the structure-MR damper system,
$\ddot{x}$: the acceleration of the structure-MR damper system,
Γ: the location vector of control forces,
Λ: the location vector of disturbance signals.

The second order differential equations can be converted into state space

$$\begin{aligned} \dot{z} &= Az + Bf_{MR}(t,z,v) - E\ddot{w}_g \\ y &= Cz + Df_{MR}(t,z,v) + n \end{aligned} \tag{1.35}$$

where

A : the system state matrix,
B : the input matrix,
C : the output matrix,
D : the feedthrough matrix,
E : the disturbance location matrix,
n : the noise,
z : the states of the structure-MR damper system,
v : the applied voltage (or current).

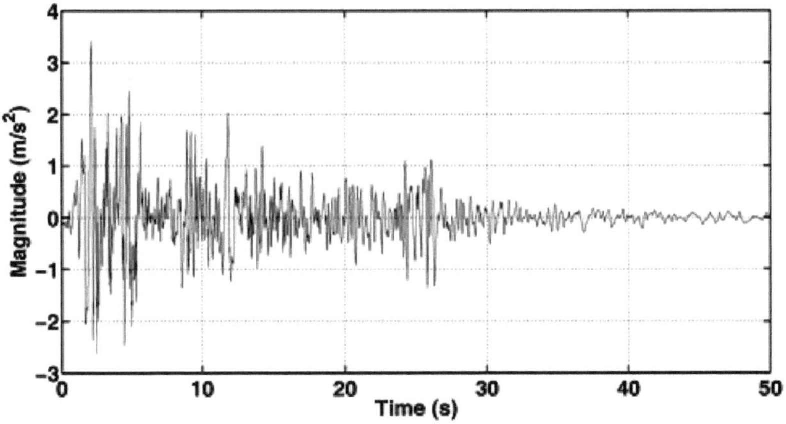

Fig. 1.2 1940 El-Centro earthquake record

In this chapter, two dynamic systems are studied, 1. a three-story building-MR damper system, 2. an twenty-story building-MR damper system.

1.5.1 Three-story Building Structure

In this section, a three-story building structure employing a SD-1000 MR damper whose parameters are given in [33] is presented. Properties of the three-story building structure are adopted from a scaled model [13] of a prototype building structure that was developed by [12]. The mass of each floor = 98.3 kg; the stiffness of each story k_1 = 516,000 N/m, k_2 = 684,000 N/m, and k_3 = 684,000 N/m; and the damping coefficients of each floor c_1 = 125 Ns/m, c_2 = 50 Ns/m, and c_3 = 50 Ns/m. In addition, the SD-1000 MR damper whose approximate capacity is 1,500 N is installed on the 1st floor using a Chevron brace, which leads to a nonlinear dynamic model, i.e., a building-MR damper system. Using (1.16) and (1.23), one can design fuzzy state feedback controllers that guarantee global asymptotical stability (GAS) of the closed-loop control system and provides the desired transient responses by constraining the closed-loop poles in a region D such that $(q_c, r_c) = (50, 45)$. This region puts a lower bound on both exponential decay rate and damping ratio of the closed-loop response.

The 1940 El-Centro earthquake input is applied as a ground motion shown in Figure 1.2. Note that the 1940 El-Centro earthquake signal is reproduced at five times the real recorded rate because the scaling factor of the laboratory scale structure model is 0.2. The time history responses that are controlled by the SNFC system at the entire floor are compared with the performance of a traditional optimal controller, i.e., H_2/LQG, while the uncontrolled system response is used as the baseline. The parameters of the linear quadratic regulator (LQR) and the Kalman filter

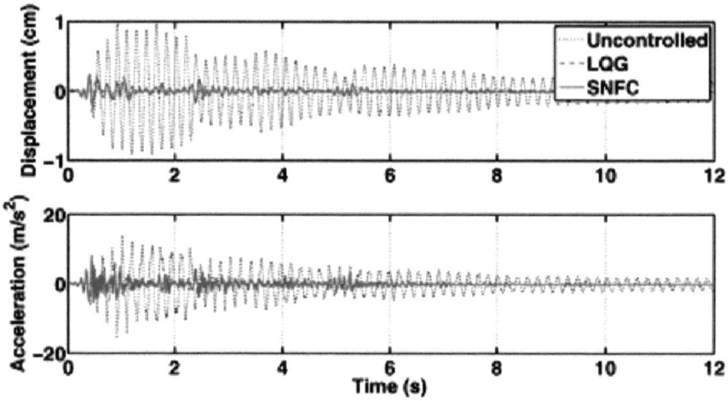

Fig. 1.3 Time history responses of a three-story building-MR damper system subjected to the 1940 El-Centro earthquake

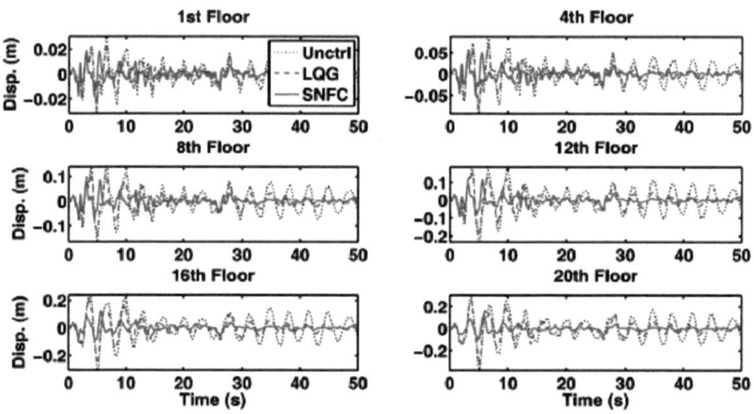

Fig. 1.4 Time history responses of a twenty-story building-MR damper system subjected to the 1940 El-Centro earthquake

are adopted from [13]. Figure 1.3 shows time history responses at the top floor of an uncontrolled, a H_2/LQG controlled, and an SNFC controlled systems, respectively. The upper graph of each figure shows the time history of displacement responses; the lower graph depicts the time history of acceleration responses. According to the time history responses, both displacement and acceleration responses are dramatically reduced when either H_2LQG or the proposed SNFC system is applied.

1.5.2 Twenty-story Building Structure

In this section, a set of SNFC are designed for vibration control of a seismically excited twenty-story benchmark building [34] equipped with 1000-kN MR dampers [20] such that each SNFC guarantees asymptotical stability and provides the desired transient response by constraining the closed-loop poles in a complex plane. Three MR dampers are located on the first eight stories and two devices are installed on the next twelve stories. The total number of MR dampers is 48. The locations of MR dampers within the building structure are determined through many trial-and-error simulations. Although the capacity and location of the MR dampers within the benchmark building can be optimized, such considerations are merely beyond the scope of the present investigation. The El-Centro earthquake record is applied to the building-MR damper system. The parameters of LQG are adopted from Spencer *et al.* (1999). There is still potential to improve the performance of the LQG controller because its parameters are not optimized for the semiactive control operation of MR dampers. Note that the objective of the current study is not to compare the performance of the SNFC system with that of the LQG system, but to investigate the 20-story benchmark building on structural control using the SNFC system, i.e., the authors offer an instance of an application of a LQG control system as a sample approach to illustrate the general effectiveness of the authors' proposed approach. The time history responses that are controlled by the SNFC system at the entire floor are compared with the performance of a traditional optimal controller, i.e., LQG, while the uncontrolled system response is used as the baseline. Figure 1.4 shows time history responses at the top floor of an uncontrolled, a LQG controlled, and an SNFC controlled systems, respectively. According to the time history responses, displacement responses are dramatically reduced when either proposed SNFC system or LQG controler is applied.

1.6 Concluding Remarks

In this chapter, an LMI-based systematic design methodology for SNFC of a class of nonlinear building-MR damper systems with multi-objective requirements (i.e., stability conditions and the performance on transient responses) in a unified framework was proposed. The design framework is based on PDC that forms a fuzzy gain scheduling scheme. It employs multiple Lyapunov-based linear controllers that correspond to local linear dynamic models with automatic scheduling performed via fuzzy rules. Each local linear controller is formulated via LMIs such that GAS is guaranteed and the performance on transient responses is also satisfied. Then, the associated Kalman observers are designed for output feedback control systems. Lastly, all of the local linear controllers and the estimators are blended via a fuzzy interpolation method. It is demonstrated from a simulation that the proposed SNFC system effectively reduces the vibration of the seismically excited building structure equipped with MR dampers. Further research is recommended to apply the proposed SNFC system to the larger-scale civil structures subjected to severe multi-hazard environments. It is also recommended that an experimental study to verify the effectiveness of the SNFC system be investigated. This might include a small-scale and a hybrid full-scale testing.

References

1. M. Abe, Rule-based Control Algorithm for Active Tuned Mass Damper, *Journal of Engineeirng Mechanics*, Vol. 122, No. 8, pp. 705-713, 1996.
2. H. Adeli & A. Saleh, *Control, Optimization, and Smart Structures: High-Performance Bridges and Buildings of the Future*, John Wiley & Sons, Inc., New York, 1999.
3. A.S. Ahlawat and A. Ramaswamy, Multiobjective Optimal FLC Driven Hybrid Mass Damper System for Torsionally Coupled, Seismically Excited Structures, *Earthquake Engineering & Structural Dynamics*, Vol. 31, pp. 1459-1479, 2000.
4. A.S. Ahlawat and A. Ramaswamy, Multi-objective Optimal Design of FLC Driven Hybrid Mass Damper for Seismically Excited Structures, *Earthquake Engineering & Structural Dynamics*, Vol. 31, pp. 2121-2139, 2002.
5. A.S. Ahlawat and A. Ramaswamy, Multiobjective Optimal Fuzzy Logic Control System for Response Control of Wind-Excited Tall Buildings, *ASCE Journal of Engineering Mechanics*, Vol. 130, pp. 524-530, 2004.
6. M. Al-Dawod, B. Samali, F. Naghdy and K.C.S. Kwok, Multiobjective Optimal Fuzzy Logic Control System for Response Control of Wind-Excited Tall Buildings, *Engineering Structures*, Vol.23, pp.1512-1522, 2001.
7. M. Al-Dawod, B. Samali, F. Naghdy, K. C. S. Kwok and F. Naghdy, Fuzzy Controller for Seismically Excited Nonlinear Buildings, *ASCE Journal of Engineering Mechanics*, Vol.130, pp. 407-415, 2004.
8. H. Alli. and O. Yakut, Fuzzy Sliding-Mode Control of Structures, *Engineering Structures*, Vol.27, pp.277-284, 2005.
9. M. Battaini, F. Casciati and L. Faravelli, Fuzzy Control of Structural Vibration. An Active Mass System Driven by a Fuzzy Controller, *Earthquake Engineering & Structural Dynamics*, Vol.27, pp.1267-1276, 1998.
10. M. Battaini, F. Casciati and L. Faravelli, Controlling Wind Response through a Fuzzy Controller, *ASCE Journal of Engineering Mechanics*, Vol. 130, pp.486-491, 2004.
11. M. Chilali and P. Gahinet, H_∞ Design with Pole Placement Constraints: An LMI Approach, *IEEE Transactions on Automatic Control*, Vol. 41, pp.358-367, 1996.
12. L.L. Chung, R.C. Lin, T.T. Soong and A.M. Reinhorn, Experiments on Active Control for MDOF Seismic Structures, *ASCE Journal of Engineering Mechanics*, Vol. 115, pp.1609-1627, 1989.
13. S.J. Dyke, B.F. Spencer, M.K. Sain and J.D. Carlson, Modeling and Control of Magnetorheological Dampers for Seismic Response Reduction, *Smart Materials and Structures*, Vol. 5, pp. 565-575, 1996.
14. L. Faravelli and R. Rossi, Adaptive Fuzzy Control: Theory versus Implementation, *Journal of Structural Control*, Vol. 9, pp. 59-73, 2002.
15. L. Faravelli and T. Yao, Use of Adaptive Networks in Fuzzy Control of Civil Structures, *Microcomputers in Civil Engineering*, Vol. 11, pp. 67-76, 1996.
16. S.S. Farinwata, D. Filev and R. Langari, *Fuzzy Control-Synthesis and Analysis*, John Wiley & Sons, New York, 2000.
17. S.K. Hong and R. Langari, An LMI-based H_∞ Fuzzy Control System Design with T-S Framework, *Information Sciences*, Vol. 123, pp. 163-179, 2000.
18. S. Hurlebaus and L. Gaul, Smart Structure Dynamics, *Mechanical Systems and Signal Processing*, Vol. 20, pp. 255-281, 2006.

19. J. Joh, R. Langari, F. T. Jeung and W. J. Chung, A New Design Method for Continuous Takagi-Sugeno Fuzzy Controller with Pole Placement Constraints: An LMI Approach, *Proceedings of the 1997 IEEE International Conference on Systems, Man, and Cybernetics*, Orlando, Florida, pp. 2969-2974, 1997.
20. H.J. Jung, B.F. Spencer Jr. and I.W. Lee, Control of Seismically Excited Cable-Stayed Bridge Employing Magnetorheological Fluid Dampers, *ASCE Journal of Structural Engineering*, Vol. 129, pp. 873-883, 2003.
21. H.S. Kim and P.N. Roschke, Design of Fuzzy Logic Controller for Smart Base Isolation System Using Genetic Algorithm, *Engineering Structures*, Vol. 28, pp. 84-96, 2006.
22. S.B. Kim, C.B. Yun and B.F. Spencer Jr., Vibration Control of Wind-Excited Tall Buildings Using Sliding Mode Fuzzy Control, *ASCE Journal of Engineering Mechanics*, Vol. 130, pp. 505-510, 2004.
23. Y. Kim and R. Langari, Nonlinear Identification and Control of a Building Structure with a Magnetorheological Damper, *Proceedings of the American Control Conference*, New York City, New York, pp. 3353-3358, 2007.
24. Y. Kim, *Nonlinear Identification and Control of Building Structures Equipped with Magneotorheological Dampers*, Ph.D. Dissertation, Zachry Department of Civil Engineering, Texas A & M University, College Station, Texas, USA, December, 2007.
25. Y. Kim, R. Langari and S. Hurlebaus, Semiactive Nonlinear Control of a Building with a Magnetorheological Damper System, *Mechanical Systems and Signal Processing*, Vol. 23, pp. 300-315, 2008.
26. R. Langari, Synthesis of Nonlinear Control Strategies via Fuzzy Logic, *Proceedings of the American Control Conference*, San Francisco, California, pp. 1855-1859, 1993.
27. R. Langari, Past, Present and Future of Fuzzy Control: A Case for Application of Fuzzy Logic in Hierarchical Control, *Proceedings of Annual Conference of the North American Fuzzy Information Processing Society*, New York City, New York, pp. 760-765, 1999.
28. S. Lei and R. Langari, Hierarchical Fuzzy Logic Control of a Double Inverted Pendulum, *Proceedings of IEEE International Conference on Fuzzy Systems*, San Antonio, Texas, pp. 1074-1077, 2000.
29. C.H. Loh, L.Y. Wu and P. Y. Lin, Displacement Control of Isolated Structures with Semi-active Control Devices, *Journal of Structural Control*, Vol. 10, pp. 77-100, 2003.
30. K.S. Park, H.M. Koh and C.W. Seo, Independent Modal Space Fuzzy Control of Earthquake-Excited Structures, *Engineering Structures*, Vol. 26, pp. 279-289, 2003.
31. K.C. Schurter and P. N. Roschke, Neuro-Fuzzy Control of Structures Using Magnetorheological Dampers, *Proceedings of the American Control Conference*, Arlington, Virginia, pp. 1097-1102, 2001.
32. T.T. Soong, *Active Structural Control: Theory and Practice*, Addison-Wesley Pub., New York, 1990.
33. B.F. Spencer, S.J. Dyke, M.K. Sain and J.D. Carlson, Phenomenological Model for Magnetorheological Dampers, *ASCE Journal of Engineering Mechanics*, Vol. 123, pp. 230-238, 1997.
34. B.F. Spencer, Jr., R.E. Christenson and S.J. Dyke, Next Generation Benchmark Control Problem for Seismically Excited Buildings, *Proceeding of the Second World Conference on Structrual Control*, Kyoto, Japan, pp. 1351-1360, 1999.
35. B.F. Spencer, Jr. and S. Nagarajaiah, State of the Art of Structural Control, *ASCE Journal of Structural Engineering*, Vol. 129, No. 7, pp. 845-856, 2003.
36. G. Strang, *Linear Algebra and Its Applications, 3rd Edition*, Harcourt Brace Jovanovich, Inc., 1988.
37. R.S. Subramaniam, A.M. Reinhorn, M.A.Riley and S. Nagarajaiah, Hybrid Control of Structures Using Fuzzy Logic, *Microcomputers in Civil Engineering*, Vol. 11, pp. 1-17, 1996.
38. M. Symans and S.W. Kelly, Fuzzy Logic Control of Bridge Structures using Intelligent Semi-active Seismic Isolation Systems, *Earthquake Engineering & Structural Dynamics*, Vol. 28, pp. 37-60, 1999.

39. T. Takagi and M. Sugeno, Fuzzy Identification of Systems and Its Applications to Modeling and Control, *IEEE Transactions on Systems, Man, and Cybernetics*, Vol. 15, pp. 116-132, 1985.
40. K. Tanaka and M. Sano, A Robust Stabilization Problem of Fuzzy Control Systems and its Application to Backing up Control of a Truck-trailer, *IEEE Transactions on Fuzzy Systems*, Vol. 2, pp. 119-134, 1994.
41. K. Tanaka and M. Sugeno, Stability Analysis and Design of Fuzzy Control Systems, *Fuzzy Sets and Systems*, Vol. 45, pp. 135-156, 1992.
42. A. Tani, H. Kawamura and S. Ryu, Intelligent Fuzzy Optimal Control of Building Structures, *Engineering Structures*, Vol. 20, pp. 184-192, 1998.
43. A.P. Wang and C.D. Lee, Fuzzy Sliding Mode Control for a Building Structure based on Genetic Algorithms, *Earthquake Engineering & Structural Dynamics*, Vol. 31, pp. 881-895, 2002.
44. H.O. Wang, K. Tanaka and M. Griffin, An Analytical Framework of Fuzzy Modeling and Control of Nonlinear Systems: Stability and Design Issues, *Proceedings of the American Control Conference*, Seattle, Washington, pp. 2272-2276, 1995.
45. G. Yan and L. L. Zhou, Integrated Fuzzy Logic and Genetic Algorithms for Multi-Objective Control of Structures Using MR Dampers, *Journal of Sound and Vibration*, Vol. 296, pp. 368-382, 2006.
46. J. Yen and R. Langari, *Fuzzy Logic-Intelligence, Control, and Information*, Prentice Hall, Upper Saddle River, New Jersey, USA, 1998.
47. L. Zhou, C.C. Chang and L.X. Wang, Adaptive Fuzzy Control for Nonlinear Building-Magnetorheological Damper System, *ASCE Journal of Structural Engineering*, Vol. 129, pp. 905-913, 2003.
48. Yoshida, O. and S. J. Dyke, Seismic Control of a Nonlinear Benchmark Building Using Smart Dampers, *Journal of Engineering Mechanics*, Vol. 130, 386-392, 1988.
49. L.A. Zadeh, Fuzzy Sets, *Information and Control*, Vol. 8, No. 3, pp. 338-353, 1965.

Chapter 2
Approaches to Robust $\mathscr{H}_\infty$ Controller Synthesis of Nonlinear Discrete-time-delay Systems via Takagi-Sugeno Fuzzy Models

Jianbin Qiu, Gang Feng and Jie Yang

Abstract This chapter investigates the problem of robust $\mathscr{H}_\infty$ piecewise state-feedback control for a class of nonlinear discrete-time-delay systems via Takagi-Sugeno (T-S) fuzzy models. The state delay is assumed to be time-varying and of an interval-like type with the lower and upper bounds. The parameter uncertainties are assumed to have a structured linear-fractional form. Based on two novel piecewise Lyapunov-Krasovskii functionals and some matrix inequality convexifying techniques, both delay-independent and delay-dependent controller design approaches are developed in terms of a set of linear matrix inequalities (LMIs). Numerical examples are also provided to illustrate the effectiveness and less conservatism of the proposed methods.

2.1 Introduction

Fuzzy logic control is a simple and effective approach to the control of many complex nonlinear systems [1–7]. During the past decades, a great number of industrial applications of fuzzy logic control have been reported in the open literature [1–4]. Among various model-based fuzzy control approaches, in particular, the method

Jianbin Qiu
Department of Manufacturing Engineering and Engineering Management, University of Science and Technology of China& City University of Hong Kong Joint Advanced Research Center, Dushu Lake Higher Education Town, Suzhou Industrial Park, 215123, P.R. China, e-mail: jbqiu@mail.ustc.edu.cn

Gang Feng
Department of Manufacturing Engineering and Engineering Management, City University of Hong Kong, 83 Tat Chee Avenue, Kowloon, Hong Kong, e-mail: megfeng@cityu.edu.hk

Jie Yang
Department of Precision Machinery and Precision Instrumentation, University of Science and Technology of China, Hefei, 230026, P.R. China, e-mail: jieyang@ustc.edu.cn

based on Takagi-Sugeno (T-S) model is well suited to model-based nonlinear control [5–7]. A T-S fuzzy model consists of a set of local linear models in different premise variable space regions which are blended through fuzzy membership functions.

During the past few years, significant research efforts have been devoted to the stability analysis and controller design of T-S fuzzy systems. The reader can refer to the survey papers [8, 9] and the references cited therein for the most recent advances on this topic. The appeal of T-S fuzzy models is that the stability analysis and controller synthesis of the overall fuzzy systems can be carried out in the Lyapunov function based framework. It has been shown in [5–11] that the stability and stabilization of the T-S fuzzy systems can be determined by searching for a common positive definite Lyapunov matrix. However, there are many fuzzy systems that do not admit a common Lyapunov function, whereas they may still be asymptotically stable [12, 14]. Recently, there have appeared some results on stability analysis and controller synthesis of fuzzy systems based on piecewise/fuzzy Lyapunov functions instead of a common Lyapunov function [12–20]. It has been shown that piecewise/fuzzy Lyapunov functions are much richer classes of Lyapunov function candidates than a common Lyapunov function candidate, and thus are able to deal with a larger class of fuzzy dynamic systems. The analysis and design results based on piecewise/fuzzy Lyapunov functions are generally less conservative than those based on a common Lyapunov function.

On the other hand, it is well-known that time-delays are frequently encountered in various complex nonlinear systems, such as chemical systems, mechanical systems, and communication networks. It has been well recognized that the presence of time-delays may result in instability, chaotic mode, and/or poor performance of control systems [21–27]. Control of dynamic systems with time-delays is a research subject of great practical and theoretical significance, which has received considerable attention in the past decades. Over the past few years, increasing attention has been drawn to the study of stability analysis and controller design for T-S fuzzy systems with time-delays [28–43]. In the context of discrete-time T-S fuzzy delay systems, some sufficient conditions for the solvability of stability analysis, state-feedback/output-feedback stabilization/$\mathscr{H}_\infty$ control problems were obtained in [28, 30, 31] in terms of LMIs. It is known that since the delay-dependent conditions include the information on the size of delays, they are usually less conservative than the delay-independent ones, especially when the size of delays is small. To reduce the design conservatism, recently, based on a piecewise Lyapunov-Krasovskii functional, the authors in [32] studied the problems of delay-dependent stability analysis and controller synthesis for a class of discrete-time T-S fuzzy systems with time-delays and it was shown that the obtained delay-dependent results are less conservative than the existing delay-independent ones. However, it is noted that the delay-dependent approaches presented in [32] suffer from a couple of drawbacks. Firstly, the delay-dependent criteria was realized by utilizing a system model transformation technique incorporating *Moon's bounding inequality* [21] to estimate the inner product of the involved crossing terms, which would lead to significant conservatism. Secondly, the results given in [32] only considered the case of time-invariant delays. Those results are unfortunately not applicable to the case of time-varying delays. In addition, the system parameter uncertainties were not taken into account in [32]. These motivate the present research.

In this chapter, we revisit the problem of robust $\mathscr{H}_\infty$ piecewise state-feedback control for a class of nonlinear discrete-time-delay systems via T-S fuzzy models. The state delay is assumed to be time-varying and of an interval-like type with the lower and upper bounds. The parameter uncertainties are assumed to have a structured linear-fractional form. Based on two novel piecewise Lyapunov-Krasovskii functionals combined with some matrix inequality convexifying techniques, both delay-independent and delay-dependent controller design approaches have been developed. It is shown that the controller gains can be obtained by solving a set of LMIs. Two simulation examples are also provided to illustrate the effectiveness and less conservatism of the proposed methods.

The rest of this chapter is structured as follows. Section 2.2 is devoted to the model description and problem formulation. The main results for robust $\mathscr{H}_\infty$ piecewise controller designs are given in Section 2.3. Two simulation examples are presented in Section 2.4 to demonstrate the applicability of the proposed approaches. In Section 2.5, some conclusions are presented.

The notations used throughout this chapter are standard. Z^+ denotes the set of nonnegative integer number. $\Re^n$ denotes the n-dimensional Euclidean space. $\Re^{n\times m}$ denotes the set of $n \times m$ real matrices. A real symmetric matrix $P > 0 (\geq 0)$ denotes P being positive definite (or being positive semi-definite). For a matrix $A \in \Re^{n\times n}$, A^{-1} and A^T are the inverse and transpose of the matrix A, respectively, and A^{-T} denotes $(A^{-1})^T$. $\mathrm{Sym}\{A\}$ is the shorthand notation for $A + A^T$. $\mathbf{I}_n$ and $\mathbf{0}_{n\times m}$ are used to denote the $n \times n$ identity matrix and $n \times m$ zero matrix, respectively. The subscripts n and $n \times m$ are omitted when the size is not relevant or can be determined from the context. $\mathrm{diag}\{\cdots\}$ denotes a block-diagonal matrix. The notation $\star$ in a symmetric matrix always denotes the symmetric block in the matrix. $l_2[0,\infty)$ refers to the space of square summable infinite vector sequences with the Euclidean norm $\|\bullet\|_2$. Matrices, if not explicitly stated, are assumed to have compatible dimensions for algebraic operations.

2.2 Model Description and Robust $\mathscr{H}_\infty$ Piecewise Control Problem

The T-S fuzzy dynamic model is described by fuzzy IF-THEN rules, each of which represents the local linear input-output relationship of nonlinear systems. Similar to [2–7], a discrete-time T-S fuzzy dynamic model with time-delay and parametric uncertainties can be described as

Plant Rule R^l: IF $\zeta_1(t)$ is F_1^l and $\zeta_2(t)$ is F_2^l and $\cdots$ and $\zeta_g(t)$ is F_g^l, THEN

$$\begin{cases} x(t+1) = A_l(t)x(t) + A_{dl}(t)x(t-\tau(t)) + B_{1l}(t)u(t) + D_{1l}(t)w(t) \\ z(t) = C_l(t)x(t) + C_{dl}(t)x(t-\tau(t)) + B_{2l}(t)u(t) + D_{2l}(t)w(t) \\ x(t) = \phi(t), -\tau_2 \leq t \leq 0, l \in \mathcal{L} := \{1,2,\cdots,r\} \end{cases} \tag{2.1}$$

where R^l denotes the lth fuzzy inference rule; r is the number of inference rules; $F_q^l(q = 1,2,\cdots,g)$ are fuzzy sets; $x(t) \in \Re^{n_x}$ is the system state; $u(t) \in \Re^{n_u}$ is the control input; $z(t) \in \Re^{n_z}$ is the regulated output; $w(t) \in \Re^{n_w}$ is the disturbance input which is assumed to belong to $l_2[0,\infty)$; $\zeta(t) := [\zeta_1(t), \zeta_2(t), \cdots, \zeta_g(t)]$ are some measurable variables of the system, for example, the state variables; $\tau(t)$ is a positive integer function representing the time-varying state delay of the system (2.1) and satisfying the following assumption

$$\tau_1 \leq \tau(t) \leq \tau_2 \tag{2.2}$$

with τ_1 and τ_2 being two constant positive integers representing the minimum and maximum time-delay, respectively. In this case, $\tau(t)$ is called an *interval-like* or *range-like* time-varying delay [24,25,37,43]. It is noted that this kind of time-delay describes the real situation in many practical engineering systems. For example, in the field of networked control systems, the network transmission induced delays (either from the sensor to the controller or from the controller to the plant) can be assumed to satisfy (2.2) without loss of generality [26,27]. $\phi(t)$ is a real-valued initial condition sequence on $[-\tau_2, 0]$. $A_l(t)$, $A_{dl}(t)$, $B_{1l}(t)$, $B_{2l}(t)$, $C_l(t)$, $C_{dl}(t)$, $D_{1l}(t)$ and $D_{2l}(t)$, $l \in \mathcal{L}$ are appropriately dimensioned system matrices with time-varying parametric uncertainties, which are assumed to be of the form

$$\begin{aligned} \begin{bmatrix} A_l(t) \; A_{dl}(t) \; B_{1l}(t) \; D_{1l}(t) \\ C_l(t) \; C_{dl}(t) \; B_{2l}(t) \; D_{2l}(t) \end{bmatrix} &= \begin{bmatrix} A_l \; A_{dl} \; B_{1l} \; D_{1l} \\ C_l \; C_{dl} \; B_{2l} \; D_{2l} \end{bmatrix} \\ &\quad + \begin{bmatrix} W_{1l} \\ W_{2l} \end{bmatrix} \Delta(t) \begin{bmatrix} E_{1l} \; E_{2l} \; E_{3l} \; E_{4l} \end{bmatrix} \end{aligned} \tag{2.3}$$

$$\Delta(t) = \Lambda(t) \left[\mathbf{I}_{s_2} - J\Lambda(t)\right]^{-1} \tag{2.4}$$

$$0 < \mathbf{I}_{s_2} - JJ^T \tag{2.5}$$

where $A_l, A_{dl}, B_{1l}, B_{2l}, C_l, C_{dl}, D_{1l}, D_{2l}, W_{1l}, W_{2l}, E_{1l}, E_{2l}, E_{3l}, E_{4l}$ and J are known real constant matrices of appropriate dimensions. $\Lambda(t) : Z^+ \rightarrow \Re^{s_1 \times s_2}$ is an unknown real-valued time-varying matrix function with Lesbesgue measurable elements satisfying

$$\Lambda^T(t)\Lambda(t) \leq \mathbf{I}_{s_2}. \tag{2.6}$$

The parameter uncertainties are said to be admissible if (2.3)-(2.6) hold. It is noted that interval bounded parameters can also be utilized to describe uncertain systems. For the discrete-time case, interval model control and applications can be found in [44] and the references therein.

Remark 2.1. The parametric uncertainties are assumed to have a structured linear-fractional form. It is noted that this kind of parameter uncertainties has been fairly investigated in the robust control theory [30,45,46]. It has been shown that every rational nonlinear system possesses a linear-fractional representation [46]. Notice that when $J = \mathbf{0}$, the linear-fractional form uncertainties reduce to the norm-bounded ones [21,25,31,37]. Notice also that the condition (2.5) guarantees that $\mathbf{I}_{s_2} - J\Lambda(t)$ is invertible for all $\Lambda(t)$ satisfying (2.6).

Let $\mu_l[\zeta(t)]$ be the normalized fuzzy-basis function of the inferred fuzzy set F^l where $F^l := \prod_{q=1}^{g} F_q^l$ and

$$\mu_l[\zeta(t)] := \frac{\prod_{q=1}^{g} \mu_{lq}[\zeta_q(t)]}{\sum_{p=1}^{r} \prod_{q=1}^{g} \mu_{pq}[\zeta_q(t)]} \geq 0, \quad \sum_{l=1}^{r} \mu_l[\zeta(t)] = 1 \tag{2.7}$$

where $\mu_{lq}[\zeta_q(t)]$ is the grade of membership of $\zeta_q(t)$ in F_q^l. In the sequel, we will drop the argument of $\mu_l[\zeta(t)]$ for clarity, i.e., denote μ_l as $\mu_l[\zeta(t)]$.

By using a center-average defuzzifier, product fuzzy inference and singleton fuzzifier, the following global T-S fuzzy dynamic model can be obtained

$$\begin{cases} x(t+1) = A(\mu,t)x(t) + A_d(\mu,t)x(t-\tau(t)) + B_1(\mu,t)u(t) + D_1(\mu,t)w(t) \\ z(t) = C(\mu,t)x(t) + C_d(\mu,t)x(t-\tau(t)) + B_2(\mu,t)u(t) + D_2(\mu,t)w(t) \\ x(t) = \phi(t), -\tau_2 \leq t \leq 0 \end{cases} \tag{2.8}$$

where

$$\begin{aligned} A(\mu,t) &:= \sum_{l=1}^{r} \mu_l A_l(t), A_d(\mu,t) := \sum_{l=1}^{r} \mu_l A_{dl}(t), B_1(\mu,t) := \sum_{l=1}^{r} \mu_l B_{1l}(t), \\ B_2(\mu,t) &:= \sum_{l=1}^{r} \mu_l B_{2l}(t), C(\mu,t) := \sum_{l=1}^{r} \mu_l C_l(t), C_d(\mu,t) := \sum_{l=1}^{r} \mu_l C_{dl}(t), \\ D_1(\mu,t) &:= \sum_{l=1}^{r} \mu_l D_{1l}(t), D_2(\mu,t) := \sum_{l=1}^{r} \mu_l D_{2l}(t). \end{aligned} \tag{2.9}$$

In this chapter, we consider the robust $\mathscr{H}_\infty$ control problem for the uncertain fuzzy dynamic model (2.8) based on piecewise Lyapunov-Krasovskii functionals. In order to facilitate the piecewise controller design, we partition the premise variable space $S \subseteq \mathbb{R}^g$ by the boundaries

$$\partial S_l^\nu := \{\zeta(t) | \mu_l[\zeta(t)] = 1, 0 \leq \mu_l[\zeta(t) + \delta] < 1, \forall 0 < |\delta| \ll 1, l \in \mathscr{L}\} \tag{2.10}$$

where ν is the set of the face indexes of the polyhedral hull satisfying $\partial S_l = \cup_\nu (\partial S_l^\nu), l \in \mathscr{L}$.

Then, based on the boundaries (2.10), we can get the induced close polyhedral regions $\bar{S}_i, i \in \mathscr{I}$ satisfying

$$\bar{S}_i \cap \bar{S}_j = \partial S_l^\nu, i \neq j, i, j \in \mathscr{I}, l \in \mathscr{L} \tag{2.11}$$

where $\mathscr{I}$ denotes the set of region indexes. The corresponding open regions are defined as $S_i, i \in \mathscr{I}$. It is noted that there are no disjointed interiors in S and $\cup_{i\in\mathscr{I}} \bar{S}_i$.

In each region $\bar{S}_i$, $i \in \mathscr{I}$, define

$$\mathscr{K}(i) := \left\{k|\mu_k[\zeta(t)] > 0, k \in \mathscr{L}, \zeta(t) \in \bar{S}_i, i \in \mathscr{I}\right\} \tag{2.12}$$

and then the global system (2.8) can be expressed by a blending of $k \in \mathscr{K}(i)$ subsystems

$$\begin{cases} x(t+1) = \mathscr{A}_i(t)x(t) + \mathscr{A}_{di}(t)x(t-\tau(t)) + \mathscr{B}_{1i}(t)u(t) + \mathscr{D}_{1i}(t)w(t) \\ z(t) = \mathscr{C}_i(t)x(t) + \mathscr{C}_{di}(t)x(t-\tau(t)) + \mathscr{B}_{2i}(t)u(t) + \mathscr{D}_{2i}(t)w(t) \\ x(t) = \phi(t), -\tau_2 \le t \le 0, \zeta(t) \in S_i, i \in \mathscr{I} \end{cases} \tag{2.13}$$

where

$$\begin{aligned} &\mathscr{A}_i(t) := \sum_{k\in\mathscr{K}(i)} \mu_k A_k(t), \mathscr{A}_{di}(t) := \sum_{k\in\mathscr{K}(i)} \mu_k A_{dk}(t), \mathscr{B}_{1i}(t) := \sum_{k\in\mathscr{K}(i)} \mu_k B_{1k}(t), \\ &\mathscr{D}_{1i}(t) := \sum_{k\in\mathscr{K}(i)} \mu_k D_{1k}(t), \mathscr{C}_i(t) := \sum_{k\in\mathscr{K}(i)} \mu_k C_k(t), \mathscr{C}_{di}(t) := \sum_{k\in\mathscr{K}(i)} \mu_k C_{dk}(t), \\ &\mathscr{B}_{2i}(t) := \sum_{k\in\mathscr{K}(i)} \mu_k B_{2k}(t), \mathscr{D}_{2i}(t) := \sum_{k\in\mathscr{K}(i)} \mu_k D_{2k}(t) \end{aligned} \tag{2.14}$$

with $0 \le \mu_k[\zeta(k)] \le 1$, $\sum_{k\in\mathscr{K}(i)} \mu_k[\zeta(k)] = 1$. For each region $\bar{S}_i$, the set $\mathscr{K}(i)$ contains the indexes for the system matrices used in the interpolation within that subspace. For a crisp subspace, $\mathscr{K}(i)$ contains a single element.

In this chapter, we address the piecewise controller design of fuzzy system (2.13), thus it is noted that when the state of the system transits from the region $\bar{S}_i$ to $\bar{S}_j$ at the time t, the dynamics of the system is governed by the dynamics of the region model $\bar{S}_i$ at that time. For future use, we also define a new set Ω to represent all possible region transitions

$$\Omega := \left\{(i,j)|\zeta(t) \in \bar{S}_i, \zeta(t+1) \in \bar{S}_j, i, j \in \mathscr{I}\right\} \tag{2.15}$$

where $j = i$ when $\zeta(t)$ stays in the same region $\bar{S}_i$ and $j \neq i$ when $\zeta(t)$ transits from the region $\bar{S}_i$ to $\bar{S}_j$.

Now, we consider the following piecewise control law for the fuzzy system (2.13)

$$u(t) = K_i x(t), \zeta(t) \in \bar{S}_i, i \in \mathscr{I}. \tag{2.16}$$

Substituting (2.16) into system (2.13) leads to the following closed-loop dynamics

$$\begin{cases} x(t+1) = \bar{\mathscr{A}}_i(t)x(t) + \mathscr{A}_{di}(t)x(t-\tau(t)) + \mathscr{D}_{1i}(t)w(t) \\ z(t) = \bar{\mathscr{C}}_i(t)x(t) + \mathscr{C}_{di}(t)x(t-\tau(t)) + \mathscr{D}_{2i}(t)w(t) \\ x(t) = \phi(t), -\tau_2 \le t \le 0, \zeta(t) \in S_i, i \in \mathscr{I} \end{cases} \tag{2.17}$$

where

$$\bar{\mathscr{A}}_i(t) := \mathscr{A}_i(t) + \mathscr{B}_{1i}(t)K_i, \bar{\mathscr{C}}_i(t) := \mathscr{C}_i(t) + \mathscr{B}_{2i}(t)K_i. \tag{2.18}$$

The robust $\mathscr{H}_\infty$ piecewise control problem to be investigated in this chapter is stated as follows.

Given an uncertain state-delayed fuzzy system (2.8) and a scalar $\gamma > 0$, design a state-feedback piecewise controller of the form (2.16) such that the closed-loop system (2.17) is robustly asymptotically stable for any fuzzy-basis functions $\mu[\zeta(t)]$ satisfying (2.7) when $w(t) = 0$, and under zero initial conditions, the l_2-gain between the exogenous input $w(t)$ and the regulated output $z(t)$ is less than γ, i.e., $\|z\|_2 \leq \gamma\|w\|_2$, for any nonzero $w \in l_2[0,\infty)$ and all admissible uncertainties. In this case, the system is said to be robustly asymptotically stable with $\mathscr{H}_\infty$ performance gamma.

Before ending this section, we introduce the following well-known lemmas, which will be used in the derivation of our main results.

Lemma 2.1. *[45] (S-procedure). Suppose that $\Delta(t)$ is given by (2.3)-(2.6), with matrices $M = M^T$, S and N of appropriate dimensions, the inequality*

$$M + Sym\{S\Delta(t)N\} < 0$$

holds if and only if for some scalar $\varepsilon > 0$

$$M + \begin{bmatrix} \varepsilon^{-1}N^T & \varepsilon S \end{bmatrix} \begin{bmatrix} I & -J \\ -J^T & I \end{bmatrix}^{-1} \begin{bmatrix} \varepsilon^{-1}N^T & \varepsilon S \end{bmatrix}^T < 0.$$

Lemma 2.2. *[45] (Schur Complements). Given constant matrices Σ_1, Σ_2, Σ_3 with $\Sigma_1 = \Sigma_1^T$ and $0 < \Sigma_2 = \Sigma_2^T$, then $\Sigma_1 + \Sigma_3^T \Sigma_2^{-1} \Sigma_3 < 0$ if and only if*

$$\begin{bmatrix} \Sigma_1 & \Sigma_3^T \\ \Sigma_3 & -\Sigma_2 \end{bmatrix} < 0 \quad or \quad \begin{bmatrix} -\Sigma_2 & \Sigma_3 \\ \Sigma_3^T & \Sigma_1 \end{bmatrix} < 0.$$

2.3 Piecewise $\mathscr{H}_\infty$ Control of T-S Fuzzy Systems with Time-delay

In this section, based on two novel piecewise Lyapunov-Krasovskii functionals combined with some matrix inequality convexifying techniques, both delay-independent and delay-dependent approaches will be developed to solve the robust $\mathscr{H}_\infty$ control problem formulated in the previous section. It is shown that the controller gains can be obtained by solving a set of LMIs.

2.3.1 Delay-independent $\mathscr{H}_\infty$ Controller Design

Theorem 2.1. *The closed-loop system (2.17) is robustly asymptotically stable with $\mathscr{H}_\infty$ performance γ if there exist matrices $0 < \underline{U}_i = \underline{U}_i^T \in \Re^{n_x \times n_x}$, $G_i \in \Re^{n_x \times n_x}$, $\bar{K}_i \in \Re^{n_u \times n_x}$, $i \in \mathscr{I}$, $0 < \bar{Q}_1 = \bar{Q}_1^T \in \Re^{n_x \times n_x}$, and a set of positive scalars $\varepsilon_{kij} > 0$, $k \in \mathscr{K}(i)$, $(i,j) \in \Omega$ such that the following LMIs hold*

$$\begin{bmatrix} -U_j & \mathbf{0} & A_kG_i+B_{1k}\bar{K}_i & A_{dk}\bar{Q}_1 & D_{1k} & \mathbf{0} & \mathbf{0} & \varepsilon_{kij}W_{1k} \\ \star & -\mathbf{I}_{n_z} & C_kG_i+B_{2k}K_i & C_{dk}Q_1 & D_{2k} & \mathbf{0} & \mathbf{0} & \varepsilon_{kij}W_{2k} \\ \star & \star & U_i-G_i-G_i^T & \mathbf{0} & \mathbf{0} & G_i^T & G_i^TE_{1k}^T+\bar{K}_i^TE_{3k}^T & \mathbf{0} \\ \star & \star & \star & -\bar{Q}_1 & \mathbf{0} & \mathbf{0} & \bar{Q}_1E_{2k}^T & \mathbf{0} \\ \star & \star & \star & \star & -\gamma^2\mathbf{I}_{n_w} & \mathbf{0} & E_{4k}^T & \mathbf{0} \\ \star & \star & \star & \star & \star & \frac{-\bar{Q}_1}{\tau_2-\tau_1+1} & \mathbf{0} & \mathbf{0} \\ \star & \star & \star & \star & \star & \star & -\varepsilon_{kij}\mathbf{I}_{s_2} & \varepsilon_{kij}J \\ \star & \star & \star & \star & \star & \star & \star & -\varepsilon_{kij}\mathbf{I}_{s_1} \end{bmatrix} < 0.$$

$$k \in \mathscr{K}(i), (i,j) \in \Omega \quad (2.19)$$

Moreover, the controller gain for each local region is given by

$$K_i = \bar{K}_i G_i^{-1}, i \in \mathscr{I}. \tag{2.20}$$

Proof. It is well-known that it suffices to find a Lyapunov function candidate $V(t,x(t)) > 0, \forall x(t) \neq 0$ satisfying the following inequality

$$V(t+1,x(t+1)) - V(t,x(t)) + \|z(t)\|_2^2 - \gamma^2\|w(t)\|_2^2 < 0 \tag{2.21}$$

to prove that the closed-loop system (2.17) is asymptotically stable with $\mathscr{H}_\infty$ performance γ under zero initial conditions for any nonzero $w \in l_2[0,\infty)$ and all admissible uncertainties.

Consider the following piecewise Lyapunov-Krasovskii functional

$$\begin{cases} V(t) := V_1(t) + V_2(t) \\ V_1(t) := x^T(t)P_ix(t), \zeta(t) \in \bar{S}_i, i \in \mathscr{I} \\ V_2(t) := \sum\limits_{s=-\tau_2}^{-\tau_1} \sum\limits_{v=t+s}^{t-1} x^T(v)Q_1x(v) \end{cases} \tag{2.22}$$

where $P_i = P_i^T > 0, i \in \mathscr{I}, Q_1 = Q_1^T > 0$ are Lyapunov matrices to be determined.

Define $\Delta V(t) := V(t+1) - V(t)$ and along the trajectory of system (2.17), one has

$$\Delta V_1(t) = x^T(t+1)P_jx(t+1) - x^T(t)P_ix(t), (i,j) \in \Omega, \tag{2.23}$$

$$\begin{aligned} \Delta V_2(t) &= (\tau_2 - \tau_1 + 1)x^T(t)Q_1x(t) - \sum_{s=t-\tau_2}^{t-\tau_1} x^T(s)Q_1x(s) \\ &\leq (\tau_2 - \tau_1 + 1)x^T(t)Q_1x(t) - x^T(t-\tau(t))Q_1x(t-\tau(t)). \end{aligned} \tag{2.24}$$

In addition, it follows from (2.17)

$$x(t+1) = \tilde{\mathscr{A}}_i(t)\xi_1(t), z(t) = \tilde{\mathscr{C}}_i(t)\xi_1(t) \tag{2.25}$$

where

$$
\begin{aligned}
\xi_1(t) &:= \left[x^T(t)\; x^T(t-\tau(t))\; w^T(t)\right]^T, \\
\tilde{\mathscr{A}_i}(t) &:= \left[\bar{\mathscr{A}_i}(t)\; \mathscr{A}_{di}(t)\; \mathscr{D}_{1i}(t)\right], \\
\tilde{\mathscr{C}_i}(t) &:= \left[\bar{\mathscr{C}_i}(t)\; \mathscr{C}_{di}(t)\; \mathscr{D}_{2i}(t)\right].
\end{aligned} \tag{2.26}
$$

Then, based on the piecewise Lyapunov-Krasovskii functional defined in (2.22) together with consideration of (2.23)-(2.26), it is easy to see that the following inequality implies (2.21),

$$
\xi_1^T \Theta_{ij}(t)\xi_1(t) < 0, \xi_1(t) \neq 0, (i,j) \in \Omega \tag{2.27}
$$

where

$$
\begin{aligned}
\Theta_{ij}(t) := {} & \tilde{\mathscr{A}_i}^T(t)P_j\tilde{\mathscr{A}_i}(t) + \tilde{\mathscr{C}_i}^T(t)\tilde{\mathscr{C}_i}(t) \\
& + \mathrm{diag}\left\{-P_i + (\tau_2 - \tau_1 + 1)Q_1, -Q_1, -\gamma^2\mathbf{I}_{n_w}\right\}.
\end{aligned} \tag{2.28}
$$

Thus, if one can show

$$
\Theta_{ij}(t) < 0, (i,j) \in \Omega \tag{2.29}
$$

then the claimed result follows. To this end, by Schur complements, (2.29) is equivalent to

$$
\begin{bmatrix}
-P_j^{-1} & \mathbf{0} & \bar{\mathscr{A}_i}(t) & \mathscr{A}_{di}(t) & \mathscr{D}_{1i}(t) & \mathbf{0} \\
\star & -\mathbf{I}_{n_z} & \bar{\mathscr{C}_i}(t) & \mathscr{C}_{di}(t) & \mathscr{D}_{2i}(t) & \mathbf{0} \\
\star & \star & -P_i & \mathbf{0} & \mathbf{0} & \mathbf{I}_{n_x} \\
\star & \star & \star & -Q_1 & \mathbf{0} & \mathbf{0} \\
\star & \star & \star & \star & -\gamma^2\mathbf{I}_{n_w} & \mathbf{0} \\
\star & \star & \star & \star & \star & \frac{-Q_1^{-1}}{\tau_2-\tau_1+1}
\end{bmatrix} < 0, (i,j) \in \Omega. \tag{2.30}
$$

Now, by introducing a nonsingular matrix $G_i \in \mathbb{R}^{n_x \times n_x}$ and pre- and post-multiplying (2.30) by $\mathrm{diag}\left\{\mathbf{I}_{(n_x+n_z)}, G_i^T, Q_1^{-1}, \mathbf{I}_{(n_w+n_x)}\right\}$ and its transpose, together with consideration of (2.14), yields

$$
\begin{bmatrix}
-U_j & \mathbf{0} & A_k(t)G_i + B_{1k}(t)\bar{K_i} & A_{dk}(t)\bar{Q}_1 & D_{1k}(t) & \mathbf{0} \\
\star & -\mathbf{I}_{n_z} & C_k(t)G_i + B_{2k}(t)\bar{K_i} & C_{dk}(t)\bar{Q}_1 & D_{2k}(t) & \mathbf{0} \\
\star & \star & -G_i^T U_i^{-1} G_i & \mathbf{0} & \mathbf{0} & G_i^T \\
\star & \star & \star & -\bar{Q}_1 & \mathbf{0} & \mathbf{0} \\
\star & \star & \star & \star & -\gamma^2\mathbf{I}_{n_w} & \mathbf{0} \\
\star & \star & \star & \star & \star & \frac{-\bar{Q}_1}{\tau_2-\tau_1+1}
\end{bmatrix} < 0,
$$

$$
k \in \mathscr{K}(i), (i,j) \in \Omega \tag{2.31}
$$

where $U_i := P_i^{-1}$, $U_j := P_j^{-1}$, $\bar{Q}_1 := Q_1^{-1}$, $\bar{K_i} := K_i G_i$.

Note that

$$
U_i - G_i - G_i^T + G_i^T U_i^{-1} G_i = (U_i - G_i)^T U_i^{-1} (U_i - G_i) \geq 0 \tag{2.32}
$$

which implies

$$-G_i^T U_i^{-1} G_i \leq U_i - G_i - G_i^T, i \in \mathscr{I}. \tag{2.33}$$

There, by (2.33), the following inequality implies (2.31)

$$\begin{bmatrix} -U_j & \mathbf{0} & A_k(t)G_i + B_{1k}(t)\bar{K}_i & A_{dk}(t)\bar{Q}_1 & D_{1k}(t) & \mathbf{0} \\ \star & -\mathbf{I}_{n_z} & C_k(t)G_i + B_{2k}(t)K_i & C_{dk}(t)Q_1 & D_{2k}(t) & \mathbf{0} \\ \star & \star & U_i - G_i - G_i^T & \mathbf{0} & \mathbf{0} & G_i^T \\ \star & \star & \star & -\bar{Q}_1 & \mathbf{0} & \mathbf{0} \\ \star & \star & \star & \star & -\gamma^2 \mathbf{I}_{n_w} & \mathbf{0} \\ \star & \star & \star & \star & \star & \frac{-\bar{Q}_1}{\tau_2 - \tau_1 + 1} \end{bmatrix} < 0,$$

$$k \in \mathscr{K}(i), (i, j) \in \Omega \tag{2.34}$$

and thus it suffices to show (2.34) instead of (2.31).

On the other hand, using the parameter uncertainty relationships (2.3)-(2.6), one has

$$\begin{bmatrix} -U_j & \mathbf{0} & A_k(t)G_i + B_{1k}(t)\bar{K}_i & A_{dk}(t)\bar{Q}_1 & D_{1k}(t) & \mathbf{0} \\ \star & -\mathbf{I}_{n_z} & C_k(t)G_i + B_{2k}(t)K_i & C_{dk}(t)Q_1 & D_{2k}(t) & \mathbf{0} \\ \star & \star & U_i - G_i - G_i^T & \mathbf{0} & \mathbf{0} & G_i^T \\ \star & \star & \star & -\bar{Q}_1 & \mathbf{0} & \mathbf{0} \\ \star & \star & \star & \star & -\gamma^2 \mathbf{I}_{n_w} & \mathbf{0} \\ \star & \star & \star & \star & \star & \frac{-\bar{Q}_1}{\tau_2 - \tau_1 + 1} \end{bmatrix}$$

$$= \begin{bmatrix} -U_j & \mathbf{0} & A_kG_i + B_{1k}\bar{K}_i & A_{dk}\bar{Q}_1 & D_{1k} & \mathbf{0} \\ \star & -\mathbf{I}_{n_z} & C_k(t)G_i + B_{2k}K_i & C_{dk}Q_1 & D_{2k} & \mathbf{0} \\ \star & \star & U_i - G_i - G_i^T & \mathbf{0} & \mathbf{0} & G_i^T \\ \star & \star & \star & -\bar{Q}_1 & \mathbf{0} & \mathbf{0} \\ \star & \star & \star & \star & -\gamma^2 \mathbf{I}_{n_w} & \mathbf{0} \\ \star & \star & \star & \star & \star & \frac{-\bar{Q}_1}{\tau_2 - \tau_1 + 1} \end{bmatrix}$$

$$+ \mathrm{Sym}\left\{ \begin{bmatrix} W_{1k} \\ W_{2k} \\ \mathbf{0} \\ \mathbf{0} \\ \mathbf{0} \\ \mathbf{0} \end{bmatrix} \Delta(t) \begin{bmatrix} \mathbf{0} & \mathbf{0} & E_{1k}G_i + E_{3k}\bar{K}_i & E_{2k}\bar{Q}_1 & E_{4k} & \mathbf{0} \end{bmatrix} \right\}. \tag{2.35}$$

Now, by Lemma 2.1, it is easy to see that (2.19) implies (2.35). The proof is thus completed. □

It is noted that in Theorem 2.1, if we set $U_i \equiv U$, $i \in \mathscr{I}$, we will get the corresponding controller design results based on a common Lyapunov-Krasovskii functional, which is summarized in the following corollary.

Corollary 2.1. *The closed-loop system (2.17) is robustly asymptotically stable with $\mathscr{H}_\infty$ performance γ if the LMIs (2.19) hold with $U_i \equiv U$, $i \in \mathscr{I}$.*

2.3.2 Delay-dependent $\mathscr{H}_\infty$ Controller Design

For the case of time-invariant delay, i.e., $\tau_1 = \tau_2$, it is noted that the results given in Theorem 2.1 are independent of the delay size. Thus, Theorem 2.1 can be applicable to the situations when no *a priori* knowledge about the size of time-delay is available. On the other hand, it is also well known that the delay-independent results for time-delay systems are usually more conservative than the delay-dependent ones especially when the size of time-delays is small. In this subsection, we consider the delay-dependent robust $\mathscr{H}_\infty$ control of fuzzy system (2.8) based on control law (2.16) by constructing a new piecewise Lyapunov-Krasovskii functional. The corresponding result is summarized in the following theorem.

Theorem 2.2. *The closed-loop system (2.17) is robustly asymptotically stable with $\mathscr{H}_\infty$ performance γ if there exist matrices $0 < U_i = U_i^T \in \Re^{n_x \times n_x}$, $\bar{M}_i = \begin{bmatrix} \bar{M}_{1i} \\ \bar{M}_{2i} \end{bmatrix} \in \Re^{2n_x \times n_x}$, $\bar{N}_i = \begin{bmatrix} \bar{N}_{1i} \\ \bar{N}_{2i} \end{bmatrix} \in \Re^{2n_x \times n_x}$, $\bar{R}_i = \begin{bmatrix} \bar{R}_{1i} \\ \bar{R}_{2i} \end{bmatrix} \in \Re^{2n_x \times n_x}$, $\bar{X}_i = \bar{X}_i^T = \begin{bmatrix} \bar{X}_{11i} & \bar{X}_{12i} \\ \star & \bar{X}_{22i} \end{bmatrix} \in \Re^{2n_x \times 2n_x}$, $\bar{Y}_i = \bar{Y}_i^T = \begin{bmatrix} \bar{Y}_{11i} & \bar{Y}_{12i} \\ \star & \bar{Y}_{22i} \end{bmatrix} \in \Re^{2n_x \times 2n_x}$, $\bar{K}_i \in \Re^{n_u \times n_x}$, $i \in \mathscr{I}$, $G \in \Re^{n_x \times n_x}$, $0 < \bar{Q}_\alpha = \bar{Q}_\alpha^T \in \Re^{n_x \times n_x}$, $\alpha \in \{1,2,3\}$, $0 < \bar{Z}_\beta = \bar{Z}_\beta^T \in \Re^{n_x \times n_x}$, $\beta \in \{1,2\}$, and a set of positive scalars $\varepsilon_{kij} > 0$, $k \in \mathscr{K}(i)$, $(i,j) \in \Omega$ such that the following LMIs hold*

$$
\begin{bmatrix}
-U_j & \mathbf{0} & \mathbf{0} & \mathbf{0} & A_kG+B_{1k}\bar{K}_i & A_{dk} & D_{1k} & \mathbf{0} & \mathbf{0} & \mathbf{0} & \varepsilon_{kij}W_{1k} \\
\star & \bar{Z}_1-G-G^T & \mathbf{0} & \mathbf{0} & \sqrt{\tau_2}\,(A_kG+B_{1k}\bar{K}_i-G) & \sqrt{\tau_2}A_{dk} & \sqrt{\tau_2}D_{1k} & \mathbf{0} & \mathbf{0} & \mathbf{0} & \sqrt{\tau_2}\varepsilon_{kij}W_{1k} \\
\star & \star & \bar{Z}_2-G-G^T & \mathbf{0} & \sqrt{\tau_2-\tau_1}\,(A_kG+B_{1k}\bar{K}_i-G) & \sqrt{\tau_2-\tau_1}A_{dk} & \sqrt{\tau_2-\tau_1}D_{1k} & \mathbf{0} & \mathbf{0} & \mathbf{0} & \sqrt{\tau_2-\tau_1}\,\varepsilon_{kij}W_{1k} \\
\star & \star & \star & -\boldsymbol{I}_{n_z} & C_kG+B_{2k}\bar{K}_i & C_{dk} & D_{2k} & \mathbf{0} & \mathbf{0} & \mathbf{0} & \mathbf{0} \\
\star & \star & \star & \star & \Pi_{55} & \Pi_{56} & \mathbf{0} & -\bar{N}_{1i} & \bar{R}_{1i} & G^TE_{1k}^T+\bar{K}_i^TE_{3k}^T & \mathbf{0} \\
\star & \star & \star & \star & \star & \Pi_{66} & \mathbf{0} & -\bar{N}_{2i} & \bar{R}_{2i} & E_{2k}^T & \mathbf{0} \\
\star & \star & \star & \star & \star & \star & -\gamma^2\boldsymbol{I}_{n_w} & \mathbf{0} & \mathbf{0} & E_{4k}^T & \mathbf{0} \\
\star & \star & \star & \star & \star & \star & \star & -\bar{Q}_2 & \mathbf{0} & \mathbf{0} & \mathbf{0} \\
\star & \star & \star & \star & \star & \star & \star & \star & -\bar{Q}_3 & \mathbf{0} & \mathbf{0} \\
\star & \star & \star & \star & \star & \star & \star & \star & \star & -\varepsilon_{kij}\boldsymbol{I}_{s_2} & \varepsilon_{kij}J \\
\star & \star & \star & \star & \star & \star & \star & \star & \star & \star & -\varepsilon_{kij}\boldsymbol{I}_{s_1}
\end{bmatrix} < 0,
$$

$$k \in \mathscr{K}(i), (i,j) \in \Omega \tag{2.36}$$

$$\begin{bmatrix} \bar{X}_i & \bar{M}_i \\ \star & \bar{Z}_1 \end{bmatrix} \geq 0, i \in \mathscr{I} \tag{2.37}$$

$$\begin{bmatrix} \bar{X}_i+\bar{Y}_i & \bar{N}_i \\ \star & \bar{Z}_1+\bar{Z}_2 \end{bmatrix} \geq 0, i \in \mathscr{I} \tag{2.38}$$

$$\begin{bmatrix} \bar{Y}_i & \bar{R}_i \\ \star & \bar{Z}_2 \end{bmatrix} \geq 0, i \in \mathscr{I} \tag{2.39}$$

where

$$
\begin{aligned}
\Pi_{55} &:= U_i - G - G^T + (\tau_2-\tau_1+1)\bar{Q}_1 + \bar{Q}_2 + \bar{Q}_3 \\
&\quad + \bar{M}_{1i} + \bar{M}_{1i}^T + \tau_2\bar{X}_{11i} + (\tau_2-\tau_1)\bar{Y}_{11i}, \\
\Pi_{56} &:= \bar{N}_{1i} - \bar{M}_{1i} - \bar{R}_{1i} + \bar{M}_{2i}^T + \tau_2\bar{X}_{12i} + (\tau_2-\tau_1)\bar{Y}_{12i}, \\
\Pi_{66} &:= -\bar{Q}_1 + \bar{N}_{2i} + \bar{N}_{2i}^T - \bar{M}_{2i} - \bar{M}_{2i}^T - \bar{R}_{2i} - \bar{R}_{2i}^T + \tau_2\bar{X}_{22i} + (\tau_2-\tau_1)\bar{Y}_{22i}.
\end{aligned}
$$

Moreover, the controller gain for each local region is given by

$$K_i = \bar{K}_i G^{-1}, i \in \mathscr{I}. \tag{2.40}$$

Proof. Similar to the proof of Theorem 2.1, it suffices to find a Lyapunov function candidate $V(t,x(t)) > 0$, $\forall x(t) \neq 0$ satisfying (2.21) to prove that the closed-loop system (2.17) is asymptotically stable with $\mathscr{H}_\infty$ performance γ under zero initial conditions for any nonzero $w \in l_2[0,\infty)$ and all admissible uncertainties.

Define $e(t) := x(t+1) - x(t)$ and consider the following piecewise Lyapunov-Krasovskii functional

$$\begin{cases} V(t) := V_1(t) + V_2(t) + V_3(t) + V_4(t) \\ V_1(t) := x^T(t)P_i x(t), \zeta(t) \in S_i, i \in \mathscr{I} \\ V_2(t) := \sum\limits_{s=-\tau_2}^{-\tau_1} \sum\limits_{v=t+s}^{t-1} x^T(v) Q_1 x(v) \\ V_3(t) := \sum\limits_{s=t-\tau_2}^{t-1} x^T(s) Q_2 x(s) + \sum\limits_{s=t-\tau_1}^{t-1} x^T(s) Q_3 x(s) \\ V_4(t) := \sum\limits_{s=-\tau_2}^{-1} \sum\limits_{m=t+s}^{t-1} e^T(m) Z_1 e(m) + \sum\limits_{s=-\tau_2}^{-\tau_1-1} \sum\limits_{m=t+s}^{t-1} e^T(m) Z_2 e(m) \end{cases} \tag{2.41}$$

where $P_i = P_i^T > 0, i \in \mathscr{I}, Q_\alpha = Q_\alpha^T > 0, \alpha \in \{1,2,3\}$ and $Z_\beta = Z_\beta^T > 0, \beta \in \{1,2\}$ are Lyapunov matrices to be determined.

Define $\Delta V(t) := V(t+1) - V(t)$ and along the trajectory of system (2.17), one has

$$\Delta V_1(t) = x^T(t+1) P_j x(t+1) - x^T(t) P_i x(t), (i,j) \in \Omega \tag{2.42}$$

$$\Delta V_2(t) \leq (\tau_2 - \tau_1 + 1) x^T(t) Q_1 x(t) - x^T(t-\tau(t)) Q_1 x(t-\tau(t)) \tag{2.43}$$

$$\begin{aligned} \Delta V_3(t) &= x^T(t)(Q_2 + Q_3)x(t) \\ &\quad - x^T(t-\tau_2) Q_2 x(t-\tau_2) - x^T(t-\tau_1) Q_3 x(t-\tau_1), \end{aligned} \tag{2.44}$$

$$\begin{aligned} \Delta V_4(t) &= \tau_2 e^T(t) Z_1 e(t) + (\tau_2 - \tau_1) e^T(t) Z_2 e(t) - \sum_{m=t-\tau(t)}^{t-1} e^T(m) Z_1 e(m) \\ &\quad - \sum_{m=t-\tau_2}^{t-\tau(t)-1} e^T(m)(Z_1 + Z_2) e(m) - \sum_{m=t-\tau(t)}^{t-\tau_1-1} e^T(m) Z_2 e(m). \end{aligned} \tag{2.45}$$

In addition, define $\xi_2(t) := \left[x^T(t) \; x^T(t-\tau(t)) \right]^T$ and from the definition of $e(t)$, for any appropriately dimensioned matrices M_i, N_i, R_i, X_i, Y_i, $i \in \mathscr{I}$ one has

$$0 \equiv 2\xi_2^T(t)M_i\left[x(t)-x(t-\tau(t))-\sum_{m=t-\tau(t)}^{t-1}e(m)\right], \tag{2.46}$$

$$0 \equiv 2\xi_2^T(t)N_i\left[x(t-\tau(t))-x(t-\tau_2)-\sum_{m=t-\tau_2}^{t-\tau(t)-1}e(m)\right], \tag{2.47}$$

$$0 \equiv 2\xi_2^T(t)R_i\left[x(t-\tau_1)-x(t-\tau(t))-\sum_{m=t-\tau(t)}^{t-\tau_1-1}e(m)\right], \tag{2.48}$$

$$\begin{aligned} 0 \equiv\ & \xi_2^T(t)\left[\tau_2 X_i+(\tau_2-\tau_1)Y_i\right]\xi_2(t)-\sum_{m=t-\tau(t)}^{t-1}\xi_2^T(t)X_i\xi_2(t) \\ &-\sum_{m=t-\tau_2}^{t-\tau(t)-1}\xi_2^T(t)\left(X_i+Y_i\right)\xi_2(t)-\sum_{m=t-\tau(t)}^{t-\tau_1-1}\xi_2^T(t)Y_i\xi_2(t). \end{aligned} \tag{2.49}$$

On the other hand, under the following bounding conditions,

$$\begin{bmatrix} X_i & M_i \\ \star & Z_1 \end{bmatrix} \geq 0, i \in \mathscr{I} \tag{2.50}$$

$$\begin{bmatrix} X_i+Y_i & N_i \\ \star & Z_1+Z_2 \end{bmatrix} \geq 0, i \in \mathscr{I} \tag{2.51}$$

$$\begin{bmatrix} Y_i & R_i \\ \star & Z_2 \end{bmatrix} \geq 0, i \in \mathscr{I} \tag{2.52}$$

one has

$$\begin{aligned} 0 &\leq \sum_{m=t-\tau(t)}^{t-1}\begin{bmatrix} \xi_2(t) \\ e(m) \end{bmatrix}^T\begin{bmatrix} X_i & M_i \\ \star & Z_1 \end{bmatrix}\begin{bmatrix} \xi_2(t) \\ e(m) \end{bmatrix} \\ &= \sum_{m=t-\tau(t)}^{t-1}\left\{\xi_2^T(t)X_i\xi_2(t)+2\xi_2^T(t)M_i e(m)+e^T(m)Z_1 e(m)\right\}, \end{aligned} \tag{2.53}$$

$$\begin{aligned} 0 &\leq \sum_{m=t-\tau_2}^{t-\tau(t)-1}\begin{bmatrix} \xi_2(t) \\ e(m) \end{bmatrix}^T\begin{bmatrix} X_i+Y_i & N_i \\ \star & Z_1+Z_2 \end{bmatrix}\begin{bmatrix} \xi_2(t) \\ e(m) \end{bmatrix} \\ &= \sum_{m=t-\tau_2}^{t-\tau(t)-1}\left\{\xi_2^T(t)\left(X_i+Y_i\right)\xi_2(t)+2\xi_2^T(t)N_i e(m)+e^T(m)(Z_1+Z_2)e(m)\right\}, \end{aligned} \tag{2.54}$$

$$\begin{aligned} 0 &\leq \sum_{m=t-\tau(t)}^{t-\tau_1-1}\begin{bmatrix} \xi_2(t) \\ e(m) \end{bmatrix}^T\begin{bmatrix} Y_i & R_i \\ \star & Z_2 \end{bmatrix}\begin{bmatrix} \xi_2(t) \\ e(m) \end{bmatrix} \\ &= \sum_{m=t-\tau(t)}^{t-\tau_1-1}\left\{\xi_2^T(t)Y_i\xi_2(t)+2\xi_2^T(t)R_i e(m)+e^T(m)Z_2 e(m)\right\}. \end{aligned} \tag{2.55}$$

Thus, based on the piecewise Lyapunov-Krasovskii functional defined in (2.41), together with consideration of the following relationships,

$$\begin{cases} x(t+1) = \hat{\mathscr{A}}_i(t)\xi_3(t) \\ z(t) = \hat{\mathscr{C}}_i(t)\xi_3(t) \\ e(t) = x(t+1) - x(t) = \check{\mathscr{A}}_i(t)\xi_3(t) \\ \xi_3(t) := \begin{bmatrix} x^T(t) \; x^T(t-\tau(t)) \; w^T(t) \; x^T(t-\tau_2) \; x^T(t-\tau_1) \end{bmatrix}^T \\ \hat{\mathscr{A}}_i(t) := \begin{bmatrix} \mathscr{A}_i(t) \; \mathscr{A}_{di}(t) \; \mathscr{D}_{1i}(t) \; \mathbf{0}_{n_x\times n_x} \; \mathbf{0}_{n_x\times n_x} \end{bmatrix} \\ \hat{\mathscr{C}}_i(t) := \begin{bmatrix} \mathscr{C}_i(t) \; \mathscr{C}_{di}(t) \; \mathscr{D}_{2i}(t) \; \mathbf{0}_{n_z\times n_x} \; \mathbf{0}_{n_z\times n_x} \end{bmatrix} \\ \check{\mathscr{A}}_i(t) := \begin{bmatrix} \mathscr{A}_i(t) - \mathbf{I}_{n_x} \; \mathscr{A}_{di}(t) \; \mathscr{D}_{1i}(t) \; \mathbf{0}_{n_x\times n_x} \; \mathbf{0}_{n_x\times n_x} \end{bmatrix} \\ M_i := \begin{bmatrix} M_{1i} \\ M_{2i} \end{bmatrix} N_i := \begin{bmatrix} N_{1i} \\ N_{2i} \end{bmatrix} R_i := \begin{bmatrix} R_{1i} \\ R_{2i} \end{bmatrix} \\ X_i := \begin{bmatrix} X_{11i} & X_{12i} \\ \star & X_{12i} \end{bmatrix} Y_i := \begin{bmatrix} Y_{11i} & Y_{12i} \\ \star & Y_{12i} \end{bmatrix} \end{cases} \tag{2.56}$$

it is easy to see that the following inequality implies (2.21)

$$\xi_3^T(t)\Sigma_{ij}(t)\xi_3(t) < 0, \xi_3(t) \neq 0, (i,j) \in \Omega \tag{2.57}$$

where

$$\begin{aligned} \Sigma_{ij}(t) &:= \hat{\mathscr{A}}_i^T(t)P_j\hat{\mathscr{A}}_i(t) + \hat{\mathscr{C}}_i^T(t)\hat{\mathscr{C}}_i(t) + \tau_2\check{\mathscr{A}}_i^T(t)Z_1\check{\mathscr{A}}_i(t) \\ &\quad + (\tau_2-\tau_1)\check{\mathscr{A}}_i^T(t)Z_2\check{\mathscr{A}}_i(t) + \Phi, \\ \Phi &:= \begin{bmatrix} \Phi_{11} & \Phi_{12} & \mathbf{0} & -N_{1i} & R_{1i} \\ \star & \Phi_{22} & \mathbf{0} & -N_{2i} & R_{2i} \\ \star & \star & -\gamma^2\mathbf{I}_{n_w} & \mathbf{0} & \mathbf{0} \\ \star & \star & \star & -Q_2 & \mathbf{0} \\ \star & \star & \star & \star & -Q_3 \end{bmatrix}, \\ \Phi_{11} &:= -P_i + (\tau_2-\tau_1+1)Q_1 + Q_2 + Q_3 + M_{1i} + M_{1i}^T + \tau_2X_{11i} + (\tau_2-\tau_1)Y_{11i}, \\ \Phi_{12} &:= N_{1i} - M_{1i} - R_{1i} + M_{2i}^T + \tau_2X_{12i} + (\tau_2-\tau_1)Y_{12i}, \\ \Phi_{22} &:= N_{2i} + N_{2i}^T - M_{2i} - M_{2i}^T - R_{2i} - R_{2i}^T + \tau_2X_{22i} + (\tau_2-\tau_1)Y_{22i}. \end{aligned} \tag{2.58}$$

Thus, if one can show (2.50)-(2.52) and

$$\Sigma_{ij}(t) < 0, (i,j) \in \Omega \tag{2.59}$$

then the claimed result follows. To this end, by Schur complements, (2.59) is equivalent to

$$\begin{bmatrix} -P_j^{-1} & \mathbf{0} & \mathbf{0} & \mathbf{0} & \hat{\mathscr{A}}_i(t) \\ \star & -Z_1^{-1} & \mathbf{0} & \mathbf{0} & \sqrt{\tau_2}\check{\mathscr{A}}_i(t) \\ \star & \star & -Z_2^{-1} & \mathbf{0} & \sqrt{\tau_2-\tau_1}\check{\mathscr{A}}_i(t) \\ \star & \star & \star & -\mathbf{I}_{n_z} & \hat{\mathscr{C}}_i(t) \\ \star & \star & \star & \star & \Phi \end{bmatrix} < 0, (i,j) \in \Omega. \tag{2.60}$$

Based on (2.14) and the parameter uncertainty relationships given in (2.3)-(2.6), by Lemma 2.1, it is easy to see that the following inequalities implies (2.60).

$$
\left[\begin{array}{ccccccccccc}
-P_j^{-1} & \mathbf{0} & \mathbf{0} & \mathbf{0} & A_k+B_{1k}K_i & A_{dk} & D_{1k} & \mathbf{0} & \mathbf{0} & \mathbf{0} & \varepsilon_{kij}W_{1k} \\
\star & -Z_1^{-1} & \mathbf{0} & \mathbf{0} & \sqrt{\tau_2}\,(A_k+B_{1k}K_i-\mathbf{I}_{n_x}) & \sqrt{\tau_2}A_{dk} & \sqrt{\tau_2}D_{1k} & \mathbf{0} & \mathbf{0} & \mathbf{0} & \sqrt{\tau_2}\varepsilon_{kij}W_{1k} \\
\star & \star & -Z_2^{-1} & \mathbf{0} & \sqrt{\tau_2-\tau_1}\,(A_k+B_{1k}K_i-\mathbf{I}_{n_x}) & \sqrt{\tau_2-\tau_1}A_{dk} & \sqrt{\tau_2-\tau_1}D_{1k} & \mathbf{0} & \mathbf{0} & \mathbf{0} & \sqrt{\tau_2-\tau_1}\varepsilon_{kij}W_{1k} \\
\star & \star & \star & -\mathbf{I}_{n_z} & C_k+B_{2k}K_i & C_{dk} & D_{2k} & \mathbf{0} & \mathbf{0} & \mathbf{0} & \mathbf{0} \\
\star & \star & \star & \star & \Phi_{11} & \Phi_{12} & \mathbf{0} & -N_{1i} & R_{1i} & E_{1k}^T+K_i^TE_{3k}^T & \mathbf{0} \\
\star & \star & \star & \star & \star & \Phi_{22} & \mathbf{0} & -N_{2i} & R_{2i} & E_{2k}^T & \mathbf{0} \\
\star & \star & \star & \star & \star & \star & -\gamma^2\mathbf{I}_{n_w} & \mathbf{0} & \mathbf{0} & E_{4k}^T & \mathbf{0} \\
\star & \star & \star & \star & \star & \star & \star & -Q_2 & \mathbf{0} & \mathbf{0} & \mathbf{0} \\
\star & \star & \star & \star & \star & \star & \star & \star & -Q_3 & \mathbf{0} & \mathbf{0} \\
\star & \star & \star & \star & \star & \star & \star & \star & \star & -\varepsilon_{kij}\mathbf{I}_{s_2} & \varepsilon_{kij}J \\
\star & \star & \star & \star & \star & \star & \star & \star & \star & \star & -\varepsilon_{kij}\mathbf{I}_{s_1}
\end{array}\right] < 0,
$$

$$k \in \mathscr{K}(i),\, (i,j) \in \Omega. \quad (2.61)$$

It is noted that the conditions given in (2.61) are nonconvex due to the simultaneous presence of the Lyapunov matrices P_j, Z_1, Z_2 and their inverses P_j^{-1}, Z_1^{-1}, Z_2^{-1}. For the matrix inequality linearization purpose, similar to the proof of Theorem 2.1, we introduce a nonsingular matrix $G \in \mathbb{R}^{n_x \times n_x}$ and define the following matrices

$$
\begin{aligned}
& U_i := P_i^{-1}, U_j := P_j^{-1}, \bar{K}_i := K_iG, \\
& \bar{\chi} := G^T\chi G, \chi \in \{Z_1, Z_2, Q_1, Q_2, Q_3, M_i, N_i, R_i, X_i, Y_i\}, \\
& \bar{M}_i = \begin{bmatrix} \bar{M}_{1i} \\ \bar{M}_{2i}^T \end{bmatrix}, \bar{N}_i = \begin{bmatrix} \bar{N}_{1i} \\ \bar{N}_{2i}^T \end{bmatrix}, \bar{R}_i = \begin{bmatrix} \bar{R}_{1i} \\ \bar{R}_{2i} \end{bmatrix}, \bar{X}_i = \begin{bmatrix} \bar{X}_{11i} & \bar{X}_{12i} \\ \star & \bar{X}_{22i} \end{bmatrix}, \bar{Y}_i = \begin{bmatrix} \bar{Y}_{11i} & \bar{Y}_{12i} \\ \star & \bar{Y}_{22i} \end{bmatrix}.
\end{aligned} \quad (2.62)
$$

Then, pre- and post-multiplying (2.61) by $\mathrm{diag}\left\{\mathbf{I}_{3n_x+n_z}, G, G, \mathbf{I}_{n_w}, G, G, \mathbf{I}_{s_1+s_2}\right\}$ and its transpose together with consideration of the following bounding inequalities leads to (2.36) exactly.

$$-G^T P_i G \le U_i - G - G^T, U_i = P_i^{-1} \tag{2.63}$$
$$-G\bar{Z}_1^{-1}G^T \le \bar{Z}_1 - G - G^T, Z_1^{-1} = G\bar{Z}_1^{-1}G^T, \tag{2.64}$$
$$-G\bar{Z}_2^{-1}G^T \le \bar{Z}_2 - G - G^T, Z_2^{-1} = G\bar{Z}_2^{-1}G^T. \tag{2.65}$$

Meanwhile, pre- and post-multiplying (2.50)-(2.52) by diag$\{G^T, G^T, G^T\}$ and its transpose yields (2.37)-(2.39) directly. The proof is thus completed. □

Remark 2.2. It is noted that from the proof of Theorem 2.2, the delay-dependent criterion is realized by using a free-weighting matrix technique [22–25], which enables one to avoid performing any model transformation to the original system and thus no bounding technique is employed to estimate the inner product of the involved crossing terms [21]. Moreover, when treating the time-varying delay and estimating the upper bound of the difference of Lyapunov functional, some useful terms such as $\sum_{m=t-\tau_2}^{t-\tau(t)-1} e^T(m)(\bullet)e(m)$ are fully utilized by introducing some additional terms into the proposed Lyapunov-Krasovskii functional. In addition, by using the relationships (2.53)-(2.55), the delay upper bound τ_2 is separated into two parts: $\tau_2 = [\tau_2 - \tau(t)] + \tau(t)$ without ignoring any useful terms. Thus, compared with the existing delay-independent and delay-dependent approaches for discrete-time T-S fuzzy delay systems [28–33, 36, 37], these features have the potential to enable one to obtain less conservative results.

From the proof of Theorem 2.2, it is easy to see that the matrix inequality liearization procedure is based on the bounding inequalities given in (2.63)-(2.65), where all three different Lyapunov matrices are constrained by a same slack variable. This inevitably brings some degree of design conservatism. Another means to solve the nonlinear matrix inequality problem given in (2.61) is by utilizing the cone complementarity linearization algorithm [47]. To this end, we define variables $U_i := P_i^{-1}$, $S_1 := Z_1^{-1}$, $S_2 := Z_2^{-1}$, and by using a cone complementarity technique [47], the nonconvex feasibility problem given in (2.50)-(2.52), (2.61) is converted to the following nonlinear minimization problem involving LMI conditions:

$$\text{Minimize Tr}\left(\sum_{i\in\mathscr{I}} P_iU_i + \sum_{\beta=1}^{2} Z_\beta S_\beta\right), \text{subject to (2.50)-(2.52) and}$$

$$\left[\begin{array}{cccccccccc} -U_j & \mathbf{0} & \mathbf{0} & \mathbf{0} & A_k+B_{1k}K_i & A_{dk} & D_{1k} & \mathbf{0} & \mathbf{0} & \mathbf{0} & \varepsilon_{kij}W_{1k} \\ \star & -S_1 & \mathbf{0} & \mathbf{0} & \sqrt{\tau_2}(A_k+B_{1k}K_i-\mathbf{I}_{n_x}) & \sqrt{\tau_2}A_{dk} & \sqrt{\tau_2}D_{1k} & \mathbf{0} & \mathbf{0} & \mathbf{0} & \sqrt{\tau_2}\varepsilon_{kij}W_{1k} \\ \star & \star & -S_2 & \mathbf{0} & \sqrt{\tau_2-\tau_1}(A_k+B_{1k}K_i-\mathbf{I}_{n_x}) & \sqrt{\tau_2-\tau_1}A_{dk} & \sqrt{\tau_2-\tau_1}D_{1k} & \mathbf{0} & \mathbf{0} & \mathbf{0} & \sqrt{\tau_2-\tau_1}\varepsilon_{kij}W_{1k} \\ \star & \star & \star & -\mathbf{I}_{n_z} & C_k+B_{2k}K_i & C_{dk} & D_{2k} & \mathbf{0} & \mathbf{0} & \mathbf{0} & \mathbf{0} \\ \star & \star & \star & \star & \Phi_{11} & \Phi_{12} & \mathbf{0} & -N_{1i} & R_{1i} & E_{1k}^T+K_i^TE_{3k}^T & \mathbf{0} \\ \star & \star & \star & \star & \star & \Phi_{22} & \mathbf{0} & -N_{2i} & R_{2i} & E_{2k}^T & \mathbf{0} \\ \star & \star & \star & \star & \star & \star & -\gamma^2\mathbf{I}_{n_w} & \mathbf{0} & \mathbf{0} & E_{4k}^T & \mathbf{0} \\ \star & \star & \star & \star & \star & \star & \star & -Q_2 & \mathbf{0} & \mathbf{0} & \mathbf{0} \\ \star & \star & \star & \star & \star & \star & \star & \star & -Q_3 & \mathbf{0} & \mathbf{0} \\ \star & \star & \star & \star & \star & \star & \star & \star & \star & -\varepsilon_{kij}\mathbf{I}_{s_2} & \varepsilon_{kij}J \\ \star & \star & \star & \star & \star & \star & \star & \star & \star & \star & -\varepsilon_{kij}\mathbf{I}_{s_1} \end{array}\right] < 0,$$

$$k\in\mathscr{K}(i), (i,j)\in\Omega. \tag{2.66}$$

$$P_i>0, U_i>0, \begin{bmatrix} P_i & \mathbf{I}_{n_x} \\ \star & U_i \end{bmatrix} \geq 0, i\in\mathscr{I} \tag{2.67}$$

$$Z_\beta>0, S_\beta>0, \begin{bmatrix} Z_\beta & \mathbf{I}_{n_x} \\ \star & S_\beta \end{bmatrix} \geq 0, \beta\in\{1,2\} \tag{2.68}$$

$$Q_\alpha \geq 0, \alpha\in\{1,2,3\}. \tag{2.69}$$

Then, for given delay bounds τ_1 and τ_2, the suboptimal performance of γ can be found by the following algorithm. The convergence of this algorithm is guaranteed in terms of similar results in [21, 47].

Algorithm 2.1 *Suboptimal performance of* γ

Step 1. *Choose a sufficiently large initial* $\gamma>0$ *such that there exists a feasible solution to (2.50)-(2.52) and (2.66)-(2.69). Set* $\gamma_0=\gamma$.

Step 2. *Find a feasible set*

$$\left\{\begin{array}{c} P_{i0}, U_{i0}, X_{i0}, Y_{i0}, M_{i0}, N_{i0}, R_{i0}, K_{i0}, i\in\mathscr{I}, \varepsilon_{kij0}, k\in\mathscr{K}(i), (i,j)\in\Omega, \\ Q_{\alpha0}, Z_{\beta0}, S_{\beta0}, \alpha\in\{1,2,3\}, \beta\in\{1,2\} \end{array}\right\}$$

satisfying (2.50)-(2.52) and (2.66)-(2.69). Set $\sigma=0$.

Step 3. *Solving the following LMIs problem for the variables* P_i, U_i, X_i, Y_i, M_i, N_i, R_i, K_i, $i\in\mathscr{I}$, Q_α, $\alpha\in\{1,2,3\}$, Z_β, S_β, $\beta\in\{1,2\}$, ε_{kij}, $k\in\mathscr{K}(i)$, $(i,j)\in\Omega$:

$$\textit{Minimize } Tr\left(\sum_{i\in\mathscr{I}}(P_{i\sigma}U_i+P_iU_{i\sigma})+\sum_{\beta=1}^{2}(Z_{\beta\sigma}S_\beta+Z_\beta S_{\beta\sigma})\right)$$
Subject to (2.58)-(2.62) and (2.65)-(2.67).

Set $P_{i(\sigma+1)}=P_i$, $U_{i(\sigma+1)}=U_i$, $i\in\mathscr{I}$, $Z_{\beta(\sigma+1)}=Z_\beta$, $S_{\beta(\sigma+1)}=S_\beta$, $\beta\in\{1,2\}$.

Step 4. *Substituting the controller gains* K_i *obtained in Step 3 into (2.66) and by some simple algebraic manipulations yield (2.70). Then, if the LMIs (2.50)-(2.52) and (2.70) are feasible with respect to the variables* $P_i=P_i^T>0$, X_i, Y_i, M_i, N_i, R_i, $\bar{\varepsilon}_{kij}$, $k\in\mathscr{K}(i)$, $(i,j)\in\Omega$, $Q_\alpha=Q_\alpha^T\ge 0$, $Z_\beta=Z_\beta^T>0$, $\alpha\in\{1,2,3\}$, $\beta\in\{1,2\}$, *then set* $\gamma_0=\gamma$ *and return to Step 2 after decreasing* γ *to some extent. If the conditions (2.50)-(2.52) and (2.70) are infeasible within the maximum number of iterations allowed, then exit. Otherwise, set* $\sigma=\sigma+1$ *and go to Step 3.*

$$\left[\begin{array}{cccccc}
-P_j & \mathbf{0} & \mathbf{0} & \mathbf{0} & P_j(A_k+B_{1k}K_i) & P_jA_{dk} \\
\star & -Z_1 & \mathbf{0} & \mathbf{0} & \sqrt{\tau_2}Z_1(A_k+B_{1k}K_i-\boldsymbol{I}_{n_x}) & \sqrt{\tau_2}Z_1A_{dk} \\
\star & \star & -Z_2 & \mathbf{0} & \sqrt{\tau_2-\tau_1}Z_2(A_k+B_{1k}K_i-\boldsymbol{I}_{n_x}) & \sqrt{\tau_2-\tau_1}Z_2A_{dk} \\
\star & \star & \star & -\boldsymbol{I}_{n_z} & C_k+B_{2k}K_i & C_{dk} \\
\star & \star & \star & \star & \Phi_{11} & \Phi_{12} \\
\star & \star & \star & \star & \star & \Phi_{22} \\
\star & \star & \star & \star & \star & \star \\
\star & \star & \star & \star & \star & \star \\
\star & \star & \star & \star & \star & \star
\end{array}\right.$$
$$\left.\begin{array}{ccccc}
D_{1k} & \mathbf{0} & \mathbf{0} & \mathbf{0} & W_{1k} \\
\sqrt{\tau_2}D_{1k} & \mathbf{0} & \mathbf{0} & \mathbf{0} & \sqrt{\tau_2}W_{1k} \\
\sqrt{\tau_2-\tau_1}D_{1k} & \mathbf{0} & \mathbf{0} & \mathbf{0} & \sqrt{\tau_2-\tau_1}W_{1k} \\
D_{2k} & \mathbf{0} & \mathbf{0} & \mathbf{0} & \mathbf{0} \\
\mathbf{0} & -N_{1i} & R_{1i} & \bar{\varepsilon}_{kij}(E_{1k}^T+K_i^TE_{3k}^T) & \mathbf{0} \\
\mathbf{0} & -N_{2i} & R_{2i} & \bar{\varepsilon}_{kij}E_{2k}^T & \mathbf{0} \\
-\gamma^2\boldsymbol{I}_{n_w} & \mathbf{0} & \mathbf{0} & \bar{\varepsilon}_{kij}E_{4k}^T & \mathbf{0} \\
\star & -Q_2 & \mathbf{0} & \mathbf{0} & \mathbf{0} \\
\star & \star & -Q_3 & \mathbf{0} & \mathbf{0} \\
\star & \star & \star & -\bar{\varepsilon}_{kij}\boldsymbol{I}_{s_2} & \bar{\varepsilon}_{kij}J \\
\star & \star & \star & \star & -\bar{\varepsilon}_{kij}\boldsymbol{I}_{s_1}
\end{array}\right]<0,$$
$$k\in\mathscr{K}(i),(i,j)\in\Omega.\quad(2.70)$$

Remark 2.3. It is noted that the results given in Theorem 2.2 and Algorithm 2.1 do not encompass each other. The results obtained in Theorem 2.2 are very nice in the sense that these conditions are convex, and thus can be readily solved with commercially available software. The design conservatism of Theorem 2.2 mainly comes from the bounding inequalities given in (2.67)-(2.69). In the iterative approach, the conditions given in (2.66)-(2.68) are equivalent to the corresponding performance analysis results presented in (2.70). This is the main advantage of Algorithm 2.1 over Theorem 2.2. However, the numerical computation cost involved in Algorithm

2.1 is much larger than that involved in Theorem 2.2, especially when the number of iterations increases.

Remark 2.4. The results presented in this chapter are obtained by using piecewise Lyapunov functions [12, 14, 15, 19, 32, 33]. However, it is noted that the extension of the proposed design techniques based on fuzzy Lyapunov functions is straightforward. A piecewise Lyapunov function is defined along the partitioned premise variable space [12, 14, 15, 19, 32, 33], while a fuzzy-basis-dependent Lyapunov function is based on a mapping from fuzzy-basis functions to a Lyapunov matrix [13, 16–18, 30]. It is hard to theoretically explain which kind of Lyapunov function is better. However, for discrete-time fuzzy systems, the controller design procedure based on these two kinds of Lyapunov functions are very similar once the Lyapunov matrices are separated from the system matrices. The controller designs for continuous-time fuzzy systems based on piecewise/fuzzy Lyapunov functions are much more complicated.

2.4 Simulation Examples

In this section, we use two examples to demonstrate the advantages and less conservatism of the controller design methods proposed in this chapter.

Example 2.1. Consider the following modified Henon mapping model with time-delay and external disturbance borrowed from [33] with some modifications.

$$\begin{cases} x_1(t+1) = -x_1^2(t) + 0.04x_1(t-\tau(t)) + 0.3x_2(t) + 1.4 + 0.5w(t) \\ x_2(t+1) = x_1(t) - 0.02x_1(t-\tau(t)) + 0.03x_2(t-\tau(t)) + 0.5w(t) \end{cases} \tag{2.71}$$

where the disturbance is given by $w(t) = 0.1\mathrm{e}^{-0.3t}\sin(0.02\pi t)$ and the time-delay $\tau(t)$ is assumed to satisfy $2 \leq \tau(t) \leq 5$. With the initial condition $\phi(t) = \begin{bmatrix} 0.1 & 0 \end{bmatrix}^T$, $-5 \leq t \leq 0$, the chaotic behavior of the above system is shown in Figure 2.1. It is observed from Figure 2.1 that the attractor region is $\{x(t) | -2 \leq x_1(t) \leq 2\}$.

Next, we consider the following time-delay T-S fuzzy system to represent the system (2.71).

Plant Rule R^l: IF $\theta_1(t)$ is F^l, THEN

$$\begin{cases} \theta(t+1) = A_l\theta(t) + A_{dl}\theta(t-\tau(t)) + B_{1l}u(t) + D_{1l}w(t) \\ z(t) = C_l\theta(t) + C_{dl}\theta(t-\tau(t)) + B_{2l}u(t) \end{cases} \tag{2.72}$$

where $z(t)$ is the regulated output, $\theta(t) := x(t) - x_f$ and $x_f = \begin{bmatrix} x_{f1} \\ x_{f2} \end{bmatrix} = \begin{bmatrix} 0.9088 \\ 0.8995 \end{bmatrix}$ is one of the fixed points of system (2.71) embedded in the attractor region. The system parameters are given as follows.

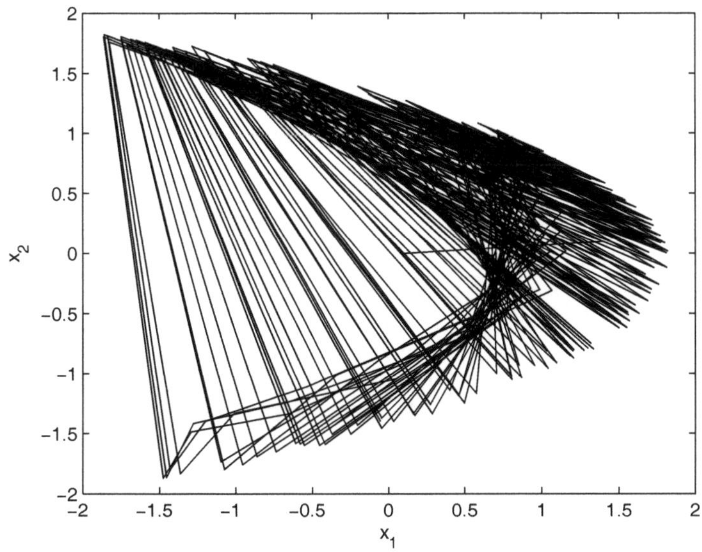

Fig. 2.1 The chaotic behavior of system (2.71)

$$
\begin{aligned}
A_1 &= \begin{bmatrix} 4 & 0.3 \\ 1 & 0 \end{bmatrix}, A_2 = \begin{bmatrix} -4 & 0.3 \\ 1 & 0 \end{bmatrix}, A_3 = \begin{bmatrix} 0 & 0.3 \\ -1 & 0 \end{bmatrix}, \\
A_{d1} &= A_{d2} = \begin{bmatrix} 0.04 & 0 \\ -0.02 & 0.03 \end{bmatrix}, A_{d3} = \begin{bmatrix} 0.009 & 0 \\ 0.6 & 0 \end{bmatrix}, \\
B_1 &= B_2 = \begin{bmatrix} 1 \\ 0 \end{bmatrix}, B_3 = \begin{bmatrix} 0.5 \\ 0.5 \end{bmatrix}, C_1 = C_2 = C_3 = \begin{bmatrix} 0.2 & -0.3 \end{bmatrix}, \\
C_{d1} &= C_{d2} = C_{d3} = \begin{bmatrix} 0.03 & 0.01 \end{bmatrix}, B_{21} = B_{22} = B_{23} = 0.1.
\end{aligned}
$$

In the region of $\{\theta(t)| - 15 \leq \theta_1(t) \leq 15\}$, we define the following membership functions, which are also shown in Figure 2.2.

$$\begin{bmatrix} \mu_1[\theta_1(t)] \\ \mu_2[\theta_1(t)] \\ \mu_3[\theta_1(t)] \end{bmatrix} = \begin{cases} \begin{bmatrix} 1 \\ 0 \\ 0 \end{bmatrix}, & -15 \le \theta_1(t) < -4 - 2x_{f1} \\ \begin{bmatrix} 0.5\left(1 - \frac{\theta_1(t)+2x_{f1}}{4}\right) \\ 0.5\left(1 + \frac{\theta_1(t)+2x_{f1}}{4}\right) \\ 0 \end{bmatrix}, & -4 - 2x_{f1} \le \theta_1(t) < 4 - 2x_{f1} \\ \begin{bmatrix} 0 \\ 1 - \left(1 + \mathrm{e}^{10-\theta_1(t)}\right)^{-1} \\ \left(1 + \mathrm{e}^{10-\theta_1(t)}\right)^{-1} \end{bmatrix}, & 4 - 2x_{f1} \le \theta_1(t) \le 15. \end{cases}$$

It is easy to see that in the region of $\{\theta(t)| -4 - 2x_{f1} \le \theta_1(t) \le 4 - 2x_{f1}\}$, system (2.72) describes the error system of (2.71) to the fixed point x_f.

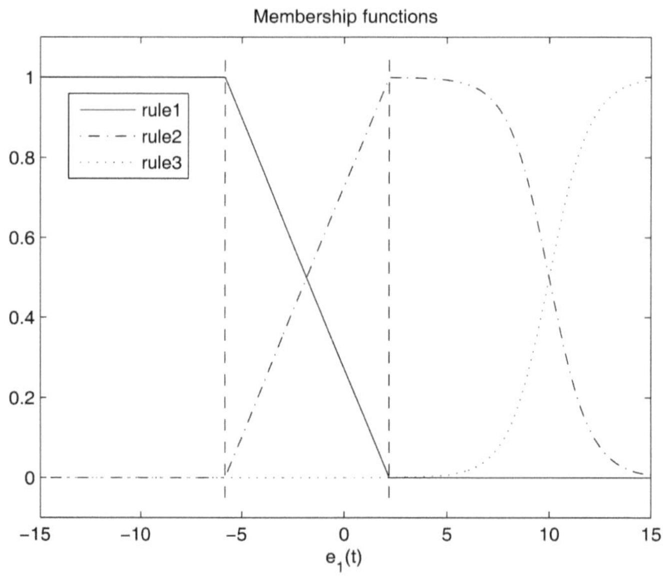

Fig. 2.2 Membership functions for the fuzzy system in Example 2.1

The objective is then to design a piecewise controller to stabilize the system (2.72) to zero with a guaranteed disturbance attenuation level γ. Based on the partition method presented in Section 2.2, there are four boundaries. Thus, the premise space is divided into three regions as shown in Figure 2.2. Let the $\mathscr{H}_\infty$ performance $\gamma = 3.5$ and it has been found that there is no feasible solution based on the delay-independent method given in Theorem 2.1, or the delay-dependent methods by [32,33,36,37] and Theorem 2.2 given in this chapter. However, by applying Algorithm 2.1 and after 37 iterations, one indeed obtains a feasible solution with controller gains given by

$$K_1 = \begin{bmatrix} -3.4795 & -0.2238 \end{bmatrix}, K_2 = \begin{bmatrix} 1.2401 & -0.2879 \end{bmatrix}, K_3 = \begin{bmatrix} 2.9863 & 0.1781 \end{bmatrix}.$$

With a randomly generated time-varying delay $\tau(t)$ between $\tau_1 = 2$ and $\tau_2 = 5$, and initial condition $\theta(t) = \begin{bmatrix} 0.1 & 0 \end{bmatrix}^T, -5 \le t \le 0$, the state trajectories of the error system are shown in Figure 2.3. It can be observed that the performance of the resulting closed-loop system is satisfactory.

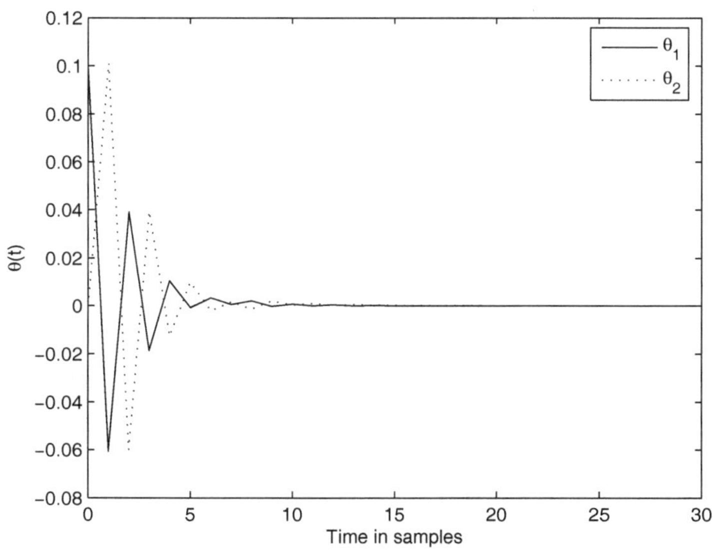

Fig. 2.3 Time response of the error system in Example 2.1

Example 2.2. Consider the following uncertain discrete-time state-delayed T-S fuzzy system of the form (2.1) with three rules.

Plant Rules R^l: IF $x_1(t)$ is F^l, THEN

$$\begin{cases} x(t+1) = A_l(t)x(t) + A_{dl}(t)x(t-\tau(t)) + B_{1l}(t)u(t) + D_{1l}(t)w(t) \\ z(t) = C_l(t)x(t) + C_{dl}(t)x(t-\tau(t)) + B_{2l}(t)u(t) + D_{2l}(t)w(t) \end{cases}$$

where

$$
\begin{aligned}
&A_1 = \rho \begin{bmatrix} 1 & 0.3 \\ 0.2 & 1 \end{bmatrix}, A_2 = \rho \begin{bmatrix} 1 & 0.3 \\ 0.4 & 1 \end{bmatrix}, A_3 = \rho \begin{bmatrix} 1.2 & 0.5 \\ 0.2 & 1 \end{bmatrix}, \\
&A_{d1} = A_{d2} = A_{d3} = \begin{bmatrix} -0.1 & 0.05 \\ 0 & 0.1 \end{bmatrix}, B_{11} = \begin{bmatrix} 0.5 \\ 1.5 \end{bmatrix}, B_{12} = \begin{bmatrix} 0.5 \\ 1 \end{bmatrix}, \\
&B_{13} = \begin{bmatrix} 0.6 \\ 1 \end{bmatrix}, D_{11} = D_{12} = D_{13} = \begin{bmatrix} 0.5 \\ 0.5 \end{bmatrix}, C_1 = C_3 = \begin{bmatrix} 0.3 & 0 \end{bmatrix}, C_2 = \begin{bmatrix} 0.4 & 0 \end{bmatrix}, \\
&C_{d1} = C_{d2} = C_{d3} = \begin{bmatrix} 0.2 & 0.2 \end{bmatrix}, B_{21} = B_{22} = B_{23} = 0.1, D_{21} = D_{22} = D_{23} = 0, \\
&W_{11} = \begin{bmatrix} 0.1 \\ -0.22 \end{bmatrix}, W_{12} = \begin{bmatrix} -0.03 \\ 0.01 \end{bmatrix}, W_{13} = \begin{bmatrix} 0.05 \\ 0.01 \end{bmatrix}, W_{21} = W_{22} = W_{23} = 0, \\
&E_{11} = \begin{bmatrix} 0.1 & 0.1 \end{bmatrix}, E_{12} = \begin{bmatrix} 0.01 & 0.03 \end{bmatrix}, E_{13} = \begin{bmatrix} 0.03 & 0.02 \end{bmatrix}, E_{21} = \begin{bmatrix} 0.01 & 0.02 \end{bmatrix}, \\
&E_{22} = \begin{bmatrix} 0.1 & 0.04 \end{bmatrix}, E_{23} = \begin{bmatrix} 0.01 & 0.02 \end{bmatrix}, E_{31} = 0.1, E_{32} = 0.2, E_{33} = 0.05, \\
&E_{41} = E_{42} = E_{43} = 0.1, J = 0.2
\end{aligned}
$$

and ρ is a scalar parameter. The normalized membership functions are shown in Figure 2.4. Then, based on the partition method presented in Section 2.2, the premise variable space can be partitioned into five regions:

$$
\begin{gathered}
\mathscr{X}_1 := \{x_1 | -\infty < x_1 < 3\}, \mathscr{X}_2 := \{x_1 | -3 \le x_1 < -1\}, \\
\mathscr{X}_3 := \{x_1 | -1 \le x_1 < 1\}, \mathscr{X}_4 := \{x_1 | 1 \le x_1 < 3\}, \mathscr{X}_5 := \{x_1 | 3 \le x_1 < \infty\}.
\end{gathered}
$$

The time-varying delay $\tau(t)$ is assumed to satisfy (2.2) with $\tau_1 = 3$, $\tau_2 = 8$. It is noted that the open loop system is unstable.

The objective is to design a piecewise controller of the form (2.16) such that the resulting closed-loop system (2.17) is robustly asymptotically stable with $\mathscr{H}_\infty$ disturbance attenuation level γ. To this end, choosing the scalar $\rho = 1.3$, it has been found that there is no feasible solution based on the delay-dependent methods by [32,33,36,37] and Theorem 2.2 given in this chapter. However, by applying Theorem 2.1 given in this chapter, one indeed obtains a set of feasible solutions with the optimal $\mathscr{H}_\infty$ performance index $\gamma_{\min} = 6.3827$. The controller gains are given by

$$
\begin{aligned}
&K_1 = \begin{bmatrix} -1.9770 & -0.7920 \end{bmatrix}, \\
&K_2 = \begin{bmatrix} -1.9593 & -0.8243 \end{bmatrix}, K_3 = \begin{bmatrix} -2.3719 & -0.8310 \end{bmatrix}, \\
&K_4 = \begin{bmatrix} -0.9968 & -1.0107 \end{bmatrix}, K_5 = \begin{bmatrix} -4.9562 & -0.7747 \end{bmatrix}.
\end{aligned}
$$

With a randomly generated time-varying delay $\tau(t)$ between $\tau_1 = 3$ and $\tau_2 = 9$ and initial condition $x(t) = \begin{bmatrix} -1.5 & 3 \end{bmatrix}^T$, $-8 \le t \le 0$, the state trajectories of the closed-loop system are shown in Figure 2.5. It can be observed that the performance of the resulting closed-loop system is satisfactory.

In addition, we also try to apply Algorithm 2.1 to solve the piecewise control problem for the above system. Unfortunately, it is observed that there is no feasible solution based on Algorithm 2.1 even when the maximum number of iterations allowed is $N_{\max} = 500$.

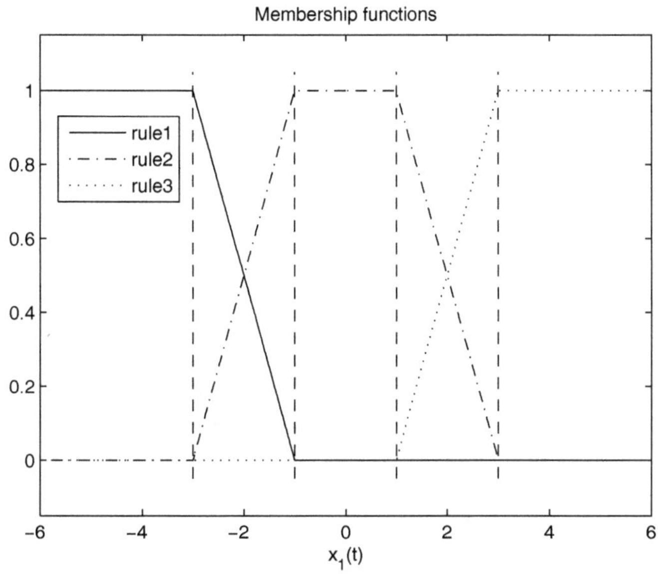

Fig. 2.4 Membership functions for the fuzzy system in Example 2.2

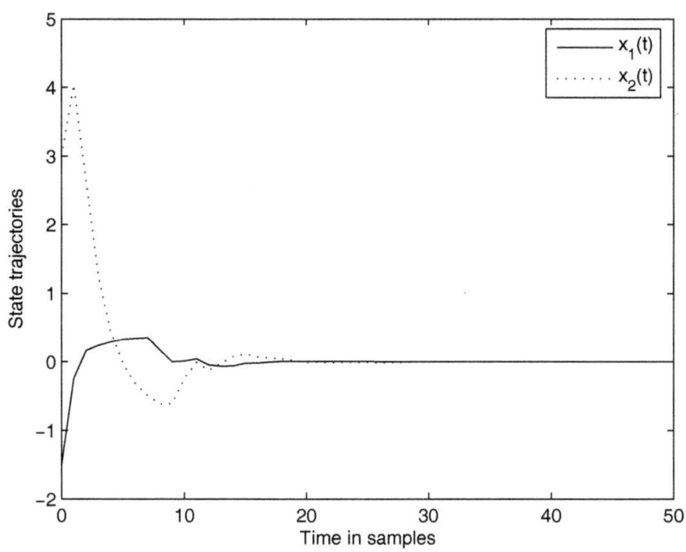

Fig. 2.5 State-trajectories of the closed-loop system in Example 2.2

Table 2.1 $\mathcal{H}_\infty$ performance for different cases with $\tau_1 = 2$, $\tau_2 = 3$ of Example 2.2

methods	$\rho = 0.9$	$\rho = 1.1$	$\rho = 1.3$	$\rho = 1.5$
[37]	1.1092	1.7595	∞	∞
Theorem 2.2	1.0577	1.6258	∞	∞
Theorem 2.1	0.5792	0.7755	1.2684	2.6634
Corollary 2.1	0.5799	0.8028	1.4156	3.5608

If the time-delay interval is reduced, feasible solutions for the several results obtained in this chapter and [37] are obtained, a more detailed comparison of the minimum robust $\mathcal{H}_\infty$ performance indexes $\gamma_{\min}$ obtained based on these methods for variety of cases is summarized in Table 2.1. The results given in this example indicate that when treating the controller design problems for time-delay systems, the delay-independent method sometimes may be more effective than the corresponding delay-dependent ones, because no matrix bounding inequality constraints and/or numerical iteration procedures are involved in the delay-independent controller synthesis case. From the table, it can also be seen that the piecewise Lyapunov-Krasovskii functional-based approach produces less conservative results than the common Lyapunov-Krasovskii functional-based approach.

2.5 Conclusions

In this chapter, based on two novel piecewise Lyapunov-Krasovskii functionals combined with some matrix inequality convexifying techniques, both delay independent and delay-dependent approaches have been developed to study the robust $\mathcal{H}_\infty$ piecewise state-feedback control problem for a class of uncertain discrete-time T-S fuzzy systems with interval-like time-varying state delay. Numerical examples are presented to demonstrate the applicability of the proposed approaches. It is also noted that the proposed methods can be extended to solve the output-feedback controller design problems for discrete-time state-delayed T-S fuzzy systems.

Acknowledgements The work described in this paper was partially supported by a grant from the Research Grants Council of the Hong Kong Special Administrative Region of China under Project CityU-112806.

References

1. L. A. Zadeh, Outline of a new approach to analysis of complex systems and decision processes, *IEEE Transactions on Systems, Man, Cybernetics*, vol. SMC-3, no. 1, pp. 28-44, Jan. 1973.
2. M. Sugeno, *Industrial Applications of Fuzzy Control*. New York: Elsevier, 1985.

3. T. Takagi and M. Sugeno, Fuzzy identification of systems and its application to modeling and control, *IEEE Transactions on Systems, Man, and Cybernetics*, vol. SMC-15, no. 1, pp. 116-132, Jan. 1985.
4. K. Tanaka and M. Sano, A robust stabilization problem of fuzzy control systems and its applications to backing up control of a truck-trailer, *IEEE Transactions on Fuzzy Systems*, vol. 2, no. 2, pp. 119-134, May 1994.
5. K. Tanaka and M. Sugeno, Stability analysis and design of fuzzy control systems, *Fuzzy Sets and Systems*, vol. 45, no. 2, pp. 135-156, Jan. 1992.
6. H. O. Wang, K. Tanaka, and M. F. Griffin, An approach to fuzzy control of nonlinear systems: stability and design issues, *IEEE Transactions on Fuzzy Systems*, vol. 4, no. 1, pp. 14-23, Jan. 1996.
7. T. Tanaka and H. O. Wang, *Fuzzy Control Systems Design and Analysis: A Linear Matrix Inequality Approach*. New York: Wiley, 2001.
8. A. Sala, T. M. Guerra, and R. Babuška, Perspectives of fuzzy systems and control, *Fuzzy Sets and Systems*, vol. 156, no. 3, pp. 432-444, Dec. 2005.
9. G. Feng, A survey on analysis and design of model-based fuzzy control systems, *IEEE Transactions on Fuzzy Systems*, vol. 14, no. 5, pp. 676-697, Oct. 2006.
10. S. G. Cao, N. W. Rees, and G. Feng, Analysis and design for a class of complex control systems, part II: fuzzy controller design, *Automatica*, vol. 33, no.6, pp. 1029-1039, June 1997.
11. H. D. Tuan, P. Apkarian, T. Narikiyo, and Y. Yamamoto, Parameterized linear matrix inequality techniques in fuzzy control system design, *IEEE Transactions on Fuzzy Systems*, vol. 9, no. 2, pp. 324-332, Apr. 2001.
12. M. Johansson, A. Rantzer, and K.-E. Årzén, Piecewise quadratic stability of fuzzy systems, *IEEE Transactions on Fuzzy Systems*, vol. 7, no. 6, pp. 713-722, Dec. 1999.
13. D. J. Choi and P. Park, $\mathscr{H}_\infty$ state-feedback controller design for discrete-time fuzzy systems using fuzzy weighting-dependent Lyapunov functions, *IEEE Transactions on Fuzzy Systems*, vol. 11, no. 2, pp. 271-278, Apr. 2003.
14. G. Feng, Stability analysis of discrete-time fuzzy dynamic systems based on piecewise Lyapunov functions, *IEEE Transactions on Fuzzy Systems*, vol. 12, no. 1, pp. 22-28, Feb. 2004.
15. L. Wang and G. Feng, Piecewise $\mathscr{H}_\infty$ controller design of discrete time fuzzy systems, *IEEE Transactions on Systems, Man, and Cybernetics-Part B: Cybernetics*, vol. 34, no. 1, pp. 682-686, Feb. 2004.
16. T. M. Guerra and L. Vermeiren, LMI-based relaxed nonquadratic stabilization conditions for nonlinear systems in the Takagi-Sugeno's form, *Automatica*, vol. 40, no. 5, pp. 823-829, May 2004.
17. S. Zhou, J. Lam, and W. X. Zheng, Control design for fuzzy systems based on relaxed nonquadratic stability and $\mathscr{H}_\infty$ performance conditions, *IEEE Transactions on Fuzzy Systems*, vol. 15, no. 2, pp. 188-199, Apr. 2007.
18. A. Kruszewski, R. Wang, and T. M. Guerra, Nonquadratic stabilization conditions for a class of uncertain nonlinear discrete time TS fuzzy models: a new approach, *IEEE Transactions on Automatic Control*, vol. 53, no. 2, pp. 606-611, Mar. 2008.
19. C.-L. Chen, G. Feng, D. Sun, and X.-P. Guan, $\mathscr{H}_\infty$ output feedback control of discrete-time fuzzy systems with application to chaos control, *IEEE Transactions on Fuzzy Systems*, vol. 13, no. 4, pp. 531-543, Aug. 2005.
20. H. Gao, Z. Wang, and C. Wang, Improved $\mathscr{H}_\infty$ control of discrete-time fuzzy systems: a cone complementarity linearization approach, *Information Sciences*, vol. 175, no. 1-2, pp. 57-77, Sep. 2005.
21. Y. S. Moon, P. Park, W. H. Kwon, and Y. S. Lee, Delay-dependent robust stabilization of uncertain state-delayed systems, *International Journal of Control*, vol. 74, no. 14, pp. 1447-1455, Sep. 2001.
22. Y. He, M. Wu, J.-H. She, and G.-P. Liu, Parameter-dependent Lyapunov functional for stability of time-delay systems with polytopic-type uncertainties, *IEEE Transactions on Automatic Control*, vol. 49, no. 5, pp. 828-832, May 2004.

23. Y. He, Q.-G Wang, L. Xie, and C. Lin, Further improvement of free-weighting matrices technique for systems with time-varying delay, *IEEE Transactions on Automatic Control*, vol. 52, no. 2, pp. 293-299, Feb. 2007.
24. Y. He, Q.-G. Wang, C. Lin, and M. Wu, Delay-range-dependent stability for systems with time-varying delay, *Automatica*, vol. 43, no. 2, pp. 371-376, Feb. 2007.
25. H. Gao and T. Chen, New results on stability of discrete-time systems with time-varying state delays, *IEEE Transactions on Automatic Control*, vol. 52, no. 2, pp. 328-334, Feb. 2007.
26. H. Gao, T. Chen, and J. Lam, A new delay system approach to network-based control, *Automatica*, vol. 44, no. 1, pp. 39-52, Jan. 2008.
27. X. He, Z. Wang, and D. Zhou, Robust $\mathscr{H}_\infty$ filtering for networked systems with multiple state delays, *International Journal of Control*, vol. 80, no. 8, pp. 1217-1232, Aug. 2007.
28. Y.-Y. Cao and P. M. Frank, Analysis and synthesis of nonlinear time-delay systems via fuzzy control approach, *IEEE Transactions on Fuzzy Systems*, vol. 8, no. 2, pp. 200-211, Apr. 2000.
29. Y.-Y. Cao and P. M. Frank, Stability analysis and synthesis of nonlinear time-delay systems via linear Takagi-Sugeno fuzzy systems, *Fuzzy Sets and Systems*, vol. 124, no. 2, pp. 213-229, Dec. 2001.
30. S. Zhou and T. Li, Robust stabilization for delayed discrete-time fuzzy systems via basis-dependent Lyapunov-Krasovskii functional, *Fuzzy Sets and Systems*, vol. 151, no. 1, pp. 139-153, Apr. 2005.
31. S. Xu and J. Lam, Robust $\mathscr{H}_\infty$ control for uncertain discrete-time-delay fuzzy systems via output feedback, *IEEE Transactions on Fuzzy Systems*, vol. 13, no. 1, pp. 82-93, Feb. 2005.
32. C.-L. Chen, G. Feng, and X.-P. Guan, Delay-dependent stability analysis and controller synthesis for discrete-time T-S Fuzzy systems with time-delays, *IEEE Transactions on Fuzzy Systems*, vol. 13, no. 5, pp. 630-643, Oct. 2005.
33. C.-L. Chen and G. Feng, Delay-dependent piecewise control for time-delay T-S fuzzy systems with application to chaos control, In *Proceedings of the 14th IEEE International Conference on Fuzzy Systems*, Reno, Nevada, USA, May 2005, pp. 1056-1061.
34. S.-S. Chen, Y.-C. Chang, S.-F. Su, S.-L. Chung, and T.-T. Lee, Robust static output-feedback stabilization for nonlinear discrete-time systems with time-delay via fuzzy control approach, *IEEE Transactions on Fuzzy Systems*, vol. 13, no. 2, pp. 263-272, Apr. 2005.
35. C.-S. Tseng, Model reference output feedback fuzzy tracking control design for nonlinear discrete-time systems with time-delay, *IEEE Transactions on Fuzzy Systems*, vol. 14, no. 1, pp. 58-70, Feb. 2006.
36. H.-N. Wu, Delay-dependent stability analysis and stabilization for discrete-time fuzzy systems with state delays: a fuzzy Lyapunov-Krsovskii functional approach, *IEEE Transactions on Systems, Man, and Cybernetics-Part B: Cybernetics*, vol. 36, no. 4, pp. 954-962, Aug. 2006.
37. B. Zhang and S. Xu, Delay-dependent robust $\mathscr{H}_\infty$ control for uncertain discrete-time fuzzy systems with time-varying delays, *IEEE Transactions on Fuzzy Systems*, to be published.
38. X.-P. Guan and C. Chen, Delay-dependent guaranteed cost control for T-S fuzzy systems with time-delays, *IEEE Transactions on Fuzzy Systems*, vol. 12, no. 2, pp. 236-249. Apr. 2004.
39. H.-N. Wu and H.-X. Li, New approach to delay-dependent stability analysis and stabilization for continuous-time fuzzy systems with time-varying delay, *IEEE Transactions on Fuzzy Systems*, vol. 15, no. 3, pp. 482-493, June 2007.
40. J. Yoneyama, New delay-dependent approach to robust stability and stabilization for Takagi-Sugeno fuzzy time-delay systems, *Fuzzy Sets and Systems*, vol. 158, no. 20, pp. 2225-2237, Oct. 2007.
41. C. Lin, Q.-G. Wang, T. H. Lee, Y. He, and B. Chen, Observer-based $\mathscr{H}_\infty$ control for T-S fuzzy systems with time-delay: delay-dependent design method, *IEEE Transactions on Systems, Man, and Cybernetics-Part B: Cybernetics*, vol. 37, no. 4, pp. 1030-1038, Aug. 2007.
42. B. Chen, X. Liu, S. Tong, and C. Lin, Guaranteed cost control of T-S fuzzy systems with state and input delays, *Fuzzy Sets and Systems*, vol. 158, no. 20, pp. 2251-2267, Oct. 2007.
43. X. Jiang and Q.-L. Han, Robust $\mathscr{H}_\infty$ control for uncertain Takagi-Sugeno fuzzy systems with interval time-varying delay, *IEEE Transactions on Fuzzy Systems*, vol. 15, no. 2, pp. 321-331, Apr. 2007.

44. C. Abdallah, P. Dorato, F. Pérez, and D. Docampo, Controller synthesis for a class of interval plants, *Automatica*, vol. 31, no. 2, pp. 341-343, Feb. 1995.
45. L. Xie, Output feedback $\mathcal{H}_\infty$ control of systems with parameter uncertainty, *International Journal of Control*, vol. 63, no. 4, pp. 741-750, Mar. 1996.
46. L. El Ghaoui and G. Scorletti, Control of rational systems using linear-fractional representations and linear matrix inequalities, *Automatica*, vol. 32, no. 9, pp. 1273-1284, Sep. 1996.
47. L. El Ghaoui, F. Oustry, and M. AitRami, A cone complementarity linearization algorithm for static output-feedback and related problems, *IEEE Transactions on Automatic Control*, vol. 42, no.8, pp. 1171-1176, Aug. 1997.

Chapter 3
$\mathscr{H}_\infty$ Fuzzy Control for Systems with Repeated Scalar Nonlinearities

Hongli Dong, Zidong Wang and Huijun Gao

Abstract This paper is concerned with the H_∞ control problem for a class of discrete-time Takagi-Sugeno (T-S) fuzzy systems with repeated scalar nonlinearities. A modified T-S fuzzy model is proposed in which the consequent parts are composed of a set of discrete-time state equations containing a repeated scalar nonlinearity. Such a model can describe some well-known nonlinear systems such as recurrent neural networks. Attention is focused on the analysis and design of $\mathscr{H}_\infty$ fuzzy controllers with the same repeated scalar nonlinearities such that the closed-loop T-S fuzzy control system is asymptotically stable and preserves a guaranteed H_∞ performance. Sufficient conditions are obtained for the existence of admissible controllers, and the cone complementarity linearization (CCL) procedure is employed to cast the controller design problem into a sequential minimization one subject to linear matrix inequalities (LMIs), which can be solved efficiently by using existing optimization techniques. A illustrative example is provided to demonstrate the effectiveness of the results proposed in this chapter.

3.1 Introduction

Over the past few decades, the nonlinear theory has been extensively studied by many researchers for the reason that nonlinearities exist universally in practice;

Hongli Dong
Space Control and Inertial Technology Research Center, Harbin Institute of Technology, Harbin 150001, P.R. China, e-mail: shiningdhl@gmail.com

Zidong Wang
Department of Information Systems and Computing, Brunel University, Uxbridge, Middlesex, UB8 3PH, United Kingdom, e-mail: zidong.wang@brunel.ac.uk

Huijun Gao
Space Control and Inertial Technology Research Center, Harbin Institute of Technology, Harbin 150001, P.R. China, e-mail: hjgao@hit.edu.cn

see [13,22–25]. It is well known that Takagi-Sugeno (T-S) fuzzy models can provide an effective representation of complex nonlinear systems in terms of fuzzy sets and fuzzy reasoning applied to a set of linear input-output submodels, which has therefore received a rapidly growing interest in the literature; see [2,9–11,20,21,29]. In particular, the control technique based on the so-called T-S fuzzy model has attracted much attention; see [1,3,8,14,17–19,26,28]. The procedure based on this technique is as follows. The nonlinear plant is represented by a T-S type fuzzy model. In this type of fuzzy model, local dynamics in different state space regions are represented by linear models. The overall fuzzy model of the system is achieved by smoothly blending these local models together through membership functions. The control design is carried out based on the fuzzy model via the so-called parallel distributed compensation scheme. The idea is that, for each local linear model, a linear feedback control is designed. The resulting overall controller, which is nonlinear in general, is again a fuzzy blending of each individual linear controller.

The T-S fuzzy model has been widely employed to approximate a nonlinear system, which is described by a family of fuzzy IF-THEN rules that represent local linear input-output relations of the system. Nevertheless, the local model is not necessarily a linear one but sometimes a "simple" or "slightly" nonlinear system whose dynamics can be thoroughly investigated. A good example of such a simple nonlinear system is the recurrent neural network that involves a nonlinear but known activation function. Therefore, there has appeared initial research interest focusing on the extended T-S model whose system dynamics is captured by a set of fuzzy implications which characterize local relations in the state space [15,16,27]. In this case, the local dynamics of each fuzzy rule is expressed by a well-studied nonlinear system, and the overall fuzzy model can be achieved by fuzzy "blending" of these simple nonlinear local systems. For example, a modified T-S fuzzy model has been proposed in [16] in which the consequent parts are composed of a set of stochastic Hopfield neural networks with time-varying delays, and a stability criterion has been derived in terms of LMIs. The results of [16] have then been extended in [15] to deal with the stability analysis problem for T-S fuzzy cellular neural networks with time-varying delays. Motivated by the works in [15,16,27], in this chapter, we will consider a more general yet well-known nonlinearity, namely, repeated scalar nonlinearity [4,6,12], which covers some typical classes of nonlinearities such as the semilinear function, the hyperbolic tangent function that has been extensively used for activation function in neural networks, and the sine function, etc.

In this chapter, we consider the $\mathscr{H}_\infty$ control problem for discrete-time T-S fuzzy systems with repeated scalar nonlinearities. The nonlinear system to be considered is described by a discrete-time state equation involving a repeated scalar nonlinearity which typically appears in recurrent neural networks. We aim to analyze and design a fuzzy controller such that the closed-loop fuzzy control system is asymptotically stable and preserves a guaranteed $\mathscr{H}_\infty$ performance. Sufficient conditions that involve matrix equalities are obtained for the existence of admissible controllers, and the CCL procedure is employed to cast the nonconvex feasibility problem into a sequential minimization problem subject to LMIs, which can then be readily solved by using existing optimization techniques. An explicit expression of the desired state feedback controller is also given. A numerical example is provided to show the effectiveness of the proposed method.

The notation used in the chapter is fairly standard. The superscript "T" stands for matrix transposition, $\mathbb{R}^n$ denotes the n-dimensional Euclidean space, $\mathbb{R}^{m\times n}$ is the set of all real matrices of dimension $m \times n$, I and 0 represent the identity matrix and zero matrix, respectively. The notation $P > 0$ means that P is real, symmetric, and positive definite; $\mathrm{tr}(M)$ refers to the trace of the matrix M; $l_2[0,\infty)$ is the space of square-integrable vector functions over $[0,\infty)$; the notation $\|A\|$ refers to the norm of a matrix A defined by $\|A\| = \sqrt{\mathrm{tr}(A^T A)}$ and $\|\cdot\|_2$ stands for the usual l_2 norm. In symmetric block matrices or complex matrix expressions, we use an asterisk $(*)$ to represent a term that is induced by symmetry, and diag $\{\dots\}$ stands for a block-diagonal matrix. Matrices, if their dimensions are not explicitly stated, are assumed to be compatible for algebraic operations.

3.2 Problem Formulation

In this section, we consider a T-S fuzzy system with repeated scalar nonlinearities. We first describe the physical plant and the associated controller in term of fuzzy rules, and then formulate the problem investigated in this chapter.

3.2.1 The Physical Plant

In this paper, we consider the following discrete-time fuzzy systems with repeated scalar nonlinearities:

Plant Rule i: IF $\theta_1(k)$ is M_{i1} and $\theta_2(k)$ is M_{i2} and $\cdots$ and $\theta_p(k)$ is M_{ip} THEN

$$\left.\begin{aligned} x_{k+1} &= A_i f(x_k) + B_{2i}u_k + B_{1i}w_k, \\ z_k &= C_i f(x_k) + D_{2i}u_k + D_{1i}w_k, \\ i &= 1,\dots,r. \end{aligned}\right\} \tag{3.1}$$

where M_{ij} is the fuzzy set; $x_k \in \mathbb{R}^n$ represents the state vector; $u_k \in \mathbb{R}^m$ is the input vector; $w_k \in \mathbb{R}^p$ is the exogenous disturbance input which belongs to $l_2[0,\infty)$; $z_k \in \mathbb{R}^q$ is the controlled output; $A_i, B_{2i}, B_{1i}, C_i, D_{2i}$ and D_{1i} are all constant matrices with compatible dimensions; r is the number of IF-THEN rules; $\theta_k = [\theta_1(k),\ \theta_2(k),\cdots,\theta_p(k)]$ is the premise variable vector. It is assumed that the premise variables do not depend on the input variable u_k, which is needed to avoid a complicated defuzzification process of fuzzy controllers. f is a nonlinear function satisfying the following assumption as in [5].

Assumption 3.1 *The nonlinear function $f : \mathbb{R} \to \mathbb{R}$ in system (3.1) satisfies*

$$\forall a, b \in \mathbb{R} \quad |f(a) + f(b)| \le |a + b|. \tag{3.2}$$

In the sequel, for the vector $x = [x_1 \; x_2 \cdots x_n]^T$, we denote

$$f(x) \triangleq [f(x_1) \; f(x_2) \; \cdots \; f(x_n)]^T.$$

Remark 3.1. The model (3.1) is called a fuzzy system with repeated scalar nonlinearity [4, 6, 12]. Note that f is odd (by putting $b = -a$) and 1-Lipschitz (by putting $b = -b$). Therefore, f encapsulates some typical classes of nonlinearities, such as:

- the semilinear function (i.e., the standard saturation $sat(s) := s$ if $|s| \leq 1$ and $sat(s) := sgn(s)$ if $|s| > 1$);
- the hyperbolic tangent function that has been extensively used for activation function in neural networks;
- the sine function, etc.

Given a pair of $(x_k, \; u_k)$, the final outputs of the fuzzy system are inferred as follows:

$$\left.\begin{aligned} x_{k+1} &= \sum_{i=1}^{r} h_i(\theta_k)[A_i f(x_k) + B_{2i} u_k + B_{1i} w_k], \\ z_k &= \sum_{i=1}^{r} h_i(\theta_k)[C_i f(x_k) + D_{2i} u_k + D_{1i} w_k], \end{aligned}\right\} \tag{3.3}$$

where the fuzzy basis functions are given by

$$h_i(\theta_k) = \frac{\vartheta_i(\theta_k)}{\sum_{i=1}^{r} \vartheta_i(\theta_k)},$$

with $\vartheta_i(\theta_k) = \prod_{j=1}^{p} M_{ij}(\theta_j(k))$. $M_{ij}(\theta_j(k))$ representing the grade of membership of $\theta_j(k)$ in M_{ij}. Here, $\vartheta_i(\theta_k)$ has the following basic property:

$$\vartheta_i(\theta_k) \geq 0, \; i = 1, 2, ..., r, \quad \sum_{i=1}^{r} \vartheta_i(\theta_k) \geq 0, \; \forall k,$$

and therefore

$$h_i(\theta_k) \geq 0, \; i = 1, 2, ..., r, \quad \sum_{i=1}^{r} h_i(\theta_k) = 1, \; \forall k.$$

3.2.2 Controller

In this paper, we consider the following fuzzy control law for the fuzzy system (3.3):

Controller Rule i: IF $\theta_1(k)$ **is** M_{i1} **and** $\theta_2(k)$ **is** M_{i2} **and** $\cdots$ **and** $\theta_p(k)$ **is** M_{ip} **THEN**

$$u_k = K_i f(x_k), \quad i = 1, 2, ..., r.$$

Here, $f(x_k) \in R^n$ is the input to the controller; $u_k \in R^m$ is the output of the controller; K_i are the gain matrices of the controller to be designed. Hence, the controller can be represented by the following input-output form:

$$u_k = \sum_{i=1}^{r} h_i(\theta_k) K_i f(x_k). \tag{3.4}$$

3.2.3 Closed-loop System

The closed-loop T-S fuzzy control system can now be obtained from (3.3) and (3.4) that

$$\left.\begin{aligned} x_{k+1} &= \sum_{i=1}^{r}\sum_{j=1}^{r} h_i(\theta_k)h_j(\theta_k)[A_{ij}f(x_k) + B_{1i}w_k], \\ z_k &= \sum_{i=1}^{r}\sum_{j=1}^{r} h_i(\theta_k)h_j(\theta_k)[C_{ij}f(x_k) + D_{1i}w_k], \end{aligned}\right\} \tag{3.5}$$

where $A_{ij} = A_i + B_{2i}K_j, C_{ij} = C_i + D_{2i}K_j$.

Before formulating the problem to be investigated, we first introduce the following definitions and lemma.

Definition 3.1. Given a scalar $\gamma > 0$, the closed-loop T-S fuzzy control system in (3.5) is said to be asymptotically stabe with a guaranteed H_∞ performance γ if it is asymptotically stable in the large when $w_k \equiv 0$, and under zero initial condition, for all nonzero$\{w_k\} \in l_2[0,\infty)$, the controlled output z_k satisfies

$$||z||_2 \leq \gamma ||w||_2. \tag{3.6}$$

Definition 3.2. A square matrix $P \triangleq [p_{ij}] \in \mathbb{R}^{n\times n}$ is called diagonally dominant if for all $i = 1,\cdots,n$

$$p_{ii} \geqslant \sum_{j\neq i} |p_{ij}|. \tag{3.7}$$

Lemma 3.1. *[5] If $P > 0$ is diagonally dominant, then for all nonlinear functions f satisfying Assumption (3.2), the following inequality holds for all $x_k \in \mathbb{R}^n$:*

$$f^T(x_k)Pf(x_k) \leq x_k^T P x_k. \tag{3.8}$$

Remark 3.2. It will be seen later that the purpose of requiring the matrix P to satisfy (3.8) is to admit the quadratic Lyapunov function $V(x_k) = x_k^T P x_k$.

The problem under consideration is to design a controller of the form (3.4) such that the closed-loop system (3.5) is asymptotically stable with $\mathscr{H}_\infty$ performance index γ.

3.3 $\mathscr{H}_\infty$ Fuzzy Control Performance Analysis

In this section, a condition guaranteeing $\mathscr{H}_\infty$ performance is established for T-S fuzzy systems via a quadratic approach described in the following theorem.

Theorem 3.1. *Given a scalar $\gamma > 0$ and supposing the gain matrices K_i $(i = 1, ..., r)$ of the controllers in (3.4) are given. The closed-loop fuzzy control system (3.5) composed of (3.3) and (3.4) is asymptotically stabe with an $\mathscr{H}_\infty$ performance γ if there exists a positive diagonally dominant matrix P satisfying*

$$G_{ii}^T P G_{ii} + H_{ii}^T H_{ii} - \bar{L} < 0, \ \ i = 1, 2, ..., r, \tag{3.9}$$

$$(G_{ij} + G_{ji})^T P (G_{ij} + G_{ji}) + (H_{ij} + H_{ji})^T (H_{ij} + H_{ji}) - 4\bar{L} < 0, \ \ 1 \leq i < j \leq r, \tag{3.10}$$

where

$$G_{ij} = \begin{bmatrix} A_{ij} & B_{1i} \end{bmatrix}, \ \ H_{ij} = \begin{bmatrix} C_{ij} & D_{1i} \end{bmatrix}, \ \ \bar{L} = \begin{bmatrix} P & 0 \\ 0 & \gamma^2 I \end{bmatrix}.$$

Proof. In order to show that the fuzzy system in (3.5) is asymptotically stable with a guaranteed $\mathscr{H}_\infty$ performance γ under conditions (3.9)-(3.10), we define the following Lyapunov function candidate

$$V(x_k) = x_k^T P x_k. \tag{3.11}$$

When $w_k \equiv 0$, the difference of the Lyapunov function is calculated as

$$\begin{aligned}
\Delta V(x_k) =& \ V(x_{k+1}) - V(x_k) \\
=& \ f^T(x_k) \left[\sum_{i=1}^{r} \sum_{j=1}^{r} h_i(\theta_k) h_j(\theta_k) A_{ij} \right]^T P \left[\sum_{i=1}^{r} \sum_{j=1}^{r} h_i(\theta_k) h_j(\theta_k) A_{ij} \right] f(x_k) - x_k^T P x_k \\
=& \ f^T(x_k) \left[\sum_{i=1}^{r} \sum_{j=1}^{r} \sum_{s=1}^{r} \sum_{t=1}^{r} h_i(\theta_k) h_j(\theta_k) h_s(\theta_k) h_t(\theta_k) A_{ij}^T P A_{st} \right] f(x_k) - x_k^T P x_k \\
\leq& \ f^T(x_k) \left[\sum_{i=1}^{r} \sum_{j=1}^{r} h_i(\theta_k) h_j(\theta_k) A_{ij}^T P A_{ij} \right] f(x_k) - x_k^T P x_k.
\end{aligned}$$

According to Lemma 3.1, we have

$$\begin{aligned}
\Delta V(x_k) \leq& \ f^T(x_k) \left[\sum_{i=1}^{r} \sum_{j=1}^{r} h_i(\theta_k) h_j(\theta_k) (A_{ij}^T P A_{ij} - P) \right] f(x_k) \\
=& \ f^T(x_k) \sum_{i=1}^{r} h_i^2(\theta_k) (A_{ii}^T P A_{ii} - P) f(x_k) \\
&+ \frac{1}{2} f^T(x_k) \sum_{i,j=1, i<j}^{r} h_i(\theta_k) h_j(\theta_k) \left[(A_{ij} + A_{ji})^T P (A_{ij} + A_{ji}) - 4P \right] f(x_k).
\end{aligned}$$

On the other hand, by the Schur complement, inequalities (3.9) and (3.10) imply that

$$A_{ii}^T P A_{ii} - P < 0, \; i = 1, 2, ..., r,$$
$$(A_{ij} + A_{ji})^T P (A_{ij} + A_{ji}) - 4P < 0, \;\; 1 \le i < j \le r.$$

Thus, we have

$$\Delta V(x_k) < 0,$$

and then the closed-loop system (3.5) is asymptotically stable in the large when $w_k \equiv 0$.

Next, the $\mathscr{H}_\infty$ performance criteria for the closed-loop system in (3.5) will be established. Assuming zero initial conditions, we consider the following index:

$$J = \; V(x_{k+1}) + z_k^T z_k - \gamma^2 w_k^T w_k - f^T(x_k) P f(x_k).$$

Defining

$$\eta_k = \left[f^T(x_k) \; w_k^T \right]^T,$$

we have

$$\begin{aligned}
J = & \left[\sum_{i=1}^r \sum_{j=1}^r h_i(\theta_k) h_j(\theta_k) G_{ij} \eta_k \right]^T P \left[\sum_{i=1}^r \sum_{j=1}^r h_i(\theta_k) h_j(\theta_k) G_{ij} \eta_k \right] \\
& + \left[\sum_{i=1}^r \sum_{j=1}^r h_i(\theta_k) h_j(\theta_k) H_{ij} \eta_k \right]^T \left[\sum_{i=1}^r \sum_{j=1}^r h_i(\theta_k) h_j(\theta_k) H_{ij} \eta_k \right] - \eta_k^T \bar{L} \eta_k \\
\le & \; \eta_k^T \sum_{i=1}^r \sum_{j=1}^r h_i(\theta_k) h_j(\theta_k) G_{ij}^T P G_{ij} \eta_k + \eta_k^T \sum_{i=1}^r \sum_{j=1}^r h_i(\theta_k) h_j(\theta_k) H_{ij}^T H_{ij} \eta_k - \eta_k^T \bar{L} \eta_k \\
= & \; \eta_k^T \sum_{i=1}^r h_i^2(\theta_k) (G_{ii}^T P G_{ii} + H_{ii}^T H_{ii} - \bar{L}) \eta_k \\
& + \frac{1}{2} \eta_k^T \sum_{i,j=1, i<j}^r h_i(\theta_k) h_j(\theta_k) [(G_{ij} + G_{ji})^T P (G_{ij} + G_{ji}) \\
& + (H_{ij} + H_{ji})^T (H_{ij} + H_{ji}) - 4\bar{L}] \eta_k.
\end{aligned}$$

From inequalities (3.9) and (3.10), we know that $J \le 0$ and, according to Lemma 3.1, we have

$$x_{k+1}^T P x_{k+1} + z_k^T z_k - \gamma^2 w_k^T w_k - x_k^T P x_k \le 0.$$

For $k = 0, 1, 2, \ldots$, summing up both sides of the above under zero initial condition and considering $x_\infty^T P x_\infty \ge 0$, we arrive at

$$\sum_{k=0}^{\infty} z_k^T z_k - \sum_{k=0}^{\infty} \gamma^2 w_k^T w_k \le 0,$$

which is equivalent to (3.6). The proof is now complete. □

Remark 3.3. In Theorem 3.1, with given controller gain and disturbance attenuation level γ, we obtain the asymptotic stability conditions of the nominal fuzzy system (3.5), which are represented via a set of matrix inequalities in (3.9)-(3.10). We will show later in the next section that such inequalities can be converted into LMIs when designing the actual controllers. Note that the feasibility of LMIs can be easily checked by using the LMI toolbox of MATLAB®.

3.4 $\mathscr{H}_\infty$ Fuzzy Controller Design

The previous section presents an $\mathscr{H}_\infty$ performance criterion for T-S fuzzy control systems. An immediate question is whether this performance condition can be further extended to cope with the $\mathscr{H}_\infty$ control problem. To this end, in this section, we aim at designing a controller in the form of (3.4) based on Theorem 3.1. That is, we are interested in determining the controller parameters such that the closed-loop fuzzy system in (3.5) is asymptotically stable with a guaranteed $\mathscr{H}_\infty$ performance. The following theorem provides sufficient conditions for the existence of such a $\mathscr{H}_\infty$ fuzzy controller for system (3.5).

Theorem 3.2. *Consider the fuzzy system in (3.3). There exists a state feedback controller in the form of (3.4) such that the closed-loop system in (3.5) is asymptotically stable with a guaranteed $\mathscr{H}_\infty$ performance γ, if there exist matrices $0 < P \triangleq [p_{ij}]$, $L > 0$, M_i, $i = 1,\dots,r$, $R = R^T \triangleq [r_{ij}]$ satisfying*

$$\begin{bmatrix} -L & * & * & * \\ 0 & -\gamma^2 I & * & * \\ A_iL + B_{2i}M_i & B_{1i} & -L & * \\ C_iL + D_{2i}M_i & D_{1i} & 0 & -I \end{bmatrix} < 0, \quad i = 1,2,\dots,r, \tag{3.12}$$

$$\begin{bmatrix} -4L & * & * & * \\ 0 & -4\gamma^2 I & * & * \\ (A_i + A_j)L + B_{2i}M_j + B_{2j}M_i & B_{1i} + B_{1j} & -L & * \\ (C_i + C_j)L + D_{2i}M_j + D_{2j}M_i & D_{1i} + D_{1j} & 0 & -I \end{bmatrix} < 0, \quad 1 \le i < j \le r, \tag{3.13}$$

$$p_{ii} - \sum_{j \neq i}(p_{ij} + 2r_{ij}) \ge 0, \tag{3.14}$$

$$r_{ij} \ge 0 \qquad \forall i \neq j, \tag{3.15}$$

$$p_{ij} + r_{ij} \ge 0 \quad \forall i \neq j, \tag{3.16}$$

$$PL = I. \tag{3.17}$$

Furthermore, if the above conditions have feasible solutions, the matrix gains for the admissible controller in (3.4) are given by

$$K_i = M_i L^{-1}. \tag{3.18}$$

Proof. From Theorem 3.1, we know that the closed-loop system in (3.5) is asymptotically stable with a guaranteed $\mathscr{H}_\infty$ performance γ if there exists a positive diagonally dominant matrix P satisfying (3.9) and (3.10). By the Schur complement, the following inequalities are obtained:

$$\begin{bmatrix} -P & * & * & * \\ 0 & -\gamma^2 I & * & * \\ A_i + B_{2i}K_i & B_{1i} & -P^{-1} & * \\ C_i + D_{2i}K_i & D_{1i} & 0 & -I \end{bmatrix} < 0, \tag{3.19}$$

$$\begin{bmatrix} -4P & * & * & * \\ 0 & -4\gamma^2 I & * & * \\ A_i + A_j + B_{2i}K_j + B_{2j}K_i & B_{1i} + B_{1j} & -P^{-1} & * \\ C_i + C_j + D_{2i}K_j + D_{2j}K_i & D_{1i} + D_{1j} & 0 & -I \end{bmatrix} < 0. \tag{3.20}$$

Performing congruence transformations to inequalities (3.19) and (3.20) by diag $\{P^{-1}, I, I, I\}$, we have

$$\begin{bmatrix} -P^{-1} & * & * & * \\ 0 & -\gamma^2 I & * & * \\ A_iP^{-1} + B_{2i}K_iP^{-1} & B_{1i} & -P^{-1} & * \\ C_iP^{-1} + D_{2i}K_iP^{-1} & D_{1i} & 0 & -I \end{bmatrix} < 0,$$

$$\begin{bmatrix} -4P^{-1} & * & * & * \\ 0 & -4\gamma^2 I & * & * \\ (A_i + A_j)P^{-1} + B_{2i}K_jP^{-1} + B_{2j}K_iP^{-1} & B_{1i} + B_{1j} & -P^{-1} & * \\ (C_i + C_j)P^{-1} + D_{2i}K_jP^{-1} + D_{2j}K_iP^{-1} & D_{1i} + D_{1j} & 0 & -I \end{bmatrix} < 0.$$

Defining $L = P^{-1}$ and $M_i = K_iP^{-1}$, we can obtain (3.12) and (3.13) readily. Furthermore, from (3.14)-(3.16), we have

$$p_{ii} \geq \sum_{j \neq i} (p_{ij} + 2r_{ij}) = \sum_{j \neq i} (|p_{ij} + r_{ij}| + |-r_{ij}|) \geq \sum_{j \neq i} |p_{ij}|,$$

which guarantees the positive definite matrix P to be diagonally dominant, and the proof is then complete. $\square$

Remark 3.4. It is worth noting that, by far, we are unable to apply the LMI approach in the design of controller because of the matrix equality in Theorem 3.2. Fortunately, this problem can be addressed with help from the CCL algorithm proposed in [7]. We can solve this nonconvex feasibility problem by formulating it into a sequential optimization problem subject to LMI constraints.

The basic idea behind the CCL algorithm is that if the LMI $\begin{bmatrix} P & I \\ I & L \end{bmatrix} \geq 0$ is feasible in the $n \times n$ matrix variables $L > 0$ and $P > 0$, then $tr(PL) \geq n$; and $tr(PL) = n$ if and only if $PL = I$. Based on this, it is likely to solve the equalities in (3.17) by applying the CCL algorithm. In view of this observation, we put forward the following nonlinear minimization problem involving LMI conditions instead of the original nonconvex feasibility problem formulated in Theorem 3.2.

The nonlinear minimization problem is, min $tr(PL)$ subject to (3.12)-(3.16) and

$$\begin{bmatrix} P & I \\ I & L \end{bmatrix} \geq 0. \tag{3.21}$$

If the solution of min $tr(PL)$ subject to (3.12)-(3.16) exists and min$tr(PL) = n$, then the conditions in Theorem 3.2 are solvable. Finally, the following algorithm is suggested to solve the above problem.

Algorithm 3.1 *(HinfFC: $\mathscr{H}_\infty$ Fuzzy Control)*

Step 1. *Find a feasible set $(P_{(0)}, L_{(0)}, M_{i(0)}, R_{(0)})$ satisfying (3.12)-(3.16) and (3.21). Set q= 0.*

Step 2. *According to (3.12)-(3.16) and (3.21), solve the LMI problem:* min $tr(PL_{(q)} + P_{(q)}L)$.

Step 3. *If the stopping criterion is satisfied, then output the feasible solutions (P, L, M_i, R) and exit. Else, set $q = q+1$ and go to Step 2.*

3.5 An Illustrative Example

In this section, a simulation example is presented to illustrate the fuzzy controller design method developed in this chapter.

Consider a T-S fuzzy model (3.1) with repeated scalar nonlinearities. The rules are given as follows:

Plant Rule 1: IF $f_1(x_k)$ **is** $h_1(f_1(x_k))$ **THEN**

$$\begin{aligned} x_{k+1} &= A_1 f(x_k) + B_{21}u_k + B_{11}w_k, \\ z_k &= C_1 f(x_k) + D_{21}u_k + D_{11}w_k. \end{aligned} \tag{3.22}$$

Plant Rule 2: IF $f_1(x_k)$ **is** $h_2(f_1(x_k))$ **THEN**

$$\begin{aligned} x_{k+1} &= A_2 f(x_k) + B_{22}u_k + B_{12}w_k, \\ z_k &= C_2 f(x_k) + D_{22}u_k + D_{12}w_k. \end{aligned} \tag{3.23}$$

Controller Rule 1: IF $f_1(x_k)$ **is** $h_1(f_1(x_k))$ **THEN** $u_k = K_1 f(x_k)$,

Controller Rule 2: IF $f_1(x_k)$ **is** $h_2(f_1(x_k))$ **THEN** $u_k = K_2 f(x_k)$.

The final outputs of the fuzzy system are inferred as follows:

$$\begin{aligned} x_{k+1} &= \sum_{i=1}^{2} h_i(f_1(x_k))[A_i f(x_k) + B_{2i}u_k + B_{1i}w_k], \\ z_k &= \sum_{i=1}^{2} h_i(f_1(x_k))[C_i f(x_k) + D_{2i}u_k + D_{1i}w_k]. \end{aligned} \tag{3.24}$$

The model parameters are given as follows:

$$A_1 = \begin{bmatrix} 1.0 & 0.31 & 0 \\ 0 & 0.33 & 0.21 \\ 0 & 0 & -0.52 \end{bmatrix}, \ B_{11} = \begin{bmatrix} 0.1 \\ 0 \\ 0 \end{bmatrix}, \ D_{11} = \begin{bmatrix} 0.15 \\ 0 \\ 0 \end{bmatrix}, \ B_{21} = \begin{bmatrix} 1 & 1 \\ 0 & 1 \\ 0 & 1 \end{bmatrix},$$

$$A_2 = \begin{bmatrix} 0.8 & -0.38 & 0 \\ -0.2 & 0 & 0.21 \\ 0.1 & 0 & -0.55 \end{bmatrix}, \ B_{12} = \begin{bmatrix} 0 \\ 0.12 \\ 0 \end{bmatrix}, \ D_{21} = \begin{bmatrix} 1 & 1 \\ 0 & 1 \\ 0 & 1 \end{bmatrix}, \ B_{22} = \begin{bmatrix} 1 & 0 \\ 0 & 1 \\ 0 & 1 \end{bmatrix},$$

$$C_1 = \begin{bmatrix} 0.2 & 0 & 0 \\ 0 & 0 & 0 \\ 0 & 0 & 0.1 \end{bmatrix}, \ D_{12} = \begin{bmatrix} 0 \\ 0 \\ 0.22 \end{bmatrix}, \ C_2 = \begin{bmatrix} -0.12 & 0 & 0.1 \\ 0 & 0 & 0 \\ 0 & 0 & 0.1 \end{bmatrix}, \ D_{22} = \begin{bmatrix} 1 & 1 \\ 0 & 1 \\ 0 & 1 \end{bmatrix}.$$

Let the $\mathscr{H}_\infty$ performance γ be specified as 0.3 and the membership function is assumed to be

$$h_1(f_1(x_k)) = \begin{cases} 1, & f_1(x_0) = 0 \\ \sin(f_1(x_0))/f_1(x_0), & else \end{cases}, \quad h_2(f_1(x_k)) = 1 - h_1(f_1(x_k)). \tag{3.25}$$

In the above parameters, the nonlinear function $f(x_k) = \sin(x_k)$ satisfies Assumption (3.2). Our aim is to design a state-feedback paralleled controller in the form of (3.4) such that the system (3.24) is asymptotically stable with a guaranteed $\mathscr{H}_\infty$ norm bound γ.

By applying Theorem 3.2 with help from Algorithm *HinfFC*, we can obtain admissible solutions as follows:

$$L = \begin{bmatrix} 0.5754 & -0.0091 & -0.0049 \\ -0.0091 & 2.7669 & 0.8980 \\ -0.0049 & 0.8980 & 0.9163 \end{bmatrix}, \quad P = \begin{bmatrix} 1.7435 & 0.0039 & 0.0061 \\ 0.0039 & 0.5314 & -0.5210 \\ 0.0061 & -0.5210 & 1.6049 \end{bmatrix},$$

$$K_1 = \begin{bmatrix} -0.3202 & -0.0223 & -0.2328 \\ -0.0167 & -0.0652 & 0.1913 \end{bmatrix}, \ K_2 = \begin{bmatrix} -0.0049 & 0.0717 & -0.2900 \\ -0.0163 & -0.0061 & 0.1992 \end{bmatrix}.$$

First, we assume the initial condition to be

$$x_0 = [0 \ 0.01 \ 0.04]^T, \tag{3.26}$$

and the external disturbance $w_k \equiv 0$. Figure 3.1 gives the state responses for the uncontrolled fuzzy systems, which are apparently unstable. Figure 3.2 gives the state simulation results of the closed-loop fuzzy system, from which we can see that the closed-loop system is asymptotically stable.

Next, to illustrate the disturbance attenuation performance, we choose the initial condition as in (3.26) and the external disturbance w_k is assumed to be

$$w_k = \begin{cases} 0.2 & 20 \le k \le 30 \\ -0.2 & 40 \le k \le 50 \\ 0 & else \end{cases}$$

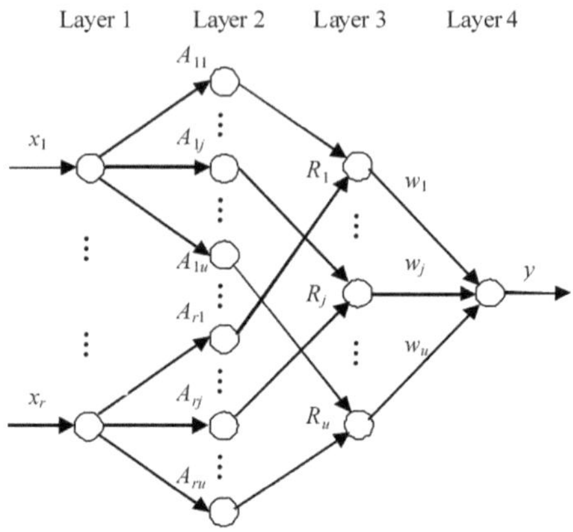

Fig. 3.1 The state evolution x_k of uncontrolled systems

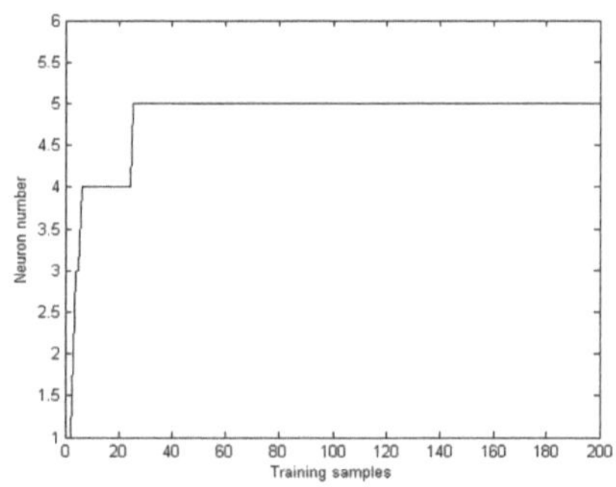

Fig. 3.2 The state evolution x_k of controlled systems

Figure 3.3 shows the controller output, and Figure 3.4 shows the evolution of the state variables. The disturbance input w_k and controlled output z_k are depicted in Figure 3.5. By simple computation, it is found that $\|z\|_2 = 0.1002$ and $\|w\|_2 = 0.9381$, which yields $\gamma = 0.1068$ (below the prescribed value $\gamma^* = 0.3$). The simulation results confirm our theoretical analysis for the problem of $\mathscr{H}_\infty$ fuzzy control for systems with repeated scalar nonlinearities.

3.6 Conclusions

The $\mathscr{H}_\infty$ control problem for discrete-time T-S fuzzy systems with repeated scalar nonlinearities has been investigated in this chapter. The nonlinear system is described by a discrete-time state equation containing a repeated scalar nonlinearity and the control strategy takes the form of parallel distributed compensation. The quadratic Lyapunov function has been used to design $\mathscr{H}_\infty$ fuzzy controllers such

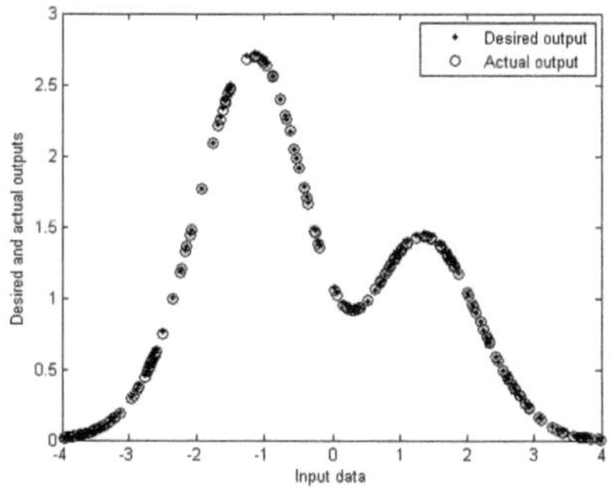

Fig. 3.3 The controllers u_k

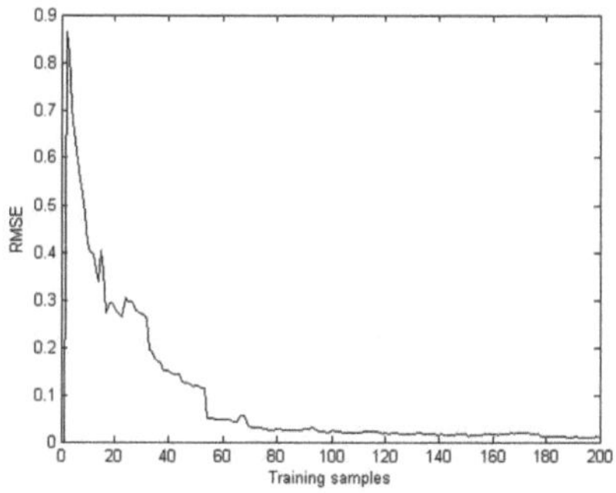

Fig. 3.4 The state evolution x_k of controlled systems when $w(k) <> 0$

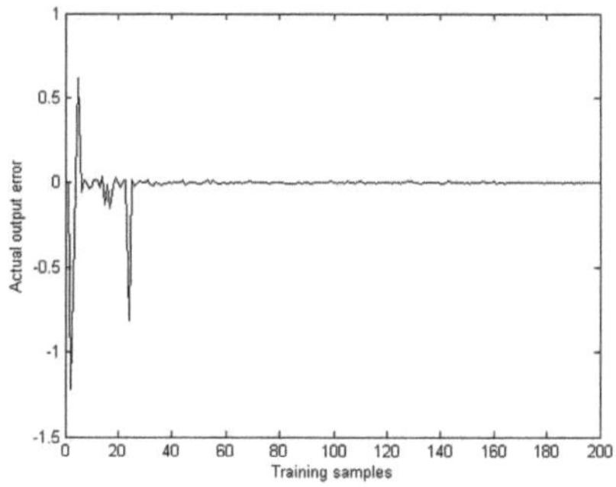

Fig. 3.5 The controlled output z_k and disturbance input w_k

that, in the presence of repeated scalar nonlinearities, the closed-loop T-S fuzzy control system is asymptotically stable and preserves a guaranteed H_∞ performance. By using the CCL algorithm, sufficient conditions have been established that ensure the asymptotic stability of the closed-loop system, and the controller gains have been obtained by the solution to a set of LMIs. An illustrative example has been provided to demonstrate the effectiveness of the approach proposed in this chapter.

References

1. W. Assawinchaichote, S. K. Nguang, P. Shi and E. Boukas, H_∞ fuzzy state-feedback control design for nonlinear systems with D-stability constraints: An LMI approach, *Mathematics and Computers in Simulation*, Vol. 78, pp. 514-531, 2008.
2. Y. Y. Cao and P. M. Frank, Analysis and synthesis of nonlinear time-delay systems via fuzzy control approach, *IEEE Trans. Fuzzy Systems*, Vol. 8, No. 2, pp. 200-211, 2000.
3. B. Chen, X. Liu and S. Tong, Robust fuzzy control of nonlinear systems with input delay, *Chaos, Solitons and Fractals*, Vol. 37, pp. 894-901, 2006.
4. Y. Chu, Further results for systems with repeated scalar nonlinearities, *IEEE Trans. Automat. Control*, Vol. 44, No. 12, pp. 2031-2035, 2001.
5. Y. Chu and K. Glover, Bounds of the induced norm and model reduction errors for systems with repeated scalar nonlinearities, *IEEE Trans. Automat. Control*, Vol. 44, No. 3, pp. 4215-4226, 1999.
6. Y. Chu and K. Glover, Stabilization and performance synthesis for systems with repeated scalar nonlinearities, *IEEE Trans. Automat. Control*, Vol. 44, No. 3, pp. 484-496, 1999.
7. L. El Ghaoui, F. Oustry and M. A. Rami, A cone complementarity linearization algorithm for static output-feedback and related problems, *IEEE Trans. Automat. Control*, Vol. 42, No. 8, pp. 1171-1176, 1997.
8. G. Feng, Controller synthesis of fuzzy dynamic systems based on piecewise Lyapunov function, *IEEE Trans. Fuzzy Systems*, Vol. 11, No. 5, pp. 605-612, 2003.
9. G. Feng, Stability analysis of discrete-time fuzzy dynamic systems based on piecewise Lyapunov functions, *IEEE Trans. Fuzzy Systems*, Vol. 12, No. 1, pp. 22-28, 2004.
10. G. Feng and J. Ma, Quadratic stabilization of uncertain discrete-time fuzzy dynamic systems, *IEEE Trans. Circuits and Systems-I: Fundamental Theory and Applications*, Vol. 48, No. 11, pp. 1337-1344, 2001.
11. H. Gao and T. Chen, Stabilization of nonlinear systems under variable sam-pling: a fuzzy control approach, *IEEE Trans. Fuzzy Systems*, Vol. 15, No. 5, pp. 972-983, 2007.
12. H. Gao, J. Lam and C. Wang, Induced l_2 and generalized H_2 filtering for systems with repeated scalar nonlinearities, *IEEE Trans. Signal Process.*, Vol. 53, No. 11, pp. 4215-4226, 2005.
13. H. Gao and C. Wang, A delay-dependent approach to robust Hinf and L2-Linf filtering for a Class of uncertain nonlinear time-delayed systems, *IEEE Trans. Automat. Control*, Vol. 48, No. 9, pp. 1661-1666, 2003.
14. H. Gao, Z. Wang and C. Wang, Improved H_∞ control of discrete-time fuzzy systems: a cone complementarity linearization approach, *Information Science*, Vol. 175, No. 1-2, pp. 57-77, 2005.
15. Y.-Y. Hou, T.-L. Liao and J.-J. Yan, Stability analysis of Takagi-Sugeno fuzzy cellular neural networks with time-varying delays, *IEEE Trans. Systems, Man, and Cybernetics: Part B*, Vol. 37, No. 3, pp. 720-726, Jun. 2007.
16. H. Huang, D.W.C. Ho and J. Lam, Stochastic stability analysis of fuzzy hopfield neural networks with time-varying delays, *IEEE Trans. Circuits and Systems II: Express Briefs*, Vol. 52, No. 5, pp. 251-255, May 2005.

17. X. Liu, Delay-dependent H_∞ control for uncertain fuzzy systems with time-varying delays, *Journal of Computational and Applied Mathematics*, Vol. 68, No. 5, pp. 1352-1361, 2008.
18. S. K. Nguang and P. Shi, H_∞ fuzzy output feedback control design for nonlinear systems: An LMI approach, *IEEE Trans. Fuzzy Systems*, Vol. 11, No. 3, pp. 331-340, 2003.
19. P. Shi and S. K. Nguang, H_∞ output feedback control of fuzzy system models under sampled measurements, *Comput. Math. Appl.*, Vol. 46, No. 5-6, pp. 705-717, 2003.
20. K. Tanaka, T. Hori and H. O. Wang, A multiple Lyapunov function approach to stabilization of fuzzy control systems, *IEEE Trans. Fuzzy Systems*, Vol. 11, No. 4, pp. 582-589, 2003.
21. Z. Wang, D. W. C. Ho and X. Liu, A note on the robust stability of uncertain stochastic fuzzy systems with time-delays, *IEEE Trans. Systems, Man and Cybernetics - Part A*, Vol. 34, No. 4, pp. 570-576, 2004.
22. Z. Wang and D. W. C. Ho, Filtering on nonlinear time-delay stochastic systems, *Automatica,* Vol. 39, No. 1, pp. 101-109, 2003.
23. Z. Wang, J. Lam and X. Liu, Nonlinear filtering for state delayed systems with Markovian switching, *IEEE Trans. Signal Processing,* Vol. 51, No. 9, pp. 2321-2328, 2003.
24. Z. Wang, J. Lam and X. Liu, Stabilization of a class of stochastic nonlinear time-delay systems, *Journal of Nonlinear Dynamics and Systems Theory,* Vol. 4, No. 3, pp. 357-368, 2004.
25. Z. Wang, Y. Liu and X. Liu, H-infinity filtering for uncertain stochastic time-delay systems with sector-bounded nonlinearities, *Automatica,* Vol. 44, No. 5, pp. 1268-1277, 2008.
26. H. Wu, Delay-dependent H_∞ fuzzy observer-based control for discrete-time nonlinear systems with state delay, *Fuzzy Sets and Systems*, Vol. 159, pp. 2696-2712, 2008.
27. K. Yuan, J. Cao and J. Deng, Exponential stability and periodic solutions of fuzzy cellular neural networks with time-varying delays, *Neurocomputing*, Vol. 69, No. 13-15, pp. 1619-1627, 2006.
28. S. Zhou and T. Li, Robust H_∞ control for discrete-time fuzzy systems via basis-dependent Lyapunov-Krasovskii functions, *Information Science*, Vol. 174, No. 3-4, pp. 197-217, 2005.
29. S. Zhou and T. Li, Robust stabilization for delayed discrete-time fuzzy systems via basis-dependent Lyapunov-Krasovskii function, *Fuzzy Sets and Systems*, Vol. 151, No. 1, pp. 139-153, 2005.

Chapter 4
Stable Adaptive Compensation with Fuzzy Cerebellar Model Articulation Controller for Overhead Cranes

Wen Yu and Xiaoou Li

Abstract This chapter proposes a novel control strategy for overhead cranes. The controller includes both position regulation and anti-swing control. Since the crane model is not exactly known, fuzzy cerebellar model articulation controller (CMAC) is used to compensate friction, and gravity, as well as the coupling between position and anti-swing control. Using a Lyapunov method and an input-to-state stability technique, the controller is proven to be robustly stable with bounded uncertainties. Real-time experiments are presented comparing this new stable control strategy with regular crane controllers.

4.1 Introduction

Although cranes are very important systems for handling heavy goods, automatic cranes are comparatively rare in industrial practice [5] [28], because of high investment costs. The need for faster cargo handling requires control of the crane motion so that its dynamic performance is optimized. Specifically, the control of overhead crane systems aims to achieve both position regulation and anti-swing control [11]. Several authors have looked at this including [4], time-optimal control was considered using boundary conditions, an idea which was further developed in [3] and [29]. Unfortunately, to increase robustness, some time optimization requirements, like zero angular velocity at the target point [23], have to be given up. Gain scheduling has been proposed as a practical method [12] to increase tracking accuracy, while observer-based feedback control was presented in [28].

Wen Yu
Departamento de Control Automatico, CINVESTAV-IPN, A.P. 14-740, Av.IPN 2508, México D.F., 07360, México, e-mail: yuw@ctrl.cinvestav.mx

Xiaoou Li
Departamento de Computación, CINVESTAV-IPN, A.P. 14-740, Av.IPN 2508, México D.F., 07360, México,e-mail: lixo@cs.cinvestav.mx

Many attempts, such as planar operation [12] and assuming the absence of friction [23], have been made to introduce simplified models for application of model-based control [28]. Thus, a self-tuning controller with a multilayer perceptron model for an overhead crane system was proposed [21], while in [10], the controller consists of a combined position servo control and a fuzzy-logic anti-swing controller.

Classical proportional and derivative (PD) control has the advantage of not requiring an overhead crane model but because of friction, gravitational forces and the other uncertainties, it cannot guarantee a zero steady-state error. While a proportional integral derivative (PID) control can remove this error, it lacks global asymptotic stability [17]. Several efforts have therefore been made to improve the performance of PD controllers. Global asymptotically stable PD control was realized by pulsing gravity compensation in [31] while in [18], a PD controller for a vertical crane-winch system was developed, which only requires the measurement of angles and their derivatives rather than a cable angle measurement. In [14], a passivity-based controller was combined with a PD control law. Here, asymptotic regulation of the gantry and payload position was proven, but unfortunately both controllers again require a crane model to compensate for the uncertainties. There is one weakness in applying PD control to this application: due to the existence of friction and gravitational forces, the steady-state error is not guaranteed to be zero [16].

Since the swing of the payload depends on the acceleration of the trolley, minimizing both the operation time and the payload swing produces partially conflicting requirements. The anti-swing control problem involves reducing the swing of the payload while moving it to the desired position as fast as possible [2]. One particular feedforward approach is input shaping [30], which is an especially practical and effective method of reducing vibrations in flexible systems. In [22] the anti-swing motion-planning problem is solved using the kinematic model in [20]. Here, anti-swing control for a three-dimensional (3D) overhead crane is proposed, which addresses the suppression of load swing. Nonlinear anti-swing control based on the singular perturbation method is presented in [35]. Unfortunately, all of these anti-swing controllers are model-based. In this paper, a PID law is used for anti-swing control which, being model-free, will affect the position control.

Therefore, there are three uncertain factors influencing the PD control for the overhead crane: friction, gravity, and errors coming from the PID anti-swing controllers. A model-free compensator is needed to reduce steady-state error. Two popular models can be used: neural networks and fuzzy systems. While neural networks are black-box models, which use input/output data to train their weights, fuzzy systems are based on fuzzy rules, which are constructed from prior knowledge [9]. Sometimes, fuzzy systems are regarded as gray-box models.

CMAC proposed by Albus [1] is an auto-associative memory feedforward neural network, it is a simplified mode of cerebellar based on the neurophysiological theory. A very important property of CMAC is that it has better convergence speed than feedforward neural networks. Many practical applications have been presented in recent literature [7], [8].

Since the data in CMAC is quantized, linguistic information cannot be dealt with. FCMAC uses fuzzy sets (fuzzy labels) as input clusters instead of crisp sets [8].

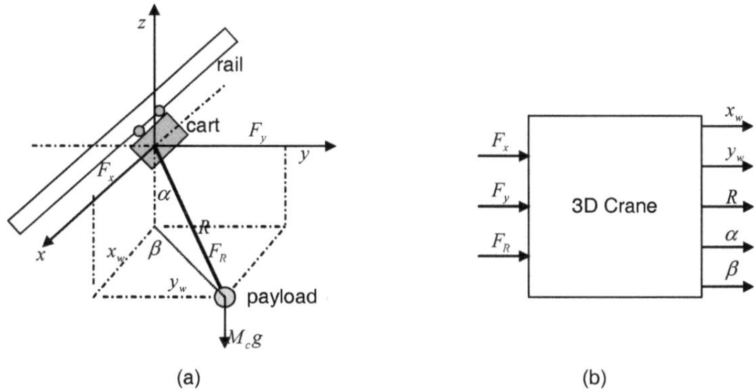

Fig. 4.1 Overhead crane

Compared with the normal CMAC, FCMAC can not only model linguistic variables based on fuzzy rules, but also is simple and highly intuitive [7]. Many ideas were realized on FCMAC extension and application. Bayesian Ying–Yang learning was introduced to determine the optimal FCMAC [24]. The Yager inference scheme was subsequently mapped onto FCMAC by [25]. In [26], a credit assignment idea was used to provide fast learning for FCMAC. Adaptation mechanisms were proposed for FCMAC learning in [34].

In this chapter, a FCMAC is used to estimate the above uncertainties. The required on-line learning rule is obtained from the tracking error analysis and there is no requirement for off-line learning. The overall closed-loop system with the FCMAC compensator is shown to be stable. Finally, results from experimental tests carried out to validate the controller are presented.

4.2 Preliminaries

The overhead crane system described schematically in Figure 4.1(a) has the system structure shown in Figure 4.1(b). Here α is the payload angle with respect to the vertical and β is the payload projection angle along the X-coordinate axis. The dynamics of the overhead crane are given by [32]:

$$M(x)\ddot{x} + C(x,\dot{x})\dot{x} + G(x) + F = \tau \tag{4.1}$$

where $x = [x_w, y_w, \alpha, \beta, R]^T$, $\tau = [F_x, F_y, 0, 0, F_R]^T$, F_x, F_y and F_R represent the control forces acting on the cart and rail and along the lift-line, (x_w, y_w, R) is position of the payload, $F = [\mu_x, \mu_y, 0, 0, \mu_R]^T \dot{x}$, μ_x, μ_y and μ_R are frictions factors, $G(x)$ is gravitational force, $C(x,\dot{x})$ is the Coriolis matrix and $M(x)$ is the dynamic matrix of the crane.

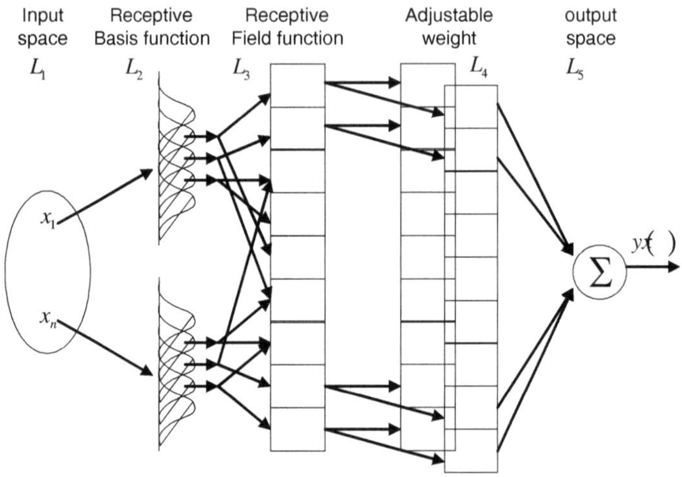

Fig. 4.2 The architecture of CMAC

In (4.1), there are some differences from other crane models in the literature. The length of the lift-line is not considered in [14], so the dimension of M is 4×4, while in [22], which also addresses anti-swing control and position control, the dimension of M is 3×3. In [19], the dimension of M is 5×5 as in this chapter. However, some uncertainties such as friction and anti-swing control coupling are not included. This overhead crane system shares one important property with robot systems: the Coriolis matrix $C(x,\dot{x})$ is skew-symmetric, *i.e.*, it satisfies the following relationship [14]

$$x^T \left[\dot{M}(x) - 2C(x,\dot{x})\right] x = 0. \tag{4.2}$$

A general FCMAC (see Figure 4.2) has five layers: input layer (L_1), fuzzified layer (L_2), fuzzy association layer (L_3), fuzzy post-association layer (L_4), and output layer (L_5).

Each input variable x_i in the $n-$dimension input space L_1 is quantized into m discrete regions (or elements) according to a resolution definition. Several elements in L_1 can be accumulated as a block. CMAC requires that the block number p is bigger than 2. By shifting an element in each input variable, different pieces are obtained. Each piece performs a basis function, which can be formulated as rectangular, triangular, Gaussian, or any continuously bounded function. If an input falls in $q-$th receptive-field, this field is active. Its neighborhoods are also activated, so it produces similar outputs near $q-$th receptive-field. This FCMAC property is called local generalization. The total shifting times is defined as l.

A fuzzified layer L_2 is also called association memory space. Each function at the fuzzified layer L_2 corresponds to a linguistic variable which is expressed by a membership functions ϕ_j^i $i = 1 \cdots n$, $j = 1 \cdots l$. The dimension of this layer is l^n. L_2 can be regarded as a fuzzification of input variables.

Fuzzy association layer L_3 is also called receptive-field space. The areas formed by the shifting units are called receptive-fields. Each location corresponds to a fuzzy association. A fuzzified layer connects and accomplishes the matching of the precondition of fuzzy logic rule. Each node at this layer completes fuzzy implication operation to obtain firing strength which is defined as α_k which is defined as

$$\alpha_j = \prod_{i=1}^{n} \lambda_q \left(\phi_j^i (x_i)\right), \quad q = 1 \cdots l, j = 1 \cdots l$$

where q is association time, l is the association number (or total shifting times), λ is the selection vector of the association memory which is defined as

$$\lambda_q \left(\phi_j^i (x_i)\right) = \phi_j^i (x_i) = [0,0 \cdots 1,0 \cdots] \begin{bmatrix} \phi_1^i \\ \vdots \\ \phi_l^i \end{bmatrix}.$$

Fuzzy post association layer L_4 is also called weight memory space, which calculates the normalization of firing strength and prepares for fuzzy inference

$$\begin{aligned} \varphi_q &= \alpha_q / \sum_{j=1}^{l} \alpha_j \\ &= \left(\prod_{i=1}^{n} \lambda_k \left(\phi_q^i (x_i)\right)\right) / \left(\sum_{j=1}^{l} \prod_{i=1}^{n} \lambda_j \left(\phi_j^i (x_i)\right)\right). \end{aligned}$$

In the output layer L_5, Takagi fuzzy inference is used, that is, the consequence of each fuzzy rule is defined as a function of input variables

$$R^j : If\ x_1\ is\ A_j^1 \cdots and\ x_n\ is\ A_j^n\ Then\ \hat{y}\ is\ f(X) \tag{4.3}$$

where $X = [x_1, \cdots, x_n]^T$. The output of the FCMAC can be expressed as

$$\hat{y} = \sum_{j=1}^{l} w_j \varphi_{kj} \tag{4.4}$$

where w_j denotes the connecting weight of j–th receptive-field. In matrix form it is

$$\hat{y} = W \varphi (X) \tag{4.5}$$

where $W = [w_1, \cdots w_l]$, $\varphi (x) = [\varphi_1, \cdots \varphi_l]^T$.

4.3 Control of an Overhead Crane

The control problem is to move the rail in such a way that the actual position of the payload reaches the desired one. The three control inputs $[F_x, F_y, F_R]$ can force the crane to the position $[x_w, y_w, R]$, but the swing angles $[\alpha, \beta]$ cannot be controlled using the dynamic model (4.1) directly. In order to design an anti-swing control, linearization models for $[\alpha, \beta]$ are analyzed. Because the acceleration of the crane is much smaller than the gravitational acceleration, the rope length is kept slowly varying and the swing is not big, giving

$$\begin{array}{c}|\ddot{x}_w| \ll g, \quad |\ddot{y}_w| \ll g, \quad |\ddot{R}| \ll g, \\ |\dot{R}| \ll R, \quad |\dot{\alpha}| \ll 1, \quad |\dot{\beta}| \ll 1, \\ s_1 = \sin\alpha \approx \alpha, \quad c_1 = \cos\alpha \approx 1.\end{array}$$

The approximated dynamics of $[\alpha, \beta]$ are then

$$\ddot{\alpha} + \ddot{x}_w + g\alpha = 0, \quad \ddot{\beta} + \ddot{y}_w + g\beta = 0.$$

Since $\ddot{x}_w = \frac{F_x}{M_r}$, $\ddot{y}_w = \frac{F_y}{M_m}$, the dynamics of the swing angles are

$$\ddot{\alpha} + g\alpha = -\frac{F_x}{M_r}, \quad \ddot{\beta} + g\beta = -\frac{F_y}{M_m}. \tag{4.6}$$

Only F_x and F_y participate the anti-swing control, F_R does not affect the swing angles α, β. The control forces F_x and F_y are assumed to have the following form

$$\begin{array}{l}F_x = A_1(x_w, \dot{x}_w) + A_2(\alpha, \dot{\alpha}), \\ F_y = B_1(y_w, \dot{y}_w) + B_2\left(\beta, \dot{\beta}\right)\end{array} \tag{4.7}$$

where $A_1(x_w, \dot{x}_w)$ and $B_1(y_w, \dot{y}_w)$ are position controllers, and $A_2(\alpha, \dot{\alpha})$ and $B_2(\beta, \dot{\beta})$ are anti-swing controllers. Substituting (4.7) into (4.6), produces the anti-swing control model

$$\begin{array}{l}\ddot{\alpha} + g\alpha + \frac{A_1}{M_r} = -\frac{A_2}{M_r}, \\ \ddot{\beta} + g\beta + \frac{B_1}{M_m} = -\frac{B_2}{M_m}.\end{array} \tag{4.8}$$

Now if $\frac{A_1}{M_r}$ and $\frac{B_1}{M_r}$ are regarded as disturbance, $\frac{A_2}{M_r}$ and $\frac{B_2}{M_m}$ as control inputs, then (4.8) is a second-order linear system with disturbances. Standard PID control can now be applied to regulate α and β thereby producing the anti-swing controllers

$$\begin{array}{l}A_2(\alpha, \dot{\alpha}) = k_{pa2}\alpha + k_{da2}\dot{\alpha} + k_{ia2}\int_0^t \alpha(\tau)\,d\tau, \\ B_2\left(\beta, \dot{\beta}\right) = k_{pb2}\beta + k_{db2}\dot{\beta} + k_{ib2}\int_0^t \beta(\tau)\,d\tau\end{array} \tag{4.9}$$

where k_{pa2}, k_{da2} and k_{ia2} are positive constants corresponding to proportional, derivative and integral gains.

Substituting (4.7) into (4.1), produces the position control model

$$M(x)\ddot{x}+C(x,\dot{x})\dot{x}+G(x)+T\dot{x}+D=\tau \tag{4.10}$$

where $D=[A_2,B_2,0,0,0]^T$, $\tau=[A_1,B_1,0,0,F_R]^T$. Using this model, a position controller will be designed in the next section.

4.4 Position Regulation with FCMAC Compensation

A PD type controller is used for position regulation, which has the following form

$$\tau=-K_p(x-x^d)-K_d(\dot{x}-\dot{x}^d)$$

where K_p and K_d are positive definite, symmetric and constant matrices, which correspond to the proportional and derivative coefficients, $x^d\in\Re^5$ is the desired position, and $\dot{x}^d\in\Re^5$ is the desired joint velocity. Here the regulation problem is discussed, so $\dot{x}^d=0$.

A filtered regulation error is defined as

$$r=(\dot{x}-\dot{x}^d)+\Lambda(x-x^d)=\tilde{x}_2+\Lambda\tilde{x}_1$$

where $\tilde{x}_1=(x-x^d)$,$\tilde{x}_2=(\dot{x}-\dot{x}^d)$, $\cdot\tilde{x}_1=\tilde{x}_2$,$\Lambda=\Lambda^T>0$. Using (4.10) and $\dot{x}^d=\ddot{x}^d=0$,

$$\begin{aligned}
M\dot{r}&=M\dot{\tilde{x}}_2+M\Lambda\dot{\tilde{x}}_1\\
&=M\ddot{x}-M\ddot{x}^d+M\Lambda\dot{\tilde{x}}_1\\
&=\tau-C\dot{x}-G-T\dot{x}-D+M\Lambda\dot{\tilde{x}}_1+C\Lambda\tilde{x}_1-C\Lambda\tilde{x}_1\\
&=\tau-Cr+f
\end{aligned} \tag{4.11}$$

where

$$f(s)=M\Lambda\dot{x}+C\Lambda\tilde{x}_1-G-T\dot{x}-D$$

where $s=\left[x^T,\dot{x}^T,\tilde{x}_1^T\right]^T$.

Because $f\left(x,\dot{x},\tilde{x}_1\right)=[f_x,f_y,f_z]^T$ is unknown, a FCMAC (4.5) is used to approximate it. The fuzzy rule (4.3) has the following form

$$\begin{array}{l} \mathrm{R}^i\text{: IF } \left(x_w \text{ is } A_{1i}^1\right) \text{ and } \left(\alpha \text{ is } A_{2i}^1\right) \text{ and } \left(\dot{x}_w \text{ is } A_{3i}^1\right) \\ \text{and } \left(\dot{\alpha} \text{ is } A_{4i}^1\right) \text{ and } \left(\tilde{x}_w \text{ is } A_{5i}^1\right) \text{THEN } \widehat{f_x} \text{ is } B_{1i} \\ \text{IF } \left(y_w \text{ is } A_{1i}^2\right) \text{ and } \left(\beta \text{ is } A_{2i}^2\right) \text{ and } \left(\dot{y}_w \text{ is } A_{3i}^2\right) \\ \text{and } \left(\dot{\beta} \text{ is } A_{4i}^2\right) \text{ and } \left(\widetilde{y}_w \text{ is } A_{5i}^2\right) \text{THEN } \widehat{f_y} \text{ is } B_{2i} \\ \text{IF } \left(R \text{ is } A_{1i}^3\right) \text{ and } \left(\dot{R} \text{ is } A_{2i}^3\right) \\ \text{and } \left(\widetilde{R} \text{ is } A_{3i}^3\right) \text{THEN } \widehat{f_z} \text{ is } B_{3i}. \end{array} \tag{4.12}$$

Here $\widehat{f_x}$, $\widehat{f_y}$ and $\widehat{f_z}$ are the uncertainties (friction, gravity and coupling errors) along the X,Y,Z -coordinate axis. $i = 1, 2 \cdots l$. A total of fuzzy IF-THEN rules are used to perform the mapping from the input vector $x = [x_w, y_w, \alpha, \beta, R]^T \in \Re^5$ to the output vector $\widehat{y}(k) = \left[\widehat{f_1}, \widehat{f_2}, \widehat{f_3}\right]^T = [\widehat{y}_1, \widehat{y}_2, \widehat{y}_3] \in R^3$. Here $A_{1i}, \cdots A_{ni}$ and $B_{1i}, \cdots B_{mi}$ are standard fuzzy sets. In this paper, some on-line learning algorithms are introduced for the membership functions A_{ji} and B_{ji} such that the PD controller with the fuzzy compensator is stable. (4.5) can be expressed as

$$\widehat{f} = \hat{W}_t \Phi(s) \tag{4.13}$$

where the parameter matrix $\hat{W} = diag\left[\hat{W}_1, \hat{W}_2, \hat{W}_3\right]$, and the data vector $\Phi(x) = [\Phi_1, \Phi_2, \Phi_3]^T$, $\hat{W}_p = \left[w_{p1} \cdots w_{pl}\right]$, $\Phi_p = \left[\phi_1^p \cdots \phi_l^p\right]^T$.

The position controller have a PD form with a fuzzy compensator

$$\begin{array}{l} \tau = \left[A_1(x_w, \dot{x}_w), B_1(y_w, \dot{y}_w), 0, 0, F_R\right]^T = -Kr - \widehat{f} \\ = -K\Lambda\left(x - x^d\right) - K\left(\dot{x} - \dot{x}^d\right) - \hat{W}_t \Phi(s) \end{array} \tag{4.14}$$

where $x = [x_w, y_w, \alpha, \beta, R]^T$, $x^d = \left[x_w^d, y_w^d, 0, 0, R^d\right]^T$, and x_w^d, y_w^d and R^d are the desired positions, $K = K_{p1}^T > 0$.

According to the Stone-Weierstrass theorem [13], a general nonlinear smooth function can be written as

$$f(s) = M\Lambda\dot{x} + C\Lambda\widetilde{x}_1 - G - T\dot{x} - D = W^* \Phi(s) + \mu(t) \tag{4.15}$$

where W^* is the optimal weight matrix, and $\mu(t)$ is the modeling error. In this chapter the fuzzy compensator (4.13) is used to approximate the unknown nonlinearity (the gravity, friction, and coupling of anti-swing control).

The coupling between anti-swing control and position control can be explained as follows. For the anti-swing control (4.8), the position control A_1 and B_1 are disturbances, which can be decreased by the integral action in PID control. Although the anti-swing model (4.8) is an approximator, the anti-swing control (4.9) does not in fact use this, as it is model-free. Hence while the anti-swing control law (4.9) cannot suppress the swing completely, it can minimize any consequent vibration.

For the position control (4.10), the anti-swing control lies in the term $D = [A_2, B_2, 0, 0, 0]^T$, which can also be regarded as a disturbance. The coupling due to anti-swing control can be compensated by the fuzzy system. For example, in order to decrease the swing, we should increase A_2 and B_2, this means increase the disturbances of the crane, so τ should be increased.

4.5 FCMAC Training and Stability Analysis

(4.15) can be rewritten as

$$M\Lambda\dot{\widehat{x}} + C\Lambda\widetilde{x}_1 - G - T\dot{x} - D = W^0\Phi(s) + \eta_g \tag{4.16}$$

where $s_1 = \left[x^T, x_2^T, \widetilde{x}_1^T\right]^T$, W^0 is a fixed bounded matrix, and η_g is the approximation error whose magnitude also depends on the value of W^0. Now, η_g is assumed to be quadratic bounded such that

$$\eta_g^T \Lambda_g \eta_g \leq \bar{\eta}_g \tag{4.17}$$

where $\bar{\eta}_g$ is a positive constant. In this paper, we use Gaussian functions in the receptive-field ϕ_i^j, which is expressed as

$$\phi_i^j = \exp\left[-\left(\frac{s_j - c_{i,j}}{\sigma_{i,j}}\right)^2\right] \tag{4.18}$$

where ϕ_i^j presents the $i-$th rule of the $j-$th input x_j, c_j^i is the mean, σ_j^i is the variance, $\Phi(x) = [\Phi_1, \Phi_2, \Phi_3]^T$, $\Phi_p = \left[\phi_1^p \cdots \phi_l^p\right]^T$.

Firstly, we assume the Gaussian functions in the receptive-field are given by prior knowledge, i.e., we only train $\hat{W}_t$. Now defining $\tilde{W}_t = W^0 - \hat{W}_t$, for the filtered regulation error $r = (\dot{x} - \dot{x}^d) + \Lambda(x - x^d)$, the following theorem holds.

Theorem 4.1. *If the updating laws for the membership functions in (4.13) are*

$$\frac{d}{dt}\hat{W}_t = K_w\Phi(s)r^T \tag{4.19}$$

where K_w is a positive definite matrix, and K_d satisfies

$$K > \frac{1}{2}\Lambda_g^{-1}, \tag{4.20}$$

then the PD control law with FCMAC compensation in (4.14) can make the tracking error r stable. In fact, the average tracking error r_1 converges to

$$\limsup_{T\to\infty} \frac{1}{T}\int_0^T \|r\|_{Q_1}^2 \, dt \leq \bar{\eta}_g \tag{4.21}$$

where $Q_1 = 2K - \Lambda_g^{-1}$.

Proof. The following Lyapunov function is proposed

$$V = r^T M r + tr\left(\widetilde{W}_t^T K_w^{-1} \widetilde{W}_t\right) \tag{4.22}$$

where K_w is a positive definite matrix. Using (4.11) and (4.14), the closed-loop system is given by

$$M\dot{r} = \tau - Cr + f = -Kr - \widetilde{W}_t \Phi(s_1) - Cr + \eta_g. \tag{4.23}$$

Now the derivative of (4.22) is

$$\dot{V} = -2r^T K r - 2r^T C r + 2r^T \eta_g + r^T \dot{M} r - 2r^T \widetilde{W}_t \Phi(s) + 2tr\left(\widetilde{W}_t^T K_w^{-1} \dot{\widetilde{W}}_t\right). \tag{4.24}$$

In view of the matrix inequality,

$$X^T Y + \left(X^T Y\right)^T \leq X^T \Lambda^{-1} X + Y^T \Lambda Y \tag{4.25}$$

which is valid for any $X, Y \in \Re^{n \times k}$ and for any positive definite matrix $0 < \Lambda = \Lambda^T \in \Re^{n \times n}$, it follows that if $X = r$, and $Y = \eta_g$, from (4.17)

$$2r^T \eta_g \leq r^T \Lambda_g^{-1} r + \eta_g^T \Lambda_g \eta_g \leq r^T \Lambda_g^{-1} r + \overline{\eta}_g. \tag{4.26}$$

Using (4.2) and (4.24), this then can be written as

$$\dot{V} \leq -r^T \left(2K - \Lambda_g^{-1}\right) r + 2tr\left[\left(K_w^{-1} \frac{d}{dt} \tilde{W}_t - \Phi(s) r^T\right) \tilde{W}\right] + \overline{\eta}_g. \tag{4.27}$$

Since $x_2^d = \dot{x}_2^d = 0$, and using the learning law (4.19), then (4.27) becomes

$$\dot{V} \leq -r^T Q r + \overline{\eta}_g \tag{4.28}$$

where $Q_1 = 2K - \Lambda_g^{-1}$. Now, from (4.20), it is known that $Q_1 > 0$, and (4.28) can then be represented as

$$\dot{V} \leq -\lambda_{\min}(Q_1) \|r\|^2 + \eta_g^T \Lambda_g \eta_g.$$

V is therefore an input-to-state stability (ISS)-Lyapunov function. Using Theorem 1 from [27], the boundedness of η_g and $\overline{\eta}_\delta$ implies that the tracking error $\|r\|$ is stable, so x and $\widehat{x}$ are bounded.

Integrating (4.28) from 0 to T yields

$$\int_0^T r^T Q_1 r dt \leq V_0 - V_T + \overline{\eta}_g T \leq V_0 + \overline{\eta}_g T.$$

(4.21) is established. □

Secondly, neither the receptive-field function nor $\hat{W}_t$ are known. pth output of the FCMAC compensator can be expressed as $\hat{W}_p \Phi_p(s)$, $p = 1, 2, 3$. Using the Taylor series,

$$\begin{aligned} &\hat{W}_p \Phi_p(s) - f(s) + \mu_p \\ &= \sum_{i=1}^{l} \left(w_{pi}^* - \hat{w}_{pi} \right) z_i^p / b_p + \sum_{i=1}^{l} \sum_{j=1}^{n_p} \frac{\partial}{\partial c_{ji}^p} \left(\frac{a_p}{b_p} \right) \left(c_{ji}^{p*} - c_{ji}^p \right) \\ &+ \sum_{i=1}^{l} \sum_{j=1}^{n_p} \frac{\partial}{\partial \sigma_{ji}^p} \left(\frac{a_p}{b_p} \right) \left(\sigma_{ji}^{p*} - \sigma_{ji}^p \right) - R_{1p} \end{aligned} \tag{4.29}$$

where R_{1p} is second order approximation error of the Taylor series, and

$$z_i^p = \prod_{i=1}^{n} \lambda_k \left(\phi_j^i \right), \quad a_p = \sum_{k=1}^{l} w_p z_p, \quad b_p = \sum_{k=1}^{l} z_p.$$

Using the chain rule, we get

$$\begin{aligned} \frac{\partial}{\partial c_{ji}^p} \left(\frac{a_p}{b_p} \right) &= \frac{\partial}{\partial z_i^p} \left(\frac{a_p}{b_p} \right) \frac{\partial z_i^p}{\partial c_{ji}^p} \\ &= \left(\frac{1}{b_p} \frac{\partial a_p}{\partial z_i^p} + \frac{\partial}{\partial z_i^p} \left(\frac{1}{b_p} \right) a_p \right) \left(2 z_i^p \frac{s_j - c_{ji}^p}{\left[\sigma_{ji}^p \right]^2} \right) \\ &= \left(\frac{w_{pi}}{b_p} - \frac{a_p}{\left[b^p \right]^2} \right) \left(2 z_i^p \frac{s_j - c_{ji}^p}{\left[\sigma_{ji}^p \right]^2} \right) \\ &= 2 z_i^p \frac{w_{pi} - \widehat{y}_p}{b_p} \frac{s_j - c_{ji}^p}{\left[\sigma_{ji}^p \right]^2} \end{aligned}$$

and

$$\begin{aligned} \frac{\partial}{\partial \sigma_{ji}^p} \left(\frac{a_p}{b_p} \right) &= \frac{\partial}{\partial z_i^p} \left(\frac{a_p}{b_p} \right) \frac{\partial z_i^p}{\partial \sigma_{ji}^p} \\ &= 2 z_i^p \frac{w_{pi} - \widehat{y}_p}{b_p} \frac{\left(s_j - c_{ji}^p \right)^2}{\left[\sigma_{ji}^p \right]^3}. \end{aligned}$$

There are three subsystems, for each one

$$\hat{W}_p \Phi_p(s) - f(s)_p = Z_p \widetilde{W}_p + \Psi_p \tilde{C}_p I + \Psi_p \tilde{B}_p I - \zeta_p \tag{4.30}$$

where $\zeta_p = \mu_p + R_{1p}$

$$
\begin{aligned}
&Z_p = \left[z_1^p/b\cdots z_l^p/b\right]^T, \quad \widetilde{W}_p = \hat{W}_p - W_p^*, \quad W_p = \left[\hat{w}_{p1}\cdots\hat{w}_{pl}\right],\\
&\Psi_p = \left[2z_1^p\tfrac{\hat{w}_{p1}-\hat{y}_p}{b_p},\cdots,2z_l^p\tfrac{\hat{w}_{pl}-\hat{y}_p}{b_p}\right], \quad I=[1,\cdots 1]^T,\\
&\tilde{C}_p = \begin{bmatrix} \frac{s_1-c_{11}^p}{\left[\sigma_{11}^p\right]^2}\left(c_{11}^p - c_{11}^{p*}\right) & & \frac{s_n-c_{n1}}{\left[\sigma_{n1}^p\right]^2}\left(c_{n1}^p - c_{n1}^{p*}\right)\\ & \ddots & \\ \frac{s_1-c_{1l}^p}{\left[\sigma_{1l}^p\right]^2}\left(c_{1l}^p - c_{1l}^{p*}\right) & & \frac{s_n-c_{nl}}{\sigma_{nl}^{p2}}\left(c_{nl}^p - c_{nl}^{p*}\right)\end{bmatrix}\\
&\tilde{B}_p = \begin{bmatrix} \frac{\left(s_1-c_{11}^{p2}\right)^2}{\sigma_{11}^{p3}}\left(\sigma_{11}^p - \sigma_{11}^{p*}\right) & & \frac{\left(s_n-c_{n1}\right)^2}{\sigma_{n1}^{p3}}\left(\sigma_{n1}^p - \sigma_{n1}^{p*}\right)\\ & \ddots & \\ \frac{\left(s_1-c_{1l}^p\right)^2}{\sigma_{1l}^{3p}}\left(\sigma_{1l}^p - \sigma_{1l}^{p*}\right) & & \frac{\left(s_n-c_{nl}\right)^2}{\sigma_{nl}^{3p}}\left(\sigma_{nl}^p - \sigma_{nl}^{p*}\right)\end{bmatrix}.
\end{aligned}
$$

In vector form

$$\hat{W}_t\Phi(s_1) - f_1(s_1) = Z\widetilde{W} + \Psi\tilde{C}I + \Psi\tilde{B}I - \zeta. \tag{4.31}$$

Now, ζ is assumed to be quadratic bounded such that

$$\zeta^T\Lambda_\zeta\zeta \le \overline{\zeta}.$$

For the filtered regulation error r, the following theorem holds.

Theorem 4.2. *If the updating laws for the membership functions in (4.13) are*

$$
\begin{aligned}
&\tfrac{d}{dt}\hat{W}_p = -K_w Z_p r^T,\\
&\tfrac{d}{dt}c_{ji}^p = -2k_c z_i^p\tfrac{\hat{w}_{pi}-\hat{y}_p}{b_p}\frac{s_j-c_{ji}^p}{\left[\sigma_{ji}^p\right]^2}r^T,\\
&\tfrac{d}{dt}\sigma_{ji}^p = -2k_b z_i^p\tfrac{\hat{w}_{pi}-\hat{y}_p}{b_p}\frac{\left(s_j-c_{ji}^p\right)^2}{\left[\sigma_{ji}^p\right]^3}r^T
\end{aligned}
\tag{4.32}
$$

where K_w is definite matrix, k_c and k_b are positive constant, and K satisfies

$$K > \frac{1}{2}\left(\Lambda_\zeta^{-1} + \Lambda_g^{-1}\right),$$

then the PD control law with fuzzy compensation in (4.14) can make the tracking error stable. The average tracking error r converges to

$$\limsup_{T\to\infty}\frac{1}{T}\int_0^T \|r\|_{Q_2}^2\,dt \le \bar{\eta}_g + \overline{\zeta}$$

where $Q_2 = 2K - \left(\Lambda_\zeta^{-1} + \Lambda_g^{-1}\right)$.

Proof. Let use define $\widetilde{c}_{ji} = \hat{c}_{ji} - c_{ji}^*$, $\widetilde{b}_{ji}(k) = \tilde{\sigma}_{ji}(k) - \sigma^*(k)$, the element of $\widetilde{C}$ is expressed as $\widetilde{c}_{ji} = \left[\widetilde{C}\right]$. We selected a positive defined scalar V_3 as

$$V_1 = r^T M r_1 + tr\left(\tilde{W}_t^T K_w^{-1} \tilde{W}_t\right) + tr\left(\tilde{C}_t^T K_c^{-1} \tilde{C}_t\right) + tr\left(\tilde{B}_t^T K_b^{-1} \tilde{B}_t\right) \tag{4.33}$$

where K_p, K_w K_c and K_b are any positive definite matrices. Using (4.30),

$$M\dot{r} = \tau - Cr + f = -Kr - \widetilde{W}_t \Phi(s) - Cr + \eta_g \tag{4.34}$$

$$M\dot{x}_2 = -Cx_2 - K_p \bar{x}_1 - K_d \tilde{x}_2 + \hat{W}_t \Phi(s) - (G(x) + F(x) + D).$$

So

$$\begin{aligned} r^T M\dot{r} &= -r^T Kr - r^T \widetilde{W}_t \Phi(s) - r^T Cr \\ &+ r^T \eta_g - r^T \left[Z\widetilde{W} + \Psi \tilde{C} I + \Psi \tilde{B} I - \zeta \right]. \end{aligned}$$

Similar as to Theorem 4.1

$$2r^T \zeta \leq r^T \Lambda_\zeta^{-1} r + \zeta^T \Lambda_\zeta \zeta \leq r^T \Lambda_g^{-1} r + \overline{\zeta},$$

$$\begin{aligned} \dot{V}_1 = &-r^T \left(2K - \Lambda_g^{-1} - \Lambda_\zeta^{-1} \right) r + \bar{\eta}_g + \overline{\zeta} + 2tr\left[\left(K_w^{-1} \tfrac{d}{dt} \tilde{W}_t - Zr^T \right) \tilde{W} \right] \\ &+ 2tr\left[\left(K_c^{-1} \tfrac{d}{dt} \tilde{C}_t - \Psi r^T \right) \tilde{C}_t \right] + 2tr\left[\left(K_b^{-1} \tfrac{d}{dt} \tilde{B}_t - \Psi r^T \right) \tilde{B}_t \right], \end{aligned} \tag{4.35}$$

using the learning law (4.32), then (4.35) becomes

$$\dot{V}_1 \leq -r^T Q_2 r + \bar{\eta}_g + \overline{\zeta}.$$

The rest of the proof is the same as the proof for Theorem 4.1. □

4.6 Experimental Comparisons

The proposed anti-swing control for overhead crane systems has been implemented on a InTeCo [15] overhead crane test-bed, see Figure 4.3. The rail is 150cm long, and the physical parameters for the system are as follows: the mass of rail is $M_r = 6.5kg$, the mass of cart is $M_c = 0.8kg$, the mass of payload is $M_m = 1.3kg$, see Figure 4.3. Here interfacing is based on a microprocessor, comprising a multifunction analog and digital I/O board dedicated to the real-time data acquisition and control in the Windows® XP environment, mounted in a PC Pentium®-III 500MHz host. Because the chip supports real-time operations without introducing latencies caused by the Windows default timing system, the control program operated in Windows XP with MATLAB® 6.5/SIMULINK®.

There are two inputs in the anti-swing model (4.14), A_1 and A_2 with A_1 from the position controller and A_2 from the anti-swing controller. When the anti-swing control A_2 is designed by (4.8), A_1 is regarded as a disturbance. The chosen parameters of the PID (4.9) control law were

$$\begin{aligned} k_{pa2} &= 2.5, \quad k_{da2} = 18, \quad k_{ia2} = 0.01, \\ k_{pb2} &= 15, \quad k_{db2} = 10, \quad k_{ib2} = 0.6. \end{aligned}$$

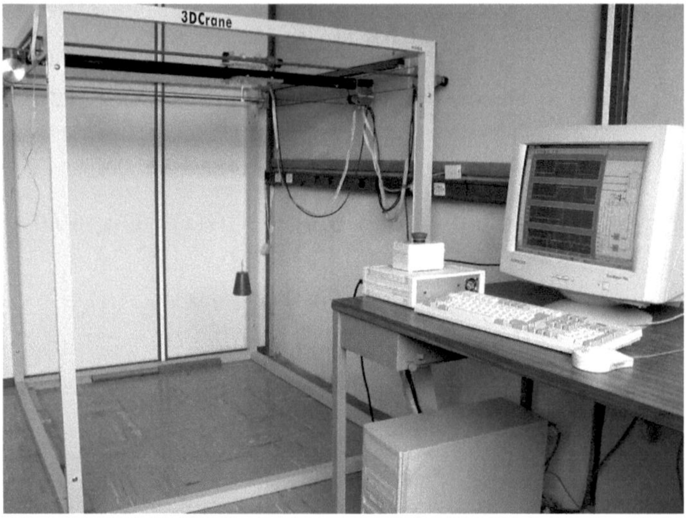

Fig. 4.3 Real-time control for an overhead crane

The position control law in (4.14) is discussed next. In this case there are two types of input to the position model (4.10), $D = [A_2, B_2, 0, 0, 0]^T$, $\tau = [A_1, B_1, 0, 0, F_R]^T$. When the position control A_1 is designed by (4.14), the anti-swing control A_2 in (4.10) is regarded as a disturbance which will be compensated for the fuzzy system (4.13). Theorem 4.2 implies that to assure stability, K_d should be large enough such that $K_d > \Lambda_g^{-1}$. Since these upper bounds are not known, $K_{d1} = diag[80, 80, 0, 0, 10]$ is selected. The position feedback gain does not effect the stability, but it should be positive, and was chosen as $K_{p1} = diag[5, 5, 0, 0, 1]$.

A total of 20 fuzzy rules in the receptive-field were used to compensate for the friction, gravity and the coupling from anti-swing control. The membership function for A_{ji} was chosen to be the Gaussian function

$$A_{ji} = \exp\left\{-(x_j - m_{ji})^2/\sigma_{ji}^2\right\}, \quad j = 1\cdots 5, \quad i = 1\cdots 20$$

where m_{ji} and σ_{ji} were selected randomly to lie in the interval $(0,1)$. Hence, $\hat{W}_t \in R^{5\times 20}$, $\Phi(x) = [\sigma_1 \cdots \sigma_{20}]^T$. The learning law took the form in (4.32) with $K_w = 10$. The desired gantry position was selected as a circle with $x_w^d = 0.5\sin(0.2t)$, $y_w^d = 0.5\cos(0.2t)$. The resulting gantry positions and angles are shown in Figure 4.4 and Figure 4.5. The control inputs are shown in Figure 4.6.

For comparison, the PID control results ($K_{d1} = diag[80, 80, 0, 0, 10]$, $K_{p1} = diag[5, 5, 0, 0, 1]$, $K_{i1} = diag[0.25, 0.25, 0, 0, 0.1]$) are shown in in Figure 4.7 and Figure 4.8.

It can be seen that the swing angles α and β are decreased a lot with the anti-swing controller. From Figure 4.6 and Figure 4.8 we see that the the improvement is

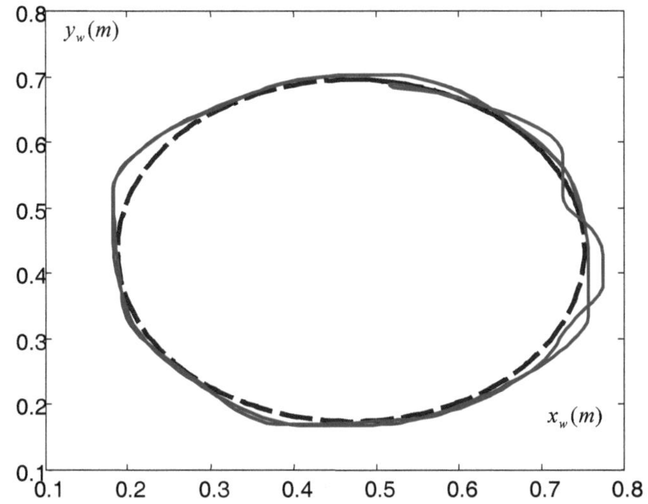

Fig. 4.4 Positions control with FCMAC compensation

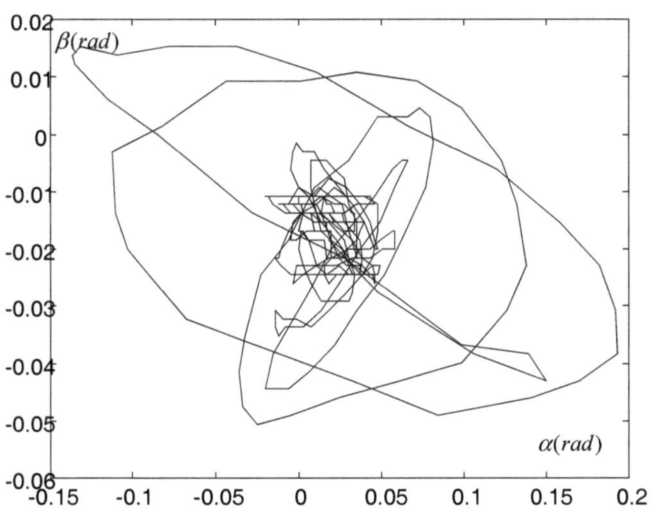

Fig. 4.5 Angles of PD control with FCMAC compensation

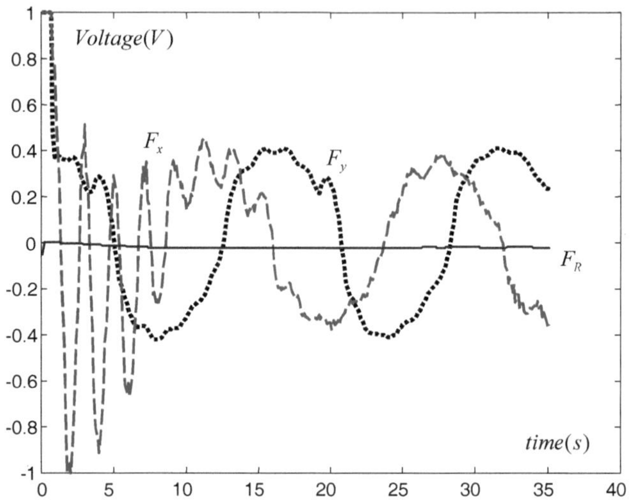

Fig. 4.6 Control inputs

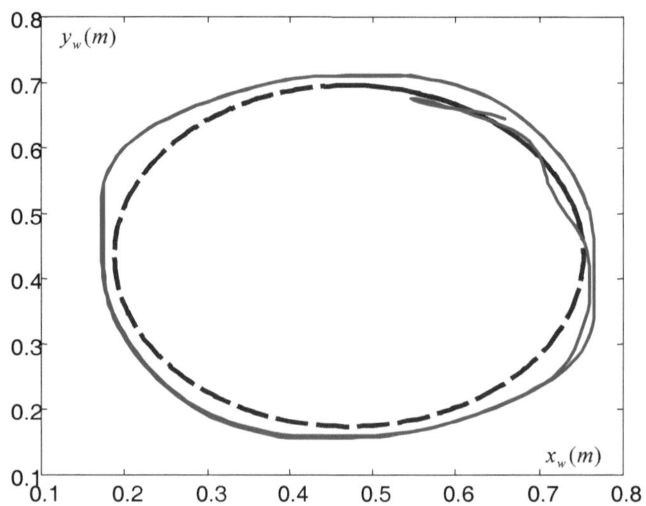

Fig. 4.7 PID position control with anti-swing control

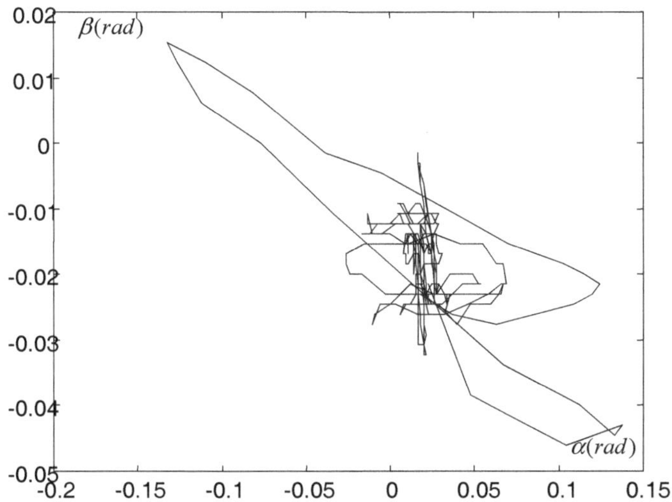

Fig. 4.8 Angles of PID position control with anti-swing control

not so good in α direction, because in this direction the inertia comes from the rail, its mass M_r is bigger than the cart M_c (β direction).

Clearly, PD control with FCMAC compensation can successfully compensate the uncertainties such as friction, gravity and anti-swing coupling. Because the PID controller has no adaptive mechanism, it does not work well for anti-swing coupling in contrast to the fuzzy compensator which can adjust its control action. On the other hand, the PID controller is faster than the PD control with fuzzy compensation in the case of small anti-swing coupling.

The structure of fuzzy compensator is very important. From fuzzy theory the form of the membership function is known not to influence the stability of the fuzzy control, but the approximation ability of fuzzy system for a particular nonlinear process depends on the membership functions selected. The number of fuzzy rules in receptive-field constitutes a structural problem. It is well known that increasing the dimension of the fuzzy rules can cause the "overlap" problem and add to the computational burden [33]. The best dimension of CMAC to use is still an open problem. In this application 20 fuzzy rules were used. Since it is difficult to obtain the fuzzy structure from prior knowledge, several fuzzy identifiers can be put in parallel and the best one selected by a switching algorithm. The learning gain K_w will influence the learning speed, so a very large gain can cause unstable learning, while a very small gain produces a slow learning process.

4.7 Conclusions

In this chapter, a FCMAC compensator is used to compensate for gravity and friction. Using Lyapunov-like analysis, the stability of the closed-loop system with the FCMAC compensation was proven. Real-time experiments were presented comparing our stable anti-swing PD control strategy with regular crane controllers. These showed that the PD control law with FCMAC compensations is effective for the overhead crane system.

References

1. J.S. Albus, A New Approach to Manipulator Control: The Cerebellar Model Articulation Controller (CMAC), *Journal of Dynamic Systems, Measurement, and Control, Transactions of ASME*, pp. 220-227, 1975.
2. E.M.Abdel-Rahman,A.H. Nayfeh, Z.N.Masoud, Dynamics and control of cranes: a review, *Journal of Vibration and Control*, Vol.9, No.7, 863-908, 2003.
3. J.W.Auernig and H.Troger, Time optimal control of overhead cranes with hoisting of the payload, *Automatica,* Vol.23, No.4, 437-447, 1987.
4. J.W.Beeston, Closed-loop time optimatial control of a suspended payload-a design study, *Proc. 4th IFAC World Congress*, 85-99, Warsaw Poland, 1969.
5. W.Blajer, K.Kotodziejczyk, Motion planning and control of gantry cranes in cluttered work environment, *IET Control Theory & Applications*, Vol.1, No.5, 1370 - 1379, 2007.
6. C.I.Byrnes, A.Isidori and J.C.Willems, Passivity, feedback equivalence, and the global stabilization of minimum phase nonlinear systems, *IEEE Trans. Automat. Contr.*, Vol.36, 1228-1240, 1991.
7. J.-Y.Chen, P.-S.Tsai and C.-C.Wong, Adaptive design of a fuzzy cerebellar model arithmetic controller neural network, *IEE Proceedings - Control Theory and Applications*, vol.152, no.2, 133-137, 2005.
8. C.-T.Chiang and C.-S. Lin, CMAC with general basis functions, *Neural Networks*, vol. 9, no. 7, pp. 1199–1211, 1996.
9. C.-Y. Chang, Adaptive Fuzzy Controller of the Overhead Cranes With Nonlinear Disturbance, *IEEE Transactions on Industrial Informatics*, Vol.3, No. 2, 164 - 172, 2007.
10. S.K.Cho, H.H.Lee, A fuzzy-logic antiswing controller for three-dimensional overhead cranes,*ISA Trans.*, Vol.41, No.2, 235-43, 2002.
11. A.H.W.Chun, R.Y.M.Wong, Improving Quality of Crane-Lorry Assignments With Constraint Programming, *IEEE Transactions on Systems, Man, and Cybernetics, Part C*, Vol.37, No. 2, 268 - 277 2007.
12. G.Corriga, A.Giua, and G.Usai, An implicit gain-scheduling controller for cranes, *IEEE Trans. Control Systems Technology,* Vol,6, No.1, 15-20, 1998.
13. G.Cybenko, Approximation by superposition of sigmoidal activation function, *Math.Control, Sig Syst*, Vol.2, 303-314, 1989.
14. Y.Fang, W.E.Dixon, D.M.Dawson and E.Zergeroglu, Nonlinear coupling control laws for an underactuated overhead crane system, *IEEE/ASME Trans. Mechatronics*, Vol.8, No.3, 418-423, 2003.
15. InTeCo, *3DCrane: Installation and Commissioning Version 1.2*, Krakow, Poland, 2000.
16. R.Kelly, Global Positioning on overhead crane manipulators via PD control plus a classs of nonlinear integral actions, *IEEE Trans. Automat. Contr.*, vol.43, No.7, 934-938, 1998.
17. R.Kelly, A tuning procedure for stable PID control of robot manipulators, *Robotica*, Vol.13, 141-148, 1995.

18. B. Kiss, J. Levine, and P. Mullhaupt, A simple output feedback PD controller for nonlinear cranes, *Proc. Conf. Decision and Control*, pp. 5097–5101, 2000.
19. H.H.Lee, Modeling and control of a three-dimensional overhead crane, *Journal of Dynamic Systems, Measurement, and Control*, Vol.120,471-476, 1998.
20. H.H.Lee, A new motion-planning scheme for overhead cranes with high-speed hoisting, *Journal of Dynamic Systems, Measurement, and Control*, Vol.126,359-364, 2004
21. J. A.Méndez, L.Acosta, L.Moreno, S.Torres, G.N.Marichal, An application of a neural self-tuning controller to an overhead crane, *Neural Computing and Applications*, Vol.8, No.2, 143-150, 1999.
22. K.A.Moustafa and A.M.Ebeid, Nonlinear modeling and control of overhead crane load sway, *Journal of Dynamic Systems, Measurement, and Control*, Vol.110, 266-271, 1988.
23. M.W.Noakes, J.F.Jansen, Generalized input for damped-vibration control of suspended payloads, *Journal of Robotics and Autonomous Systems*, Vol.10, No.2, 199-205, 1992.
24. M.N.Nguyen, D.Shi, and C.Quek, FCMAC-BYY: Fuzzy CMAC Using Bayesian Ying–Yang Learning, *IEEE Trans. Syst., Man, Cybern. B*, vol.36, no.5, 1180-1190, 2006.
25. J.Sim, W.L.Tung and C.Quek, CMAC-Yager: A Novel Yager-Inference-Scheme-Based Fuzzy CMAC, *IEEE Trans. Neural Networks*, vol.17, no.6, 1394-1410, 2006.
26. S.-F.Su, Z.-J.Lee and Y.-P.Wang, Robust and Fast Learning for Fuzzy Cerebellar Model Articulation Controllers, *IEEE Trans. Syst., Man, Cybern. B*, vol.36, no.1, 203-208, 2006.
27. E.D.Sontag and Y.Wang, On characterization of the input-to-state stability property, *System & Control Letters*, Vol.24, 351-359, 1995.
28. O.Sawodny, H.Aschemann and S.Lahres, An automated gantry crane as a large workspace robot, *Control Engineering Practice*, Vol.10, No.12, 1323-1338, 2002.
29. Y.Sakawa and Y.Shindo, Optimal control of container cranes, *Automatica,* Vol.18, No.3, 257-266, 1982.
30. W. Singhose, W. Seering and N. Singer, Residual vibration reduction using vector diagrams to generate shaped inputs, *Journal of Dynamic Systems, Measurement, and Control*, Vol.116, 654 -659,1994.
31. M.Takegaki and S.Arimoto, A new feedback method for dynamic control of manipulator, *ASME J. Dynamic Syst. Measurement, and Contr.*, Vol.103, 119-125, 1981.
32. R.Toxqui, W.Yu, and X.Li, PD control of overhead crane systems with neural compensation, *Advances in Neural Networks -ISNN 2006, Springer-Verlag, Lecture Notes in Computer Science, LNCS 3972*, 1110-1115, 2006
33. L.X.Wang, *Adaptive Fuzzy Systems and Control*, Englewood Cliffs NJ: Prentice-Hall, 1994.
34. S.Wu and M.J.Er, Dynamic fuzzy neural networks- a novel approach to function approximation, *IEEE Trans. Syst., Man, Cybern. B*, Vol.30, 358-364, 2000.
35. J.Yu, F.L.Lewis and T.Huang, Nonlinear feedback control of a gantry crane, *Proc. 1995 American Control Conference,* Seattle, 4310-4315, USA, 1995.

Part II
Neural Control

Chapter 5
Estimation and Control of Nonlinear Discrete-time Systems

Balaje T. Thumati and Jagannathan Sarangapani

Abstract Real world systems with unknown dynamics, uncertainties and disturbances pose a great challenge in developing control schemes. As a consequence, conventional controllers cannot be used for such nonlinear systems acted upon uncertainties. Hence, more advanced techniques such as robust control to mitigate disturbances and uncertainties, adaptive control using neural networks to learn unknown dynamics, and robust adaptive control to mitigate unknown dynamics and uncertainties have been introduced in the past decade. As a first step, the unknown system dynamics have to be constructed using online approximates such as neural networks and then controllers have to be designed. In this chapter, estimation or system identification and control using neural networks is thoroughly investigated. To begin with, some basic concepts on the neural networks are introduced, and then some stability properties are proposed for a general class of nonlinear discrete-time systems. Later, the development of nonlinear estimation techniques for a class of nonlinear discrete-time systems is introduced. Finally, neural networks control strategies are presented for controlling a class of nonlinear discrete-time system. Rigorous mathematical results for asymptotic stability on the estimation and control scheme are demonstrated. Additionally, some simulation examples are used to illustrate the performance of the estimation and control schemes, respectively.

Balaje T. Thumati
Department of Electrical and Computer Engineering, Missouri University of Science and Technology, Rolla, MO, USA, e-mail: sarangap@umr.edu

Jagannathan Sarangapani
Department of Electrical and Computer Engineering, Missouri University of Science and Technology, Rolla, MO, USA, e-mail: sarangap@umr.edu

5.1 Background

Neural networks are used for learning any unknown nonlinear function, which in turn helps in the estimation and control of nonlinear systems. Typically, there exists a wide range of neural network architectures for approximating any unknown function. In the following discussion, the basic concepts on neural networks are explained. Subsequently, some definitions on the stability of a general class nonlinear system are introduced.

5.1.1 Neural Networks

A general nonlinear continuous function $f(x) \in C^p(s)$ which maps $f : S \rightarrow R^p$ where S is a simply-connected set of R^n and $C^p(s)$ is the space where f is continous can be written as

$$f(x) = w^T \sigma(V^T x) + \varepsilon_1(k), \tag{5.1}$$

where V and W represent input-to-the-hidden layer and hidden-to-the-output layer weights respectively and $\varepsilon_1(k)$ is a neural net functional reconstruction error vector such that $||\varepsilon_1|| \leq \varepsilon_{1_N}$ [10] for all $x \in R^n$. Additionally the activation functions $\sigma(.)$ are bounded, measurable, non-decreasing functions from the real numbers onto [0, 1] which include for instance sigmoid etc. We define the output of a neural network as

$$\hat{y}(k) = \hat{W}^T(k)\varphi(x(k)), \tag{5.2}$$

where $\hat{W}(k)$ is the actual weight matrix and the input-to-the-hidden layer weights are considered identity and $\varphi(x(k))$ is the activation function, which is selected as a basis function [7] that guarantees function approximation. (5.2) represents a single layer network and Figure 5.1 shows a single layer neural network. Usually the input layer weights are considered fixed, and the weights of the output layer are tuned. Later in this chapter, different ways of tuning the neural network weights are discussed in detail. Similar to the case of a single layer neural network, a given continuous function *f(.)* could be written using a three layer neural network as [16]

$$f(x) = W_3^T \varphi_3(W_2^T \varphi_2(W_1^T \varphi_1(x(k)))) + \varepsilon_1(k), \tag{5.3}$$

where W_1, W_2 and W_3 are the ideal weights of the multilayer neural network respectively. Additionally, the ideal weights are considered to be bounded such that $||W_1|| \leq W_{1max}$, $||W_2|| \leq W_{2max}$ and $||W_3|| \leq W_{3max}$ with $\varphi_1(.)$, $\varphi_2(.)$ and $\varphi_3(.)$ are the activation functions of the first, second and third layers of the neural network respectively. Next we define the output of a three layer neural network as

$$\hat{y}(k) = \hat{W}_3^T(k)\hat{\varphi}_3(\hat{W}_2^T(k)\hat{\varphi}_2(\hat{W}_1^T(k)\hat{\varphi}_1(x(k)))), \tag{5.4}$$

where $\hat{W}_3(k), \hat{W}_2(k), \hat{W}_1(k)$ are the actual neural network weights of the third, second and first layer respectively and $\hat{\varphi}_1(x(k))$ represent the activation function vector of the input layer. Then $\hat{\varphi}_2(\hat{W}_1^T(k)\hat{\varphi}_1(x(k)))$, $\hat{\varphi}_3(\hat{W}_2^T(k)\hat{\varphi}_2(\hat{W}_1^T(k)\hat{\varphi}_1(x(k))))$ denote the hidden layer and output layers activation function respectively at the kth instant. For a multilayer function approximation, the activation function vector need not form a basis function [16]. Figure 5.2 shows a three layer network where the activations functions of the three layer network are shown. Here again the network weights of the first, second, third are tuned online. As a note, a three layer network renders a better function approximation rather than a single layer network.

With this basic understanding about the neural networks, we introduce some stability concepts that are required to understand the theoretical development in the following section.

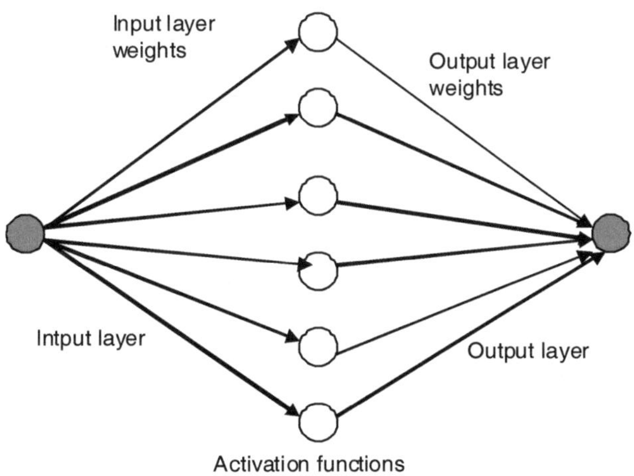

Fig. 5.1 A single layer neural network architecture

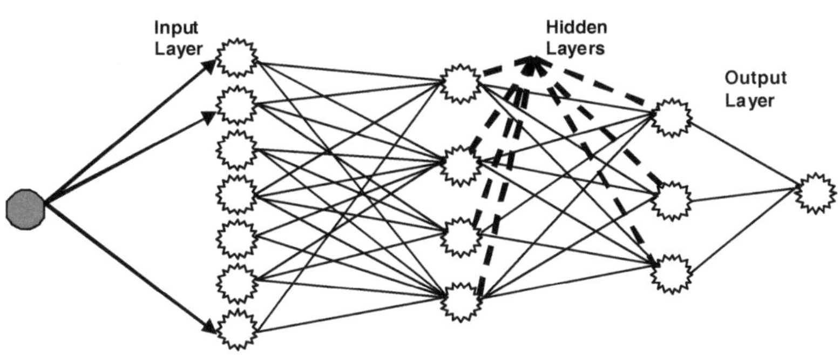

Fig. 5.2 A three layer neural network architecture

5.1.2 Stability Concepts

For a nonlinear system described by

$$x(k+1) = f(x(k)), \tag{5.5}$$

the equilibrium point $x = 0$ is locally asymptotically stable if the following definition is satisfied.

Definition 5.1. The solution for (5.5) is locally asymptotically stable if for any Ω, a compact subset of S, where $S \in R^n$ and for all $x(k_0) \in \Omega$, then for all $k \geq k_0, x(k) \to 0$.

This definition shows the criteria for the asymptotic convergence of the solution for a general class of nonlinear systems. Further definitions and explanations on stability please refer to [9].

Next, the development of the estimation scheme using neural networks is proposed in the following section. Initally, some previous developments on the estimation or identification scheme is presented. Later in the section, the estimation scheme is developed and some simulation examples are used to illustrate the estimation scheme.

5.2 Estimation of an Unknown Nonlinear Discrete-time System

5.2.1 Introduction

Identifying unknown system dynamics is an important research interest across the widespread control scientific community. This is possibly due to a large number of real-time applications wherein the system dynamics is unavailable *a priori*. Typical applications include aerospace, automotive, power system and process industries. With this widespread need for a reliable identification or estimation scheme, there has been a significant amount of research to identify nonlinear system dynamics. The need for understanding the behavior of unknown nonlinear systems began in the late 1960s, when the most prominent Kalman filtering technique was introduced to identify linear systems with unknown parameters [1]. An extended Kalman filtering technique [2] was introduced for nonlinear systems where the nonlinear system is linearized at each time instant and is identified by using the well known linear estimation approach [2]. Other techniques are available for estimating linear and nonlinear systems [3]. However, the Kalman filter technique is preferred or proven reliable over others.

Other techniques that are widely used for identifying nonlinear systems include autoregressive (AR) and autoregressive moving average (ARMA) based modeling schemes, and nonlinear and linear and autoregressive exogenous (ARX) modeling schemes [4]. Additionally, a nonlinear variant of the above listed techniques are available for predicting nonlinear series using only input-output measurements [5]. Yet, another technique that has become most popular in recent years is the use of neural networks for the approximation of nonlinear functions in the abovementioned models. Commonly used neural network architectures include feed-forward neural networks, radial basis function neural networks, recurrent neural networks, recurrent higher order neural networks, and so on [6].

In short, there are many schemes available for identifying linear and nonlinear systems. However, a few schemes [2, 8] address stability. Detailed stability analysis is available for the estimation and control of linear systems. For instance the stability analysis of Kalman filter and extended Kalman filtering technique is available [2]. However, the stability analysis of nonlinear neural network based schemes for the estimation and control is beginning to appear in the literature [22]. Due to online learning capabilities, neural networks are now utilized for estimation and control of nonlinear continuous-time and discrete-time systems.

Recurrent higher order networks were used for identifying a class of nonlinear systems in [8] and stability analysis is also demonstrated using Lyapunov theory. Additionally, the stability proof is much easier in continuous-time since the first derivative of the Lyapunov function is linear with respect to the states whereas in discrete-time the first difference is quadratic making it difficult to prove stability. In [7], a suite of neural network-based schemes were developed for identifying a class of nonlinear discrete-time systems. However, only boundedness of the identification error is demonstrated. Hence, in this chapter, the main objective is to develop a stable neural network based estimator for identifying an unknown nonlinear discrete-time system with guaranteed performance. In the next section, asymptotic stability of the identification system is demonstrated.

In this chapter, a nonlinear estimator consisting of an neural network-based online approximator with a robust term is utilized to identify unknown nonlinear discrete-time systems subjected to noise and bounded disturbances. To learn the unknown nonlinear system dynamics, a stable adaptive weight update law is proposed for tuning the nonlinear estimator. The significance of introducing a robust term, which is a function of the estimation error and an additional tunable parameter, is to guarantee asymptotic stability. The additional parameter of the robust term is also tuned online using a novel parameter update law for the general case whereas this parameter can be held fixed for ideal scenarios. By using Lyapunov theory, the asymptotic stability of the proposed nonlinear estimation scheme is proven by using mild assumptions such as an upper bound on the system uncertainties and the approximation error [12, 13]. In the next section, the system under investigation is explained in detail and later the estimation scheme is introduced.

5.2.2 Nonlinear Dynamical System

Consider the following estimators utilized for system identification commonly used in the literature [7], and are used for a wide class of nonlinear discrete-time system

[6] as

$$x_e(k+1) = \sum_{i=0}^{n-1} \alpha_i x_e(k-i) + g(u(k), u(k-1), ..., u(k-m+1)) + \eta(k) \quad (5.6)$$

$$x_e(k+1) = f(x_e(k), x_e(k-1), ..., x_e(k-n+1)) + \sum_{i=0}^{m-1} \beta_i u(k-i) + \eta(k) \quad (5.7)$$

$$\begin{aligned} x_e(k+1) =& f(x_e(k), x_e(k-1), .., x_e(k-n+1)) + g(u(k), u(k-1), .., \\ & u(k-m+1) + \eta(k), \end{aligned} \quad (5.8)$$

$$\begin{aligned} x_e(k+1) =& f(x_e(k), x_e(k-1), .., x_e(k-n+1); u(k), u(k-1), .., \\ & u(k-m+1)) + \eta(k), \end{aligned} \quad (5.9)$$

where $x_e(k) \in R^n$ is the system state, $f(.) \in R^n$, $g(.) \in R^n$ are the unknown nonlinear functions, $\alpha_i \in R^{n\times n}$, $\beta_i \in R^{n\times n}$ are unknown constant coefficient matrices, $u(k) \in R^n$ is the input vector, and $\eta(k) \in R^n$ is the unknown disturbance vector with an upper bound, i.e., $||\eta(k)|| \leq \eta_M$, where η_M a known constant. The initial state vector is considered available, i.e., $x(0) = x_0$. The system representation discussed in (5.6)-(5.9) present a suite of nonlinear autoregressive moving average (NARMA) models that are commonly deployed in the literature. For more information on these models, please refer to [6]. In the next section, a suite of nonlinear estimators are proposed,where the estimator learns the unknown system dynamics while guaranteeing aymptotic stability.

5.2.3 Identification Scheme

The identification scheme using the neural network has a similar structure as that of the unknown system in (5.6)-(5.9)

$$\hat{x}_e(k+1) = \sum_{i=0}^{n-1} \alpha_i x_e(k-i) + \hat{g}(u(k), u(k-1), ..., u(k-m+1)) - v(k), \quad (5.10)$$

$$\hat{x}_e(k+1) = \hat{f}(x_e(k), x_e(k-1), ..., x_e(k-n+1)) + \sum_{i=0}^{m-1} \beta_i u(k-i) - v(k), \quad (5.11)$$

$$\begin{aligned} \hat{x}_e(k+1) =& \hat{f}(x_e(k), x_e(k-1), ..., x_e(k-n+1)) \\ & + \hat{g}(u(k), u(k-1), ..., u(k-m+1) - v(k), \end{aligned} \quad (5.12)$$

$$\begin{aligned} \hat{x}_e(k+1) =& \hat{f}(x_e(k), x_e(k-1), ..., x_e(k-n+1); \\ & u(k), u(k-1), ..., u(k-m+1)) - v(k), \end{aligned} \quad (5.13)$$

where $\hat{f}(.)$ is an estimate of $f(.)$ and $\hat{g}(.)$ is an estimate of $g(.)$, and $v(k)$ is a robust term, which will be explained later. In general, as presented in the above identification model, the idea is to develop a suitable model, which could be tuned online by using the available states and inputs, the model must estimate the entire state vector $x_e(k)$. With this understanding of the problem definition, next neural networks are employed to provide $\hat{f}(.)$ and $\hat{g}(.)$. To tune the neural network online, we define the state estimation error as

$$e_e(k) = x_e(k) - \hat{x}_e(k). \tag{5.14}$$

Therefore, the estimation error dynamics for the system discussed in (5.6)-(5.9) are given by

$$e_e(k+1) = \tilde{g}(.) + \eta(k) + v(k), \tag{5.15}$$
$$e_e(k+1) = \tilde{f}(.) + \eta(k) + v(k), \tag{5.16}$$
$$e_e(k+1) = \tilde{f}(.) + \tilde{g}(.) + \eta(k) + v(k), \tag{5.17}$$
$$e_e(k+1) = \tilde{f}(.) + \eta(k) + v(k), \tag{5.18}$$

where $\tilde{f}(.) = f(.) - \hat{f}(.)$, $\tilde{g}(.) = g(.) - \hat{g}(.)$ represent the functional approximation errors. It is important to note that (5.16) and (5.18) are similar expect the fact that the $\tilde{f}(.)$ in (5.18) has the current and delayed values of the states $x_e(k)$ and the input $u(k)$. Then both equations could be rewritten as

$$e_e(k+1) = \tilde{f}(\bar{x_e}(k)) + \eta(k) + v(k), \tag{5.19}$$

where $\bar{x_e}(k)$ is the argument of $\tilde{f}(.)$, which consists of $x_e(k)$ and its previous values in (5.16) and also includes $u(k)$ and its previous values in (5.18). This error system could be a result of either (5.7) or (5.9). Similarly, by considering a general case, the error system in (5.15) and (5.17) could be taken as

$$e_e(k+1) = \tilde{f}(\bar{x_e}(k)) + \tilde{g}(\bar{u}(k)) + \eta(k) + v(k), \tag{5.20}$$

where $\bar{u}(k)$ denotes $u(k)$ and its previous values. For the following discussion, the error system given in (5.19) and (5.20) will be referred. Next, the robust term is defined based on the error system (5.19) and (5.20) as

$$v(k) = \frac{\hat{\lambda}_e(k) e_e(k)}{e_e^T(k) e_e(k) + c_e}, \tag{5.21}$$

where $\hat{\lambda}_e(k) \in R$ is an additional tunable parameter and $c_e > 0$ is a constant. The parameter $\hat{\lambda}_e(k)$ could be considered either tunable or a constant. Later in this chapter, we highlight the importance of the parameter $\hat{\lambda}_e(k)$. Additionally depending upon the identification model, the error dynamics used in (5.21) would vary.

5.2.4 Neural Network Structure Design

A single layer neural network is used for identifying the unknown nonlinear system dynamics. For the error system in (5.19), a single layer neural network will be utilized, but for the error system in (5.20), two single layer neural network will be utilized. Therefore, the nonlinear function $f(.)$ in (5.19) and (5.20), and the nonlinear function $g(.)$ in (5.20) could be written on a compact set S as

$$f(\bar{x_e}(k)) = w_f^T \varphi_f(\bar{x_e}(k)) + \varepsilon_{1f}(k), \tag{5.22}$$

$$g(\bar{u}(k)) = w_g^T \varphi_g(\bar{u}(k)) + \varepsilon_{1g}(k), \tag{5.23}$$

where $w_f \in R^{l\times n}$, $w_g \in R^{q\times n}$ are the ideal weights such that $||w_f|| \leq w_{fmax}$, $||w_g|| \leq w_{gmax}$ are bounded. The term $\varepsilon_{1f}(k)$ and $\varepsilon_{1g}(k)$ are defined as the neural network approximation errors, φ_f and φ_g are the neural networks basis functions, which could be a sigmoid function, radial basis function (RBF) etc.

Additionally, the basis functions are selected such that they are bounded, i.e., $\varphi_f \leq ||\varphi_{fmax}||$ and $\varphi_g \leq ||\varphi_{gmax}||$. Next, the estimate $\hat{f}(.)$ and $\hat{g}(.)$ could be written as

$$\hat{f}(\bar{x_e}(k)) = \hat{w}_f^T(k)\varphi_f(\bar{x_e}(k)), \tag{5.24}$$

$$\hat{g}(\bar{u}(k)) = \hat{w}_g^T(k)\varphi_g(\bar{u}(k)), \tag{5.25}$$

where $\hat{w}_f$ and $\hat{w}_g$ are the estimated weights of the neural networks. The following weight update laws are proposed for tuning the neural network weights as

$$\hat{w}_f(k+1) = \hat{w}_f(k) + \alpha_f \varphi_f(\bar{x_e}(k)) e_e^T(k+1), \tag{5.26}$$

$$\hat{w}_g(k+1) = \hat{w}_g(k) + \alpha_g \varphi_g(\bar{u}(k)) e_e^T(k+1), \tag{5.27}$$

where $\alpha_f > 0$ and $\alpha_g > 0$ are the learning rates. The update law for tuning $\hat{\lambda}_e(k)$ is defined as

$$\hat{\lambda}_e(k+1) = \hat{\lambda}_e(k) - \gamma_e \frac{e_e^T(k+1)e_e(k)}{e_e^T(k)e_e(k)+c_e}, \tag{5.28}$$

where $\gamma_e > 0$ is a design parameter. Alternatively a standard delta-based parameter tuning algorithm can be utilized whereas it is slow in convergence [7]. Next, we present the following stability analysis for the error system in (5.19), whereas the analysis is quite similar for the error system (5.20) and therefore omitted. The error system in (5.19) can be rewritten as

$$e_e(k+1) = \tilde{w}_f^T(k)\varphi_f(\bar{x_e}(k)) + \frac{\hat{\lambda}_e(k)e_e(k)}{e_e^T(k)e_e(k)+c_e} + \varepsilon_f(k), \tag{5.29}$$

where $\varepsilon_f(k) = \varepsilon_{1f}(k) + \eta(k)$, and the neural network estimation error is defined as $\tilde{w}_f(k) = w_f - \hat{w}_f(k)$. Next, the following assumption is required in order to proceed further.

Assumption 5.1 *The term $(\varepsilon_f(k))$ comprising of the approximation error $(\varepsilon_{1f}(k))$ and the system uncertainty $(\eta(k))$ are assumed to be upper bounded by a function of the estimation error [12, 13], i.e.,*

$$\varepsilon_f^T(k)\varepsilon_f(k) \leq \varepsilon_{1f}^T(k)\varepsilon_{1f}(k) + \eta^T(k)\eta(k) \leq \varepsilon_M = \lambda_e^* e_e^T(k)e_e(k), \tag{5.30}$$

where λ_e^* is the target value with an upper bound $||\lambda_e^*|| \leq \lambda_{e_{max}}$.

This assumption is stated by a number of control researchers [12, 13] and is considered mild in comparison with the approximation error being bound by a known constant. Additionally, the persistence of excitation condition is relaxed without the addition of an extra term which is typically referred as "ε -modification" [7].

Next by adding and subtracting $\frac{\lambda_e^* e_e(k)}{e_e^T(k)e_e(k)+c_e}$ in (5.30) and rewriting it as

$$e_e(k+1) = \psi_{1e}(k) - \psi_{2e}(k) + \varepsilon_f(k) + \frac{\lambda_e^* e_e(k)}{e_e^T(k)e_e(k)+c_e}, \tag{5.31}$$

where the parameter estimation error of the robust term is defined as $\tilde{\lambda}_e(k) = \lambda_e^* - \hat{\lambda}_e(k)$. Denote $\psi_{1e}(k) = \tilde{w}_f^T(k)\varphi_f(\bar{x_e}(k))$, and $\psi_{2e}(k) = \frac{\tilde{\lambda}_e(k)e_e(k)}{e_e^T(k)e_e(k)+c_e}$ for convenience.

Note the update laws in (5.26)-(5.28) guarantee local asymptotic stability of the proposed estimator scheme, which will be discussed next when compared to boundedness results from other works [5, 8]. Additionally, no prior offline training is needed for tuning the networks. This makes the proposed estimator scheme unique when compared to others and also could be used for learning any unknown system dynamics. In the following text, the stability analysis of the estimation scheme is presented by tuning the additional parameter in the robust term.

Theorem 5.1. *Let the proposed identification scheme in (5.11) and (5.13) be used to identify the system in (5.7) and (5.9) respectively, and the parameter update law given in (5.27) and (5.28) be used for tuning the neural network and the robust term respectively. In the presence of bounded uncertainties and disturbances, the state estimation error $e_e(k)$ is asymptotically stable while the neural network weight estimation errors $\tilde{w}_f(k)$ and parameter estimation errors $\tilde{\lambda}_e(k)$ respectively are bounded.*

Proof. Consider the Lyapunov function candidate as

$$V = e_e^T(k)e_e(k) + \frac{1}{\alpha_f}tr[\tilde{w}_f^T(k)\tilde{w}_f(k)] + \frac{1}{\gamma_e}\tilde{\lambda}_e^2(k),$$

whose first difference is given by

$$\Delta V = \underbrace{e_e^T(k+1)e_e(k+1) - e_e^T(k)e_e(k)}_{\Delta V_1} + \underbrace{\frac{1}{\alpha_f} tr[\tilde{w}_f^T(k+1)\tilde{w}_f(k+1) - \tilde{w}_f^T(k)\tilde{w}_f(k)]}_{\Delta V_2}$$

$$+ \underbrace{\frac{1}{\gamma_e}[\tilde{\lambda}_e^2(k+1) - \tilde{\lambda}_e^2(k)]}_{\Delta V_3}. \tag{5.32}$$

Substituting the error dynamics (5.31) in ΔV_1 of (5.32),we obtain

$$\Delta V_1 = (\psi_{1e}(k) - \psi_{2e}(k) + \varepsilon_f(k) + \frac{\lambda_e^* e_e(k)}{e_e^T(k)e_e(k) + c_e})^T(\psi_{1e}(k) - \psi_{2e}(k) +$$
$$\varepsilon_f(k) + \frac{\lambda_e^* e_e(k)}{e_e^T(k)e_e(k) + c_e}) - e_e^T(k)e_e(k).$$

By performing some mathematical manipulation, the above equation could be rewritten as

$$\Delta V_1 = \Psi_{1e}^T(k)\psi_{1e}(k) - 2\psi_{1e}^T\psi_{2e}(k) + \frac{2\lambda_e^*\psi_{1e}^T e_e(k)}{e_e^T(k)e_e(k) + c_e} +$$
$$+ 2\psi_{1e}^T(k)\varepsilon_f(k) + \psi_{2e}^T(k)\psi_{2e}(k) - \frac{2\lambda_e^*\psi_{2e}^T(k)e_e(k)}{(e_e^T(k)e_e(k) + c_e)} - 2\psi_{2e}^T(k)\varepsilon_f(k)$$
$$+ \frac{2\lambda_e^*\varepsilon_f^T(k)e_e(k)}{e_e^T(k)e_e(k) + c_e} + \frac{\lambda_e^{*2}e_e^T(k)e_e(k)}{(e_e^T(k)e_e(k) + c_e)^2} + \varepsilon_f^T(k)\varepsilon_f(k) - e_e^T(k)e_e(k).$$

Next upon the substitution of the weight update (5.26) law in ΔV_2 of (5.32) to get

$$\Delta V_2 = \frac{1}{\alpha_f} tr[(\Delta\tilde{w}_f(k) + \tilde{w}_f(k))^T(\Delta\tilde{w}_f(k) + \tilde{w}_f(k)) - \tilde{w}_f^T(k)\tilde{w}_f(k)],$$

where after some manipulation we get

$$\Delta V_2 = \frac{1}{\alpha_f} tr[\Delta\tilde{w}_f^T(k)\Delta\tilde{w}_f(k) + 2\Delta\tilde{w}_f^T(k)\tilde{w}_f(k)]$$
$$= \frac{1}{\alpha_f} tr[\alpha_f^2\varphi_f(e_e^T(k+1))(e_e(k+1)\varphi_f^T) - 2\alpha_f e_e(k+1)\varphi_f^T\tilde{w}_f(k)]$$
$$= \frac{1}{\alpha_f} tr[\alpha_f^2\varphi_f(\psi_{1e}(k) - \psi_{2e}(k) + \frac{\lambda_e^* e_e(k)}{e_e^T(k)e_e(k) + c_e})^T(\psi_{1e}(k)$$
$$- \psi_{2e}(k) + \varepsilon_f(k) + \frac{\lambda_e^* e_e(k)}{e_e^T(k)e_e(k) + c_e}\varphi_f^T - 2\alpha_f(\psi_{1e}(k) - \psi_{2e}(k) + \varepsilon_f(k)$$
$$+ \frac{\lambda_e^* e_e(k)}{e_e^T(k)e_e(k) + c_e}\varphi_f^T\tilde{w}_f(k)].$$

Apply the Cauchy-Schwarz inequality $((a_1+a_2+....+a_n)^T(a_1+a_2+....+a_n) \leq n(a_1^Ta_1+a_2^Ta_2+....+a_n^Ta_n))$ for the first term in the above equation, and apply the trace operator (given a vector $x \in R^n$, $tr(xx^T)=x^Tx$), to get

$$\begin{aligned}
&\leq 4\alpha_f\varphi_f^T\varphi_f(\psi_{1e}^T(k)\psi_{1e}(k)+\psi_{2e}^T(k)\psi_{2e}(k)+\frac{\lambda_e^{*2}e_e^T(k)e_e(k)}{(e_e^T(k)e_e(k)+c_e)^2}\\
&\quad+\varepsilon_f^T(k)\varepsilon_f(k))-2\psi_{1e}^T(k)(\psi_{1e}(k)-\psi_{2e}(k)+\varepsilon_f(k)+\frac{\lambda_e^*e_e(k)}{e_e^T(k)e_e(k)+c_e}).
\end{aligned}$$

Next substitute the parameter update (5.6) law in ΔV_3 of (5.32) to get

$$\begin{aligned}
\Delta V_3 =&\frac{1}{\gamma_e}(\tilde{\lambda}_e^2(k+1)-\tilde{\lambda}_e^2(k))\\
=&\frac{1}{\gamma_e}((\tilde{\lambda}_e(k)+\gamma_e\frac{e_e^T(k+1)e_e(k)}{e_e^T(k)e_e(k)+c_e})^2-\tilde{\lambda}_e^2(k))\\
=&2\psi_{1e}^T(k)\psi_{2e}(k)-2\psi_{2e}^T(k)\psi_{2e}(k)+\frac{2\lambda_e^*e_e(k)\psi_{2e}(k)}{e_e^T(k)e_e(k)+c_e}+2\varepsilon_f^T(k)\psi_{2e}(k)\\
&+\frac{\gamma_e(e_e^T(k+1)e_e(k))^2}{(e_e^T(k)e_e(k)+c_e)^2}.
\end{aligned}$$

Now observing that $\Delta V=\Delta V_1+\Delta V_2+\Delta V_3$, and combining the terms to get

$$\begin{aligned}
\Delta V \leq&\psi_{1e}^T(k)\psi_{1e}(k)-2\psi_{1e}^T(k)\psi_{2e}(k)+\frac{2\lambda_e^*\psi_{1e}^T(k)e_e(k)}{e_e^T(k)e_e(k)+c_e}\\
&+2\psi_{1e}^T(k)\varepsilon_f(k)+\psi_{2e}^T(k)\psi_{2e}(k)-\frac{2\lambda_e^*\psi_{2e}^T(k)e_e(k)}{e_e^T(k)e_e(k)+c_e}-2\psi_{2e}^T(k)\varepsilon_f(k)\\
&+\frac{2\lambda_e^*\varepsilon_f^T(k)e_e(k)}{e_e^T(k)e_e(k)+c_e}+\frac{\lambda_e^{*2}e_e^T(k)e_e(k)}{(e_e^T(k)e_e(k)+c_e)^2}+\varepsilon_f^T(k)\varepsilon_f(k)\\
&-e_e^T(k)e_e(k)+4\alpha_f\varphi_f^T\varphi_f\psi_{1e}^T(k)\psi_{1e}(k)+4\alpha_f\varphi_f^T\varphi_f\psi_{2e}^T(k)\psi_{2e}(k)\\
&+4\alpha_f\varphi_f^T\varphi_f\frac{\lambda_e^{*2}e_e^T(k)e_e(k)}{(e_e^T(k)e_e(k)+c_e)^2}+4\alpha_f\varphi_f^T\varphi_f\varepsilon_f^T(k)\varepsilon_f(k)-2\psi_{1e}^T(k)\psi_{1e}(k)\\
&+2\psi_{1e}^T(k)\psi_{2e}(k)-2\psi_{1e}^T(k)\varepsilon_f(k)-\frac{2\lambda_e^*\psi_{1e}^T(k)e_e(k)}{e_e^T(k)e_e(k)+c_e}+2\psi_{1e}^T(k)\psi_{2e}(k)\\
&-2\psi_{2e}^T(k)\psi_{2e}(k)+\frac{2\lambda_e^*\psi_{2e}^T(k)e_e(k)}{e_e^T(k)e_e(k)+c_e}+2\varepsilon_f^T(k)\psi_{2e}(k)+\frac{\gamma_e(e_e^T(k+1)e_e(k))^2}{(e_e^T(k)e_e(k)+c_e)^2}.
\end{aligned}$$

Cancelling similar terms in the above equation we get

$$\Delta V \leq -\psi_{1e}^T(k)\psi_{1e}(k) - \psi_{2e}^T(k)\psi_{2e}(k) + \underbrace{2\lambda_e^*\varepsilon_f^T(k)e_e(k)}_{1} + \underbrace{\frac{\lambda_e^{*2}e_e^T(k)e_e(k)}{(e_e^T(k)e_e(k)+c_e)^2}}_{1}$$
$$+\varepsilon_f^T(k)\varepsilon_f(k) - e_e^T(k)e_e(k) + 4\alpha_f\varphi_f^T\varphi_f\psi_{1e}^T(k)\psi_{1e}(k)$$
$$+4\alpha_f\varphi_f^T\varphi_f\psi_{2e}^T(k)\psi_{2e}(k) + 4\alpha_f\varphi_f^T\varphi_f\frac{\lambda_e^{*2}e_e^T(k)e_e(k)}{(e_e^T(k)e_e(k)+c_e)^2}$$
$$+4\alpha_f\varphi_f^T\varphi_f\varepsilon_f^T(k)\varepsilon_f(k) + 2\psi_{1e}^T(k)\psi_{2e}(k) + \frac{\gamma_e(e_e^T(k+1)e_e(k))^2}{(e_e^T(k)e_e(k)+c_e)^2}.$$

Again apply the Cauchy-Schwarz inequality ($2ab \leq a^2+b^2$) to terms numbered as 1 in the above equation to get

$$\Delta V \leq -\psi_{1e}^T(k)\psi_{1e}(k) - \psi_{2e}^T(k)\psi_{2e}(k) + \varepsilon_f^T(k)\varepsilon_f(k) + \frac{\lambda_e^{*2}e_e^T(k)e_e(k)}{(e_e^T(k)e_e(k)+c_e)^2}$$
$$+\frac{\lambda_e^{*2}e_e^T(k)e_e(k)}{(e^T(k)e(k)+c_e)^2} + \varepsilon_f^T(k)\varepsilon_f(k) - e_e^T(k)e_e(k) + 2\psi_{1e}^T(k)\psi_{2e}(k)$$
$$+4\alpha_f\varphi_f^T\varphi_f\psi_{1e}^T(k)\psi_{1e}(k) + 4\alpha_f\varphi_f^T\varphi_f\psi_{2e}^T(k)\psi_{2e}(k)$$
$$+4\alpha_f\varphi_f^T\varphi_f\frac{\lambda_e^{*2}e_e^T(k)e_e(k)}{(e_e^T(k)e_e(k)+c_e)^2} + 4\alpha_f\varphi_f^T\varphi_f\varepsilon_f^T(k)\varepsilon_f(k)$$
$$+\frac{\gamma_e((\psi_{1e}(k)-\psi_{2e}(k)+\varepsilon_f(k)+\frac{\lambda_e^*e_e(k)}{e_e^T(k)e_e(k)+c_e})^Te_e(k))^2}{(e_e^T(k)e_e(k)+c_e)^2}.$$

It is important to observe that the last term in the above equation could be written as

$$\frac{\gamma_e((\psi_{1e}(k)-\psi_{2e}(k)+\varepsilon_f(k)+\frac{\lambda_e^*e_e(k)}{e_e^T(k)e_e(k)+c_e})^Te_e(k))^2}{(e_e^T(k)e_e(k)+c_e)^2} \leq$$
$$\frac{4\gamma_e(\psi_{1e}^T(k)\psi_{1e}(k)+\psi_{2e}^T(k)\psi_{2e}(k)+\frac{\lambda_e^{*2}e_e^T(k)e_e(k)}{(e_e^T(k)e_e(k)+c_e)^2}+\varepsilon_f^T(k)\varepsilon_f(k))e_e^T(k)e_e(k)}{(e_e^T(k)e_e(k)+c_e)^2} \leq$$
$$4\gamma_e(\psi_{1e}^T(k)\psi_{1e}(k)+\psi_{2e}^T(k)\psi_{2e}(k)+\frac{\lambda_e^{*2}e_e^T(k)e_e(k)}{(e_e^T(k)e_e(k)+c_e)^2}+\varepsilon_f^T(k)\varepsilon_f(k)).$$

The above modification is possible by using the Cauchy-Schwarz inequality ($(a_1+a_2+....+a_n)^T(a_1+a_2+....+a_n) \leq n(a_1^Ta_1+a_2^Ta_2+....+a_n^Ta_n)$). After the application of this inequality, the first difference of the Lyapunov function is written as

$$\begin{aligned}\Delta V \leq & 2\varepsilon_f^T(k)\varepsilon_f(k) + 2\frac{\lambda_e^{*2} e_e^T(k)e_e(k)}{(e_e^T(k)e_e(k)+c_e)^2} - \psi_{1e}^T(k)\psi_{1e}(k) - \psi_{2e}^T(k)\psi_{2e}(k) \\ & + 4\gamma_e\psi_{1e}^T(k)\psi_{1e}(k) + 4\gamma_e\psi_{2e}^T(k)\psi_{2e}(k) + 4\gamma_e\frac{\lambda_e^{*2} e_e^T(k)e_e(k)}{(e_e^T(k)e_e(k)+c_e)^2} \\ & + 4\gamma_e\varepsilon_f^T(k)\varepsilon_f(k) - e_e^T(k)e_e(k) + 2\psi_{1e}^T(k)\psi_{2e}(k) + 4\alpha_f\varphi_f^T\varphi_f\psi_{1e}^T(k)\psi_{1e}(k) \\ & + 4\alpha_f\varphi_f^T\varphi_f\psi_{2e}^T(k)\psi_{2e}(k) + 4\alpha_f\varphi_f^T\varphi_f\frac{\lambda_e^{*2} e_e^T(k)e_e(k)}{(e_e^T(k)e_e(k)+c_e)^2} \\ & + 4\alpha_f\varphi_f^T\varphi_f\varepsilon_f^T(k)\varepsilon_f(k).\end{aligned}$$

Use the fact $\varepsilon_f^T(k)\varepsilon_f(k) \leq \lambda_e^* e_e^T(k)e_e(k)$ in the above equation results to get

$$\begin{aligned}\Delta V \leq & 2\lambda_e^* e_e^T(k)e_e(k) + 2\frac{\lambda_e^{*2} e_e^T(k)e_e(k)}{(e_e^T(k)e_e(k)+c_e)^2} - \psi_{1e}^T(k)\psi_{1e}(k) - \psi_{2e}^T(k)\psi_{2e}(k) + \\ & 2\psi_{1e}^T(k)\psi_{2e}(k) - e_e^T(k)e_e(k) + 4\alpha_f\varphi_f^T\varphi_f\psi_{1e}^T(k)\psi_{1e}(k) \\ & + 4\alpha_f\varphi_f^T\varphi_f\psi_{2e}^T(k)\psi_{2e}(k) + 4\alpha_f\varphi_f^T\varphi_f\lambda_e^* e_e^T(k)e_e(k) \\ & + 4\alpha_f\varphi_f^T\varphi_f\frac{\lambda_e^{*2} e_e^T(k)e_e(k)}{(e_e^T(k)e_e(k)+c_e)^2} + 4\gamma_e\psi_{1e}^T(k)\psi_{1e}(k) \\ & + 4\gamma_e\psi_{2e}^T(k)\psi_{2e}(k) + 4\gamma_e\frac{\lambda_e^{*2} e_e^T(k)e_e(k)}{(e_e^T(k)e_e(k)+c_e)^2} + 4\gamma_e\lambda_e^* e_e^T(k)e_e(k).\end{aligned}$$

Using the facts $\frac{e_e^T(k)e_e(k)}{(e_e^T(k)e_e(k)+c_e)^2} < e_e^T(k)e_e(k)$, $||\lambda_e^*|| \leq \lambda_{e_{max}}$, and taking the Frobenius norm, we get

$$\begin{aligned}\Delta V \leq & -(1 - 2\lambda_{e_{max}} - 2\lambda_{e_{max}}^2 - 4\alpha_f\varphi_{f_{max}}^2\lambda_{e_{max}} - 4\alpha_f\varphi_{f_{max}}^2\lambda_{e_{max}}^2 - 4\gamma_e\lambda_{e_{max}} \\ & - 4\gamma_e\lambda_{e_{max}}^2)||e_e(k)||^2 - (1 - 4\alpha_f\varphi_{f_{max}}^2 - 4\gamma_e)||\psi_{1e}(k)||^2 - (1 - 4\alpha_f\varphi_{f_{max}}^2 - \\ & 4\gamma_e)||\psi_{2e}(k)||^2 + 2||\psi_{1e}(k)||||\psi_{2e}(k)||.\end{aligned}$$

Next,$\psi_{1e}(k)$ is multiplied and divided by a constant k_1 , to get $\frac{k_1}{k_1}\psi_{1e}(k) = \frac{k_1}{k_1}\tilde{w}^T(k)\varphi_f(\bar{x_e}(k)) = k_1\psi_{1es}(k)$,where $\psi_{1es}(k) = \frac{\tilde{w}^T(k)\varphi_f(\bar{x_e}(k))}{k_1}$. Similarly, $\psi_{2e}(k)$ is multiplied and divided by a constant k_2, to get $\psi_{2e}(k) = k_2\psi_{2es}(k)$ and $\psi_{2es}(k) = \frac{\tilde{\lambda}_e(k)e_e(k)}{k_2(e_e^T(k)e_e(k)+c_e)}$. After performing these operations, the first difference is given by

$$\begin{aligned}\Delta V \leq & -(1 - 2\lambda_{e_{max}} - 2\lambda_{e_{max}}^2 - 4\alpha_f\varphi_{f_{max}}^2\lambda_{e_{max}} - 4\alpha_f\varphi_{f_{max}}^2\lambda_{e_{max}}^2 - 4\gamma_e\lambda_{e_{max}} - \\ & 4\gamma_e\lambda_{e_{max}}^2)||e_e(k)||^2 - k_1^2(1 - 4\alpha_f\varphi_{f_{max}}^2 - 4\gamma_e)||\psi_{1es}(k)||^2 - k_2^2(1 - 4\alpha_f\varphi_{f_{max}}^2 \\ & - 4\gamma_e)||\psi_{2es}(k)||^2 + 2k_1k_2||\psi_{1es}(k)||||\psi_{2es}(k)||.\end{aligned}$$

Define $\gamma_e = \alpha_f\varphi_{f_{max}}^2$, $\alpha_f \leq \frac{k_1^2}{8\varphi_{f_{max}}^2}$, $k_2 = \frac{\delta_e}{k_1}$; where $\delta_e > 0$, a constant. Using these in the above equation results in

$$\Delta V \leq -(1-2\lambda_{e_{max}}-2\lambda_{e_{max}}^2-k_1^2\lambda_{e_{max}}-k_1^2\lambda_{e_{max}}^2)||e_e(k)||^2-k_1^2(1-k_1^2)||\psi_{1es}(k)||^2$$
$$-k_2^2(1-k_2^2)||\psi_{2es}(k)||^2+2\delta_e||\psi_{1es}(k)||||\psi_{2es}(k)||.$$

Again applying the Cauchy-Schwartz inequality ($2ab \leq a^2+b^2$) to the last term in the above equation to get the first difference as

$$\Delta V \leq -(1-2\lambda_{e_{max}}-2\lambda_{e_{max}}^2-k_1^2\lambda_{e_{max}}-k_1^2\lambda_{e_{max}}^2)||e_e(k)||^2-k_1^2(1-k_1^2)||\psi_{1es}(k)||^2$$
$$-k_2^2(1-k_2^2)||\psi_{2es}(k)||^2+\delta_e||\psi_{1es}(k)||^2+\delta_e||\psi_{2es}(k)||^2.$$

Combining terms, the first difference is rewritten as

$$\Delta V \leq -(1-2\lambda_{e_{max}}-2\lambda_{e_{max}}^2-k_1^2\lambda_{e_{max}}-k_1^2\lambda_{e_{max}}^2)||e_e(k)||^2-$$
$$[(1-k_1^2)-\frac{\delta_e}{k_1^2}]||\tilde{w}_f^T(k)\varphi_f||^2-[(1-k_1^2)-\frac{k_1^2}{\delta_e}]||\tilde{\lambda}_e(k)||^2||\frac{e_e(k)}{(e_e^T(k)e_e(k)+c_e)}||^2. \tag{5.33}$$

Hence $\Delta V \leq 0$ provided the gains are taken as $k_1 \leq min(a_s,b_s,c_s,d_s)$ and $\delta_e < 1/4$, where $a_s = \sqrt{\frac{1+\sqrt{1-4\delta_e}}{2}}$, $b_s = \sqrt{\frac{1-\sqrt{1-4\delta_e}}{2}}$, $c_s = \sqrt{\frac{\delta_e}{1+\delta_e}}$, and $d_s = \sqrt{\frac{1-2\lambda_{e_{max}}-2\lambda_{e_{max}}^2}{\lambda_{e_{max}}+\lambda_{e_{max}}^2}}$. As long as the gains are selected as discussed above, $\Delta V \leq 0$ in (5.33) which shows stability in the sense of Lyapunov. Therefore $e_e(k)$, $\tilde{w}_f(k)$ and $\tilde{\lambda}_e(k)$ are bounded, provided $e_e(k_0)$, $\tilde{w}_f(k_0)$ and $\tilde{\lambda}_e(k_0)$ are bounded in the compact set S. Additionally by using [7], and summing both sides to infinity and taking limits when $k \to \infty$, it can be concluded that the estimation error $||e_e(k)||$ approaches zero with $k \to \infty$. □

In the above proof, the parameter, $\hat{\lambda}_e(k)$, of the robust term was assumed to be a tunable parameter. This could be used for a general class of nonlinear systems satisfying Assumption 5.1. However in an ideal case the parameter could be assumed to be a constant. This implies that the approximation error and the system uncertainty would be restricted to a more confined region rather than the region defined in Assumption 5.1. To show that the proposed estimation scheme would be stable with the new assumption on the parameter of the robust term, the following corollary is proposed. In this corollary the asymptotic convergence of the estimation error and the weights of the neural network are shown. Before understanding the stability of the revised estimation scheme, the error dynamics in (5.31) is now given by

$$e_e(k+1) = \psi_{1e}(k)+\varepsilon_f(k)+\frac{\lambda_e e_e(k)}{e_e^T(k)e_e(k)+c_e} \tag{5.34}$$

where $v(k) = \frac{\lambda_e e_e(k)}{e_e^T(k)e_e(k)+c_e}$ is the robust term and λ_e a constant. Next, the following corollary presents the stability results for the ideal case.

Corollary 5.1. *Consider the hypothesis given in Theorem 5.1 and the parameter update law given in (5.26) be used for tuning the neural network. In the presence* *of bounded uncertainties, the state estimation error $e_e(k)$ and the neural network estimation error $\tilde{w}_f(k)$ respectively are locally asymptotically stable.*

Proof. Define the Lyapunov function

$$V = e_e^T(k)e_e(k) + \frac{1}{\alpha_f} tr[\tilde{w}_f^T(k)\tilde{w}_f(k)].$$

After performing procedure similar to Theorem 5.1, the last step of the proof will be obtained as

$$\begin{aligned}\Delta V \leq & -(1-2\lambda_{e_{max}} - 2\lambda_{e_{max}}^2 - 3\alpha_f\varphi_{f_{max}}^2\lambda_{e_{max}} - 3\alpha_f\varphi_{f_{max}}^2\lambda_{e_{max}}^2)||e_e(k)||^2 \\ & -(1-3\alpha_f\varphi_{f_{max}}^2)||\psi_{1e}(k)||^2.\end{aligned} \tag{5.35}$$

Hence $\Delta V < 0$ provided the gains are selected as $\alpha_f \leq 1/3\varphi_{f_{max}}^2$ and $\lambda_{e_{max}} \leq 0.264$ This implies that the first difference is negative definite, i.e., $\Delta V < 0$ in (5.35) and the state estimation error $e_e(k)$ and the neural network weight estimation error, $\tilde{w}_f(k)$, respectively are locally asymptotically stable.

Mathematically, the stability of the identification scheme was demonstrated through the addition of the robust term. This term minimizes the effects of the approximation error and the system uncertainty. Obviously, one could always show similar stability results for the error system in (5.20) and using the weight update law (5.27). Till now, the four different identification models were introduced, their error dynamics were generated, and the stability was proven. To verify the mathematical results, in the next section, simulation results are used to illustrate the performance of the proposed estimation scheme. □

5.2.5 Simulation Results

Example 5.1. (Identification of an Unknown Nonlinear Dynamic System) Consider a single link robot system described by the following nonlinear discrete-time system [9] as

$$\begin{aligned}x_1(k+1) =& T(x_2(k)) + x_1(k) \\ x_2(k+1) =& T((-k_s/j_l)x_1(k) + (-f_l/j_l)x_2(k) + (k_s/j_l)x_3(k) \\ & + (-mglsin(x_1(k))/j_l) + sin(x_1(k))) + x_2(k) \\ x_3(k+1) =& T(x_4(k)) + x_3(k) \\ x_4(k+1) =& T((k_s/j_m)x_1(k) + (-k_s/j_m)x_3(k) + ((-f_m/j_m)x_4(k)) + (u(k)/j_m) + \\ & 0.05(mgl/j_l)sin(x_2(k))) + x_4(k)\end{aligned} \tag{5.36}$$

where $x(k) = [x_1(k), x_2(k), x_3(k), x_4(k)]^T$ denote system states. It was assumed that the above system is considered partially unknown, i.e., the system dynamics associated with the states $x_2(k)$ and $x_4(k)$ are unknown. Hence, we estimate the above defined system using the proposed neural network based estimator, as given next.

The nonlinear estimator is defined as

$$\begin{aligned}
\hat{x}_1(k+1) =& T(x_2(k)) + x_1(k) + 0.1(x_1(k) - \hat{x}_1(k)) \\
\hat{x}_2(k+1) =& f(x_1(k)) + 0.1(x_2(k) - \hat{x}_2(k)) + v_1(k) \\
\hat{x}_3(k+1) =& T(x_4(k)) + x_3(k) + 0.1(x_3(k) - \hat{x}_3(k)) \\
\hat{x}_4(k+1) =& f_2(x_4(k)) + 0.5(x_4(k) - \hat{x}_4(k)) + v_2(k)
\end{aligned} \tag{5.37}$$

where $\hat{x}(k) = [\hat{x}_1(k), \hat{x}_2(k), \hat{x}_3(k), \hat{x}_4(k)]^T$ are the estimated states. Here two single layer neural networks with sigmoid activation functions are employed.

However the size of $f_1(.)$ has eighteen neurons whereas the network that approximates $f_2(.)$ has sixteen neurons. The networks are tuned using the parameter update law defined in (5.26). The learning rate in (5.26) for the neural network $f_1(.)$ is taken as $\alpha_f = 0.09$ whereas for $f_2(.)$ is taken as $\alpha_f = 0.03$. Additionally, the two robust terms are taken as

$$v_1(k) = \frac{\hat{\lambda}_e(k) e_{e_2}(k)}{e_e^T(k) e_e(k) + c_e} \tag{5.38}$$

$$v_2(k) = \frac{\hat{\lambda}_e(k) e_{e_4}(k)}{e_e^T(k) e_e(k) + c_e}. \tag{5.39}$$

The parameter $(\hat{\lambda}_e(k))$ is tuned online using the update law in (5.28) with $\gamma_e = 0.0027$. The values of the system parameter used in this simulation are taken as $k_s = 2$, $j_l = 2$, $j_m = 1$, $f_l = 0.5$, $m = 4$, $g = 9.8$, $l = 0.5$, $f_m = 1$, $x_1(0) = 0$, $x_2(0) = 0$, $x_3(0) = 0$, $x_4(0) = 0$, $\hat{x}_1(0) = 0$, $\hat{x}_2(0) = 0$, $\hat{x}_3(0) = 0$, $\hat{x}_4(0) = 0$, and $c_e = 0.75$. Figures 5.3 to 5.5 display the response of the the nonlinear estimator for the two states $x_2(k)$ and $x_4(k)$ along with the Frobenius norm on the neural network weights respectively. From Figures 5.3 and 5.4, it could be observed that the state estimation error is close to zero which indicates that the unknown system dynamics are being approximated accurately. Figure 5.5 illustrates the Frobenius norm of the representative neural network weights, where it is observed that the neural network weights are bounded. Hence the proposed estimation scheme guarantees a highly satisfactory learning of the unknown system dynamics.

Example 5.2. (Fault Detection) A problem similar to system identification is the fault detection and learning. In this case, conventional residuals and threshold based techniques are used for detecting faults in a system [20, 21]. However, the fault detection will be more accurate if the fault dynamics are better approximated. Here the above proposed identification scheme would be readily be used for the online approximation of the fault dynamics. Briefly, for fault detection, a nonlinear estimator is developed with the knowledge of the given system, then by generating residuals (residuals are obtained by comparing the output of the estimator with that of the given system), faults in the system are detected. Subsequently, a neural network based online approximator is used for approximating the fault dynamics online. With this understanding, the following example is used to illustrate the fault detection (FD) scheme, which is based on the above presented system identification technique.

Consider a fourth order translational oscillator with rotating actuator (TORA) system to test the FD scheme. In this simulation, system uncertainty is assumed to be present. The modified discrete-time state space model of the TORA system is given below [9] as

$$x(k+1) = p(x(k),u(k)) + \eta(x(k),u(k)) + h(x(k),u(k)) \tag{5.40}$$

where the system states are given by the state vector $x(k) = [x_1(k), x_2(k), x_3(k), x_4(k)]^T$, $\eta(x(k),u(k))$ is the system uncertainity,$h(x(k),u(k)) = [0,0,0, (1000000000(sin(\omega_f x_1(k))))]^T$ is the induced spring stiffness abrupt fault, where $\omega_f = 10^{-9}$ and $k = 25sec$ is the time instant when the fault occurs. The nominal dynamics of the system is given by

$$p(x(k),u(k)) = \begin{vmatrix} t_s(x_2(k)) \\ t_s(\frac{1}{d}).((m+M)u(k) - mLcos(x_1(k)).(mLx_2^2(k)sin(x_1(k)) - k_s x_3(k))) \\ t_s(x_4(k)) \\ t_s(\frac{1}{d})(-mLcos(x_1(k))u(k) + (I+mL^2).(mL.x_2^2(k)sin(x_1(k)) - k_s x_3(k))) \end{vmatrix}$$

$d = (I+mL^2)(m+M) - m^2L^2cos^2(x_1(k))$, and input $u(k) = 0.5sin(kt_s)$.

To monitor the chosen system in (5.40) for detecting and learning the unknown fault that occurs in the system, the following nonlinear estimator is used as

$$\hat{x}(k+1) = A(x(k) - \hat{x}(k)) + p(x(k),u(k)) + \hat{h}(x(k),u(k);\hat{w}(k)) - v(k) \tag{5.41}$$

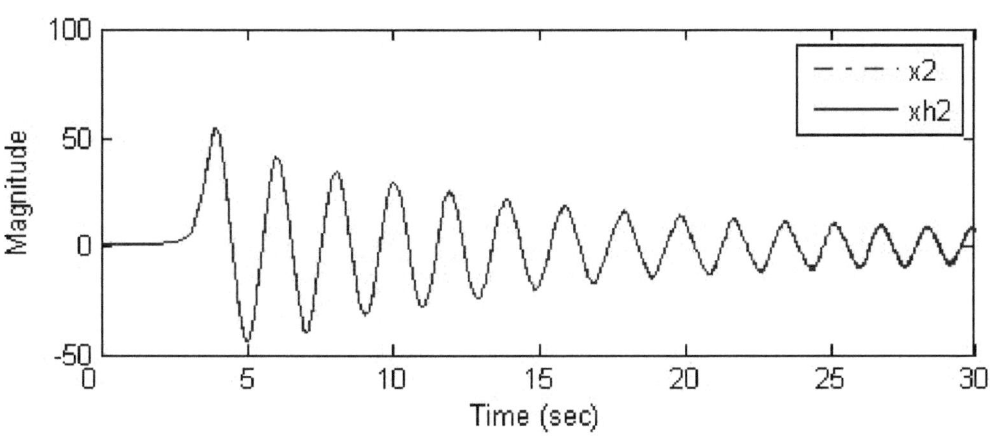

Fig. 5.3 Response of the estimated state $\hat{x}_2(k)$ and the actual state $x_2(k)$

where the estimated states are $\hat{x}(k) = [\hat{x}_1(k), \hat{x}_2(k), \hat{x}_3(k), \hat{x}_4(k)]^T$, $A = 0.25I_m$, with I_m is an appropriate dimensioned identity matrix. The online approximator in discrete-time (OLAD) $\hat{h}(x(k), u(k); \hat{w}(k)) = \hat{w}^T(k)\varphi(x(k), u(k))$ approximates the unknown fault dynamics, but is not triggered until the fault is detected. In this example, it is observed that the problem of fault detection is similar to system identification. However, prior to the fault detection, there is no need for any online approximation of the unknown system dynamics, which in this case would be the unknown fault dynamics. Additionally, the robust term is taken as $v(k) = \frac{\hat{\lambda}_e(k)e(k)}{e^T(k)e(k)+c_e}$, but is not initiated until a fault is detected. The values of the system parameter used for this simulation are taken as $m = 2$, $M = 10$, $k_s = 1$, $L = 0.5$, $I = 2$, and $t_s = 0.01$. In this simulation, assume that the system uncertainty exists from the initial time of the system operation, i.e., $\eta(x(k), u(k)) = [0, 0, 0, sin(0.005)]^T$ for all $k > 0$. Because of the system uncertainty, there is a need of a threshold to improve the robustness of the FD scheme. To detect the fault, the estimation error is monitored as shown in Figure 5.6. Note the threshold is derived based on $\rho_i = \frac{\beta_{c_i}\eta_{i_M}}{(1-\mu_i)}$ [21], where $\eta_{i_M} = 1$, $\beta_{c_i} = 0.45$, $\mu_i = 0.1$, and $i = 4$, thus rendering a constant threshold value, i.e., $\rho_4 = 0.5$ [20]. Note that the fault is detected quickly, in 0.11 seconds after its occurrence. Therefore, even in the presence of some system uncertainty, the scheme is able to detect the fault quickly. Next to understand the learning capabilities of the FD scheme, the response of the OLAD and the unknown fault dynamics are shown in Figure 5.7 where the OLAD used is a single layer network with 12 sigmoid activation functions.The design constant of the parameter update law in (5.26) is chosen as $\alpha_e = 0.0231$. Additionally, for the update law in (5.28), $\gamma_e = 1.0231$.

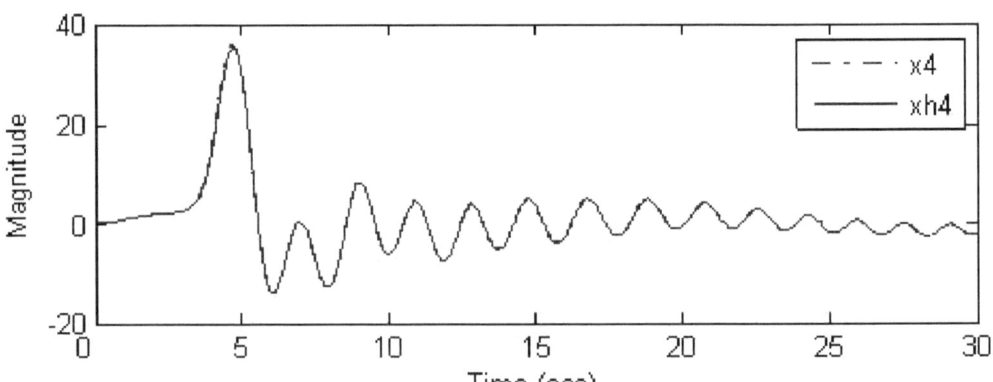

Fig. 5.4 Response of the estimated state $\hat{x}_4(k)$ and the actual state $x_4(k)$

5.3 Neural Network Control Design for Nonlinear Discrete-time Systems

5.3.1 Introduction

In the previous section, a new nonlinear estimator design was proposed and the mathematical stability was analyzed rigorously. In this section, a novel neural network based direct adaptive control strategy is proposed for a class of nonlinear discrete-time systems with guarantees of asymptotic stability. The neural network controller allows the system to track a reference trajectory, but could certainly be used for a set point regulation. Although there are significant attempts made in the neural network control of nonlinear discrete-time systems [7, 11], most of them assure uniformly ultimately bounded (UUB) stability. Next, an asymptotically stable neural network control strategy for nonlinear discrete-time systems is presented.

Significant research has been performed in the past decade in the area of neural network control of nonlinear systems. The neural networks became popular [14] due to their function approximation capabilities, which are utilized to learn system uncertainties. The neural network controller designs were first introduced for continuous-time systems [11, 15] and later extended to control nonlinear discrete-time systems [16,17]. Development of stable controllers for discrete-time systems is rather difficult since the first difference of a Lyapunov function candidate is not linear with respect to its states in contrast to a first derivative of a Lyapunov candidate for continuous-time systems. Additionally, recently developed neural network controllers relax the requirement of the persistence of excitation condition. The neural network controller designs were then extended to a more general class of nonlinear

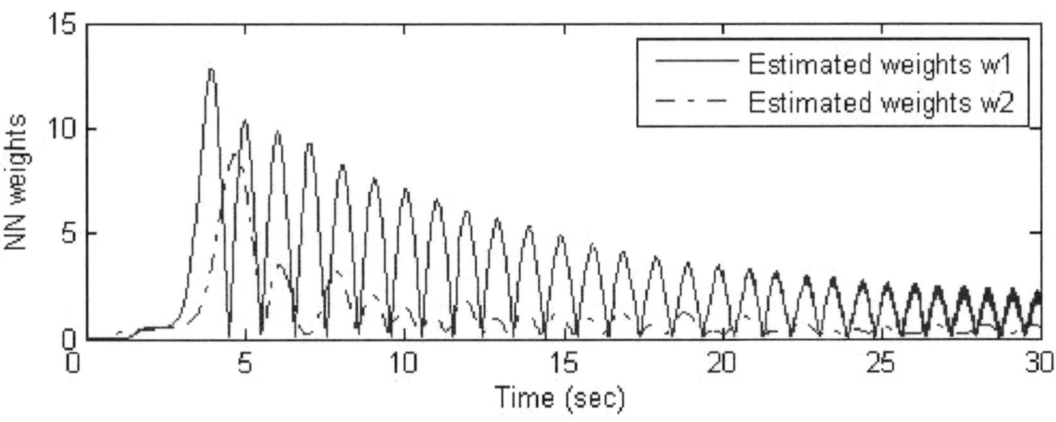

Fig. 5.5 Frobenius norm of the representative weights of the neural networks

systems with state and output feedback [17] and for nonlinear discrete-time systems [17].

It is important to note that due to the presence of functional reconstruction errors with any neural network approach [15], typically the controller designs render a UUB stability result since the functional reconstruction errors are assumed to be upper bounded by a known constant [11,14]. Recently, in another approach, a robust integral of the sign of the error (RISE) feedback method is used to show semi-global asymptotic tracking of continuous-time nonlinear systems [18] in the presence of reconstruction errors and bounded disturbances. To ensure asymptotic performance of the neural network controller, an attempt has been made in [12] for a class of continuous-time and discrete-time nonlinear systems by using a sector boundedness assumption on the uncertainties [12]. A single layer neural network is utilized in this controller design.

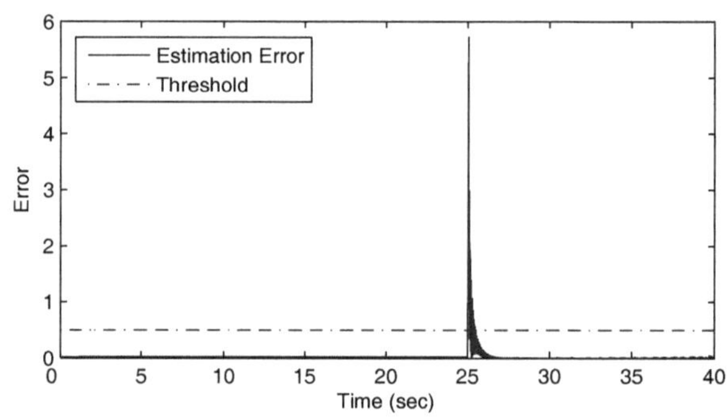

Fig. 5.6 Detection of the fault in the presence of the system uncertainty

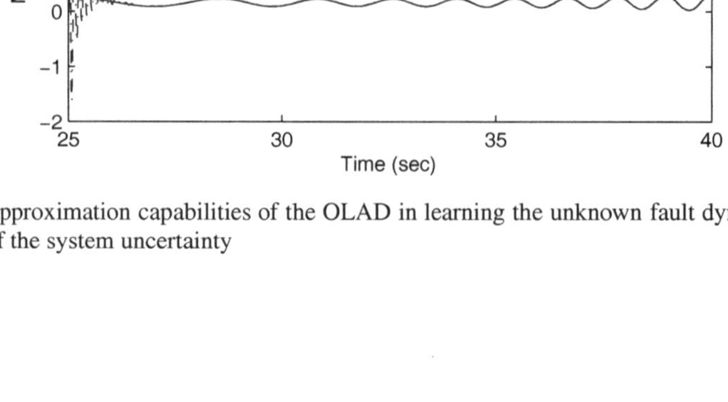

Fig. 5.7 Approximation capabilities of the OLAD in learning the unknown fault dynamics in the presence of the system uncertainty

By contrast, in this chapter, a suite of neural network controllers for a class of nonlinear discrete-time systems that guarantee local asymptotic stability under a mild assumption on the uncertainties [12, 13, 19, 20] is introduced. The controllers utilize the filter tracking error notion and a robust modifying term that was introduced in the estimator design except this new robust modifying term now is a function of the filtered tracking error. Initially, a linearly parameterized neural network is utilized in the controller design and later extended to multilayer neural networks. The stability is shown using the Lyapunov theory. Finally, we use a simulation example to illustrate the performance of the proposed neural network controllers.

5.3.2 Dynamics of MIMO Systems

Consider a mnth-order multi-input multi-output (MIMO) discrete-time nonlinear system given by

$$\begin{aligned} x_1(k+1) &= x_2(k) \\ &\vdots \\ x_{n-1}(k+1) &= x_n(k) \\ x_n(k+1) &= h(x(k)) + u(k) + d(k) \end{aligned} \tag{5.42}$$

where $x(k) = [x_1(k), ..., x_n(k)]^T$ with $x_i(k) \in R^n, i = 1, ..., n$, $u(k) \in R^n$ is the input vector, $h(k)$ is the unknown system dynamics, and $d(k)$denotes a disturbance vector at k^{th} instant with $||d(k)|| \leq d_M$ a known constant. Given a desired trajectory $x_{nd}(k)$ and its delayed values, define the tracking error as $e_n(k) = x_n(k) - x_{nd}(k)$

Next define the filtered tracking error $r(k) \in R^m$ [14] as

$$r(k) = e_n(k) + \lambda_{c_1} e_{n-1}(k) + ... + \lambda_{c_{n-1}} e_1(k) \tag{5.43}$$

where $e_{n-1}(k), ..., e_1(k)$ are the delayed values of the error $e_n(k)$, and $\lambda_{c_1}, ..., \lambda_{c_{n-1}}$ are constant matrices selected such that $|z^{n-1} + \lambda_{c_1} z^{n-2} + ... + \lambda_{c_{n-1}}|$ is within a unit disc. Subsequently, (5.43) is written as

$$r(k+1) = e_n(k+1) + \lambda_{c_1} e_{n-1}(k+1) + ... + \lambda_{c_{n-1}} e_1(k+1). \tag{5.44}$$

By substituting (5.42) in (5.44), (5.44) is rewritten as

$$r(k+1) = h(x(k)) - x_{nd}(k+1) + \lambda_{c_1} e_n(k) + ... + \lambda_{c_{n-1}} e_2(k) + u(k) + d(k). \tag{5.45}$$

Now define the control input $u(k)$ as

$$u(k) = x_{nd}(k+1) - \hat{h}(x(k)) + k_v r(k) + v(k) - \lambda_{c_1} e_n(k) - .... - \lambda_{c_{n-1}} e_2(k) \quad (5.46)$$

where k_v is a user selectable diagonal matrix, $v(k)$ is the robust control vector which is defined later and $\hat{h}(x(k))$ is an estimate of $h(x(k))$. The closed-loop error system becomes

$$r(k+1) = k_v r(k) + \tilde{h}(x(k)) + v(k) + d(k) \quad (5.47)$$

where $\tilde{h}(x(k)) = h(x(k)) - \hat{h}(x(k))$ is the functional estimation error. In the next subsection, the neural network weight update law and the robust control term are introduced. Additionally, the stability of the proposed neural network controller is demonstrated.

5.3.3 *Neural Network Controller Design*

A single and three layers neural network based controllers are proposed in this section. Additionally, the controllers are tuned online using novel weight update laws, which relax the persistency of excitation condition and the certainty equivalence principle [15].

5.3.3.1 Single Layer Neural Network Controller

Initially, we consider a single layer neural network, by considering constant ideal weights, the nonlinear function in (5.42) could be written as

$$h(x) = W^T \varphi(x(k)) + \varepsilon_1(k)$$

where the ideal weight matrix in the above equation is assumed bounded above, such that $||W|| \leq W_{max}$. Define the neural network functional estimate as

$$\hat{h}(x) = \hat{W}^T(k)\varphi(x(k)) \quad (5.48)$$

and the weights estimation error as

$$\hat{W}(k) = W - \hat{W}(k). \quad (5.49)$$

Thus the control input (5.46) is given by

$$u(k) = x_{nd}(k+1) - \hat{W}^T(k)\varphi(x(k)) + k_v r(k) + v(k) - \lambda_{c_1} e_n(k) - ... - \lambda_{c_{n-1}} e_2(k).$$

Substituting the above equation in (5.45) results in the following closed-loop filtered error dynamics as

$$r(k+1) = k_v r(k) + v(k) + \Psi_{1c}(k) + d(k) + \varepsilon_1(k) \quad (5.50)$$

where $\Psi_{1c}(k) = \hat{W}^T(k)\varphi(x(k))$. Select the robust control term as $v(k) = \frac{\hat{\lambda}_m(k)r(k)}{r^T(k)r(k)+c_c}$, where $\hat{\lambda}_m \in R$ is a tunable parameter and $c_c > 0$ is a constant. Next, substituting the robust control input into (5.50) renders

$$r(k+1) = k_v r(k) + \frac{\hat{\lambda}_m(k)r(k)}{r^T(k)r(k)+c_c} + \Psi_{1c}(k) + \varepsilon(k) \tag{5.51}$$

where $\varepsilon(k) = d(k) + \varepsilon_1(k)$. In order to proceed, the following assumption on the modeling uncertainty and bounded disturbances is needed.

Assumption 5.2 *The term (ε) comprising of the approximation error (ε_1) and bounded disturbance ($d(k)$) is assumed to be upper bounded by a smooth nonlinear function of filter tracking error [13, 20], such that*

$$\varepsilon^T(k)\varepsilon(k) \le \varepsilon_1^T\varepsilon_1 + d^T d \le \varepsilon_1^T\varepsilon_1 + d_M \le \varepsilon_M = \lambda_m^* r^T(k)r(k) \tag{5.52}$$

where λ^* is the ideal weight and has a bound, i.e., $||\lambda_m^*|| \le \lambda_{m_{max}}$.

Remark 5.1. This assumption which is stated by a number of researchers both in continuous [13, 20] and discrete-time [12], is considered mild in comparison with the assumption that the functional reconstruction error and disturbances are bounded above by a known constant.

Next by adding and subtracting $\frac{\lambda_m^*(k)r(k)}{r^T(k)r(k)+c_c}$ in (5.51) to get

$$r(k+1) = k_v r(k) - \Psi_{2c}(k) + \Psi_{1c}(k) + \varepsilon(k) + \frac{\lambda_m^*(k)r(k)}{r^T(k)r(k)+c_c} \tag{5.53}$$

where $\Psi_{2c}(k) = \frac{\tilde{\lambda}_m(k)r(k)}{r^T(k)r(k)+c_c}$, and $\tilde{\lambda}_m(k) = \lambda_m^* - \hat{\lambda}_m(k)$ is the parameter estimation error.

Note the parameter, $\hat{\lambda}_m(k)$ of the robust modifying term is considered tunable in the first case similar to the estimator design and later assumed constant in the corollary, which would be a more restrictive case. Next to show that the proposed control law using the filtered tracking error and the robust modifying term is stable, we propose the following theorem.

Theorem 5.2. *Let $x_{nd}(k)$ be the desired trajectory and the initial conditions be bounded in a compact set S. Assume the unknown nonlinear discrete-time system uncertainity is bounded above and the control law (5.46) be applied on the system. Let the neural network weight update law be selected as*

$$\hat{w}(k+1) = \hat{w}(k) + \alpha_c\varphi(k)r^T(k+1) \tag{5.54}$$

where $\alpha_c > 0$ is the learning rate. Now the parameter of the robust modifying term is tuned online using

$$\hat{\lambda}_m(k+1) = \hat{\lambda}_m(k) - \gamma_c \frac{r^T(k+1)r(k)}{r^T(k)r(k)+c_c} \tag{5.55}$$

where $\gamma_c > 0$ is the learning rate. Then, the filter tracking error $r(k)$ is locally asymptotically stable, and the neural network weight estimates $\hat{w}(k)$, the parameter estimate $\hat{\lambda}_m(k)$ are bounded.

Proof. Consider a Lyapunov function candidate as

$$V = r^T(k)r(k) + \frac{1}{\alpha_c} tr[\tilde{w}^T(k)\tilde{w}(k)]) + \frac{1}{\gamma_c}\tilde{\lambda}_m^2(k),$$

whose first difference is given by

$$\Delta V = \underbrace{r^T(k+1)r(k+1) - r^T(k)r(k)}_{\Delta V_1} + \underbrace{\frac{1}{\alpha_c} tr[\tilde{w}^T(k+1)\tilde{w}(k+1) - \tilde{w}^T(k)\tilde{w}(k)]}_{\Delta V_2}$$
$$+ \underbrace{\frac{1}{\gamma_c}[\tilde{\lambda}_m^2(k+1) - \tilde{\lambda}_m^2(k)]}_{\Delta V_3}. \tag{5.56}$$

Substitute the filter tracking error (5.53) in ΔV_1 of (5.56) to get

$$\Delta V_1 = (k_v e(k) + \Psi_{1c}(k) - \Psi_{2c}(k) + \varepsilon(k) + \frac{\lambda_m^* r(k)}{r^T(k)r(k)+c_c})^T (k_v e(k) + \Psi_{1c}(k)$$
$$- \Psi_{2c}(k) + \varepsilon(k) + \frac{\lambda_m^* r(k)}{r^T(k)r(k)+c_c}) - r^T(k)r(k).$$

After some mathematical manipulations, the first difference of the first term is given by

$$\Delta V_1 = r^T(k){k_v}^T k_v r(k) + 2r^T(k){k_v}^T \Psi_{1c}(k) - 2r^T(k){k_v}^T \Psi_{2c}(k) + 2r^T(k){k_v}^T \varepsilon(k)$$
$$+ \frac{2\lambda_m^* r^T(k){k_v}^T r(k)}{r^T(k)r(k)+c_c} + \Psi_{1c}^T(k)\Psi_{1c}(k) - 2\Psi_{1c}^T(k)\Psi_{2c}(k) + \frac{2\lambda_m^* \Psi_{1c}^T(k)r(k)}{r^T(k)r(k)+c_c}$$
$$+ 2\Psi_{1c}^T(k)\varepsilon(k) + \Psi_{2c}^T(k)\Psi_{2c}(k) - \frac{2\lambda_m^* \Psi_{2c}^T(k)r(k)}{r^T(k)r(k)+c_c} - 2\Psi_{2c}^T(k)\varepsilon(k)$$
$$+ \frac{2\lambda_m^* \varepsilon^T(k)r(k)}{r^T(k)r(k)+c_c}$$
$$+ \frac{{\lambda_m^*}^2 r^T(k)r(k)}{(r^T(k)r(k)+c_c)^2} + \varepsilon^T(k)\varepsilon(k) - r^T(k)r(k). \tag{5.57}$$

Next the weight update law (5.54) is substituted in ΔV_2 of (5.56), and using (5.49), the second term is written as

$$\Delta V_2 = \frac{1}{\alpha_c} tr[(\Delta\tilde{w}(k)+\tilde{w}(k))^T(\Delta\tilde{w}(k)+\tilde{w}(k)) - \tilde{w}^T(k)\tilde{w}(k)]$$

$$\begin{aligned}
\Delta V_2 &= \frac{1}{\alpha_c} tr[\Delta\tilde{w}^T(k)\Delta\tilde{w}(k) + 2\Delta\tilde{w}^T(k)\tilde{w}(k)] \\
&= \frac{1}{\alpha_c} tr[\alpha_c^2(r(k+1)\varphi^T)(\varphi r^T(k+1)) - 2\alpha_c r(k+1)\varphi^T\tilde{w}(k)] \\
&= \frac{1}{\alpha_c} tr[\alpha_c^2(k_v r(k) + \Psi_{1c}(k) - \Psi_{2c}(k) + \varepsilon(k) + \frac{\lambda_m^* r(k)}{r^T(k)r(k)+c_c})\varphi^T \\
&\varphi(k_v r(k) + \Psi_{1c}(k) - \Psi_{2c}(k) + \varepsilon(k) + \frac{\lambda_m^* r(k)}{r^T(k)r(k)+c_c})^T \\
&- 2\alpha_c(k_v e(k) + \Psi_{1c}(k) - \Psi_{2c}(k) + \varepsilon(k) + \frac{\lambda_m^* r(k)}{r^T(k)r(k)+c_c})\varphi^T\tilde{w}(k)].
\end{aligned}$$

Apply the Cauchy-Schwarz inequality($(a_1+a_2+...+a_n)^T(a_1+a_2+...+a_n) \le n({a_1}^T a_1 + {a_2}^T a_2 + ... + {a_n}^T a_n)$) for the first term in the above equation, and applying the trace operator (given a vector $x \in R^n$, $tr(xx^T) = x^T x$) to obtain

$$\begin{aligned}
&\le 5\alpha_c\varphi^T\varphi(r^T(k)k_v^T k_v r(k) + \Psi_{1c}^T(k)\Psi_{1c}(k) + \Psi_{2c}^T(k)\Psi_{2c}(k) + \frac{\lambda_m^{*2} r^T(k)r(k)}{(r^T(k)r(k)+c_c)^2}) \\
&+ 5\alpha_c\varphi^T\varphi\varepsilon^T(k)\varepsilon(k) - 2{\Psi_{1c}}^T(k_v r(k) + \Psi_{1c}(k) - \Psi_{2c}(k) + \varepsilon(k) + \frac{\lambda_m^* r(k)}{r^T(k)r(k)+c_c}).
\end{aligned} \tag{5.58}$$

Substitution of the parameter update law (5.55) for ΔV_3 in (5.56) yields

$$\begin{aligned}
\Delta V_3 &= \frac{1}{\gamma_c}(\tilde{\lambda}_m^2(k+1) - \tilde{\lambda}_m^2(k)) \\
&= \frac{1}{\gamma_c}((\tilde{\lambda}_m(k) + \gamma_c\frac{r^T(k+1)r(k)}{r^T(k)r(k)+c_c}) - \tilde{\lambda}_m^2(k)) \\
&= 2r^T(k)k_v^T\Psi_{2c}(k) + 2\Psi_{1c}^T(k)\Psi_{2c}(k) - 2\Psi_{2c}^T(k)\Psi_{2c}(k) + \frac{2\lambda_m^* r(k)\Psi_{2c}(k)}{r^T(k)r(k)+c_c} \\
&+ 2\varepsilon^T(k)\Psi_{2c}(k) + \frac{\gamma_c(r^T(k+1)r(k))^2}{(r^T(k)r(k)+c_c)^2}.
\end{aligned} \tag{5.59}$$

Next noting that $\Delta V = \Delta V_1 + \Delta V_2 + \Delta V_3$ from (5.57)-(5.59), the first difference of the Lyapunov function is given by

$$\begin{aligned}
\Delta V \le\, & r^T(k){k_v}^T k_v r(k) + 2r^T(k){k_v}^T \Psi_{1c}(k) - 2r^T(k){k_v}^T \Psi_{2c}(k) + 2r^T(k){k_v}^T \varepsilon(k) \\
& + \frac{2\lambda_m^* r^T(k){k_v}^T r(k)}{r^T(k)r(k)+c_c} + \Psi_{1c}^T(k)\Psi_{1c}(k) - 2\Psi_{1c}^T(k)\Psi_{2c}(k) + \frac{2\lambda_m^* \Psi_{1c}^T(k) r(k)}{r^T(k)r(k)+c_c} \\
& + 2\Psi_{1c}^T(k)\varepsilon(k) + \Psi_{2c}^T(k)\Psi_{2c}(k) - \frac{2\lambda_m^* \Psi_{2c}^T(k) r(k)}{r^T(k)r(k)+c_c} - 2\Psi_{2c}^T(k)\varepsilon(k) \\
& + \frac{2\lambda_m^* \varepsilon^T(k) r(k)}{r^T(k)r(k)+c_c} + \frac{{\lambda_m^*}^2 r^T(k) r(k)}{(r^T(k)r(k)+c_c)^2} + \varepsilon^T(k)\varepsilon(k) - r^T(k)r(k) \\
& + 5\alpha_c \varphi^T \varphi r^T(k){k_v}^T k_v r(k) + 5\alpha_c \varphi^T \varphi {\Psi_{1c}}^T(k)\Psi_{1c}(k) + 5\alpha_c \varphi^T \varphi {\Psi_{2c}}^T(k)\Psi_{2c}(k) \\
& + 5\alpha_c \varphi^T \varphi \frac{{\lambda_m^*}^2 r^T(k) r(k)}{(r^T(k)r(k)+c_c)^2} + 5\alpha_c \varphi^T \varphi \varepsilon^T(k)\varepsilon(k) - 2{\Psi_{1c}}^T(k) k_v r(k) \\
& - 2{\Psi_{1c}}^T(k)\Psi_{1c}(k) + 2{\Psi_{1c}}^T(k)\Psi_{2c}(k) - 2{\Psi_{1c}}^T(k)\varepsilon(k) - \frac{2{\Psi_{1c}}^T \lambda_m^* r(k)}{r^T(k)r(k)+c_c} \\
& + \frac{\gamma_c (r^T(k+1)r(k))^2}{(r^T(k)r(k)+c_c)^2} + 2r^T(k){k_v}^T \Psi_{2c}(k) + 2{\Psi_{1c}}^T(k)\Psi_{2c}(k) - 2\Psi_{2c}^T(k)\Psi_{2c}(k) \\
& + \frac{2\lambda_m^* r(k)\Psi_{2c}(k)}{r^T(k)r(k)+c_c} + 2\varepsilon^T(k)\Psi_{2c}(k).
\end{aligned}$$

Cancelling terms to get

$$\begin{aligned}
\Delta V \le\, & r^T(k){k_v}^T k_v r(k) + \underbrace{2r^T(k)k_v\varepsilon(k)}_{1} + \underbrace{\frac{2\lambda_m^* r^T(k){k_v}^T r(k)}{r^T(k)r(k)+c_c}}_{1} \\
& - {\Psi_{1c}}^T(k)\Psi_{1c}(k) - {\Psi_{2c}}^T(k)\Psi_{2c}(k) + \underbrace{\frac{2\lambda_m^* \varepsilon^T(k) r(k)}{r^T(k)r(k)+c_c}}_{1} + \frac{{\lambda_m^*}^2 r^T(k) r(k)}{(r^T(k)r(k)+c_c)^2} \\
& + \varepsilon^T(k)\varepsilon(k) - r^T(k)r(k) + 5\alpha_c \varphi^T \varphi r^T(k){k_v}^T k_v r(k) \\
& + 5\alpha_c \varphi^T \varphi {\Psi_{1c}}^T(k)\Psi_{1c}(k) + 5\alpha_c \varphi^T \varphi {\Psi_{2c}}^T(k)\Psi_{2c}(k) + 5\alpha_c \varphi^T \varphi \varepsilon^T(k)\varepsilon(k) \\
& + 5\alpha_c \varphi^T \varphi \frac{{\lambda_m^*}^2 r^T(k) r(k)}{(r^T(k)r(k)+c_c)^2} + 2{\Psi_{1c}}^T(k)\Psi_{2c}(k) + \frac{\gamma_c (r^T(k+1)r(k))^2}{(r^T(k)r(k)+c_c)^2}.
\end{aligned}$$

Now apply the Cauchy-Schwarz inequality ($2ab \le a^2 + b^2$) to terms numbered as 1 in the above equation yields

$$\begin{aligned}
\Delta V \leq & \, r^T(k){k_v}^T k_v r(k) + r^T(k){k_v}^T k_v r(k) + \varepsilon^T(k)\varepsilon(k) + r^T(k){k_v}^T k_v r(k) \\
& + \frac{{\lambda_m^*}^2 r^T(k)r(k)}{(r^T(k)r(k)+c_c)^2} - {\Psi_{1c}}^T(k)\Psi_{1c}(k) - {\Psi_{2c}}^T(k)\Psi_{2c}(k) \\
& + \frac{{\lambda_m^*}^2 r^T(k)r(k)}{(r^T(k)r(k)+c_c)^2} + \varepsilon^T(k)\varepsilon(k) \\
& + \frac{{\lambda_m^*}^2 r^T(k)r(k)}{(r^T(k)r(k)+c_c)^2} + \varepsilon^T(k)\varepsilon(k) - r^T(k)r(k) + 2{\Psi_{1c}}^T(k)\Psi_{2c}(k) \\
& + 5\alpha_c\varphi^T\varphi r^T(k){k_v}^T k_v r(k) + 5\alpha_c\varphi^T\varphi{\Psi_{1c}}^T(k)\Psi_{1c}(k) + 5\alpha_c\varphi^T\varphi{\Psi_{2c}}^T(k)\Psi_{2c}(k) \\
& + 5\alpha_c\varphi^T\varphi\frac{{\lambda_m^*}^2 r^T(k)r(k)}{(r^T(k)r(k)+c_c)^2} + 5\alpha_c\varphi^T\varphi\varepsilon^T(k)\varepsilon(k) \\
& + \frac{\gamma_c((k_v r(k) + \Psi_{1c}(k) - \Psi_{2c}(k) + \varepsilon(k) + \frac{\lambda_m^* r(k)}{r^T(k)r(k)+c_c})^T r(k))^2}{(r^T(k)r(k)+c_c)^2}.
\end{aligned}$$

The last term in the above equation could be written as

$$\begin{aligned}
& \frac{\gamma_c((k_v r(k) + \Psi_{1c}(k) - \Psi_{2c}(k) + \varepsilon(k) + \frac{\lambda_m^* r(k)}{r^T(k)r(k)+c_c})^T r(k))^2}{(r^T(k)r(k)+c_c)^2} \leq \\
& \leq 5\gamma_c(r^T(k){k_v}^T k_v r(k) + {\Psi_{1c}}^T(k)\Psi_{1c}(k) + \Psi_{2c}^T(k)\Psi_{2c}(k) \\
& + \frac{{\lambda_m^*}^2 r^T(k)r(k)}{(r^T(k)r(k)+c_c)^2} + \varepsilon^T(k)\varepsilon(k)).
\end{aligned}$$

The above modification is possible by using the results of the Cauchy-Schwarz inequality ($(a_1+...+a_n)^T(a_1+...+a_n) \leq n({a_1}^T a_1 + ... + {a_n}^T a_n)$). Next by using the fact from (5.52) (i.e., $\varepsilon^T(k)\varepsilon(k) \leq \lambda_m^* r^T(k)r(k)$), to get

$$\begin{aligned}
\Delta V \leq & \, 3r^T(k){k_v}^T k_v r(k) + 3\lambda_m^* r^T(k)r(k) + 3\frac{{\lambda_m^*}^2 r^T(k)r(k)}{(r^T(k)r(k)+c_c)^2} - {\Psi_{1c}}^T(k)\Psi_{1c}(k) \\
& - {\Psi_{2c}}^T(k)\Psi_{2c}(k) + 2{\Psi_{1c}}^T(k)\Psi_{2c}(k) + 5\alpha_c\varphi^T\varphi r^T(k){k_v}^T k_v r(k) \\
& + 5\alpha_c\varphi^T\varphi{\Psi_{1c}}^T(k)\Psi_{1c}(k) + 5\alpha_c\varphi^T\varphi{\Psi_{2c}}^T(k)\Psi_{2c}(k) \\
& + 5\alpha_c\varphi^T\varphi\lambda_m^* r^T(k)r(k) - r^T(k)r(k) + 5\alpha_c\varphi^T\varphi\frac{{\lambda_m^*}^2 r^T(k)r(k)}{(r^T(k)r(k)+c_c)^2} \\
& + 5\gamma_c r^T(k){k_v}^T k_v r(k) + 5\gamma_c{\Psi_{1c}}^T(k)\Psi_{1c}(k) + 5\gamma_c{\Psi_{2c}}^T(k)\Psi_{2c}(k) \\
& + 5\frac{\gamma_c{\lambda_m^*}^2 r^T(k)r(k)}{(r^T(k)r(k)+c_c)^2} + 5\gamma_c\varepsilon^T(k)\varepsilon(k).
\end{aligned}$$

Additionally, noting that $\frac{r^T(k)r(k)}{(r^T(k)r(k)+c_c)^2} < r^T(k)r(k)$, $||\lambda_m^*|| \leq \lambda_{m_{max}}$, and taking the Frobenius norm, the first difference is given by

$$\begin{aligned}\Delta V \leq &-(1-3k_{vmax}^2-3\lambda_{mmax}-3\lambda_{mmax}^2-5\alpha_c\varphi_{max}^2k_{vmax}^2-5\alpha_c\varphi_{max}^2\lambda_{mmax}\\&-5\alpha\varphi_{max}^2\lambda_{mmax}^2-5\gamma_c k_{vmax}^2-5\gamma_c\lambda_{mmax}-5\gamma_c\lambda_{mmax}^2)||r(k)||^2\\&-(1-5\alpha_c\varphi_{max}^2-5\gamma_c)||\Psi_{1c}(k)||^2-(1-5\alpha_c\varphi_{max}^2-5\gamma_c)||\Psi_{2c}(k)||^2\\&+2||\Psi_{1c}(k)||||\Psi_{2c}(k)||\frac{k_{1c}}{k_{1c}}\Psi_{1c}(k)=\frac{k_{1c}}{k_{1c}}\tilde{w}^T(k)\varphi(x,u)=k_{1c}\Psi_{1cs}(k)\end{aligned}$$

where $\Psi_{1cs}(k)=\frac{\tilde{w}^T(k)\varphi(x,u)}{k_{1c}}$. Similarly, $\Psi_{2c}(k)$ is multiplied and divided by a constant k_{2c}, to get $\Psi_{2c}(k)=k_{2c}\Psi_{2cs}(k)$ and $\Psi_{2cs}(k)=\frac{\tilde{\lambda}_m(k)r(k)}{k_2(r^T(k)r(k)+c_c)}$. Hence the first difference of the Lyapunov function is given by

$$\begin{aligned}\Delta V \leq &-(1-3k_{vmax}^2-3\lambda_{mmax}-3\lambda_{mmax}^2-5\alpha_c\varphi_{max}^2k_{vmax}^2-5\alpha_c\varphi_{max}^2\lambda_{mmax}\\&-5\alpha_c\varphi_{max}^2\lambda_{mmax}^2-5\gamma_c k_{vmax}^2-5\gamma_c\lambda_{mmax}-5\gamma_c\lambda_{mmax}^2)||r(k)||^2\\&-k_{1c}^2(1-5\alpha_c\varphi_{max}^2-5\gamma_c)||\Psi_{1cs}||^2+2k_{1c}k_{2c}||\Psi_{1cs}||||\Psi_{2cs}||\\&-k_{2c}^2(1-5\alpha_c\varphi_{max}^2-5\gamma_c)||\Psi_{2cs}||^2.\end{aligned}\tag{5.60}$$

Now take $\gamma_c=\alpha_c\varphi_{max}^2$, $\alpha_c\leq\frac{k_{1c}^2}{10\varphi_{max}^2}$, $k_{2c}=\frac{\delta_c}{k_{1c}}$, where $\delta_c>0$, a constant and using these results in (5.60) to obtain the first difference as

$$\begin{aligned}\Delta V \leq &-(1-3k_{vmax}^2-3\lambda_{mmax}-3\lambda_{mmax}^2-k_{1c}^2k_{vmax}^2-k_{1c}^2\lambda_{mmax}-k_{1c}^2\lambda_{mmax}^2)||r(k)||^2\\&-k_{1c}^2(1-k_{1c}^2)||\Psi_{1cs}(k)||^2-k_{2c}^2(1-k_{1c}^2)||\Psi_{2cs}(k)||^2+2\delta_c||\Psi_{1cs}(k)||||\Psi_{2cs}(k)||.\end{aligned}$$

Applying the Cauchy-Schwarz inequality ($2ab\leq a^2+b^2$) to the last term in the above equation to get

$$\begin{aligned}\Delta V \leq &-(1-3k_{vmax}^2-3\lambda_{mmax}-3\lambda_{mmax}^2-k_{1c}^2k_{vmax}^2-k_{1c}^2\lambda_{mmax}-k_{1c}^2\lambda_{mmax}^2)||r(k)||^2\\&-k_{1c}^2(1-k_{1c}^2)||\Psi_{1cs}(k)||^2-k_{2c}^2(1-k_{1c}^2)||\Psi_{2cs}(k)||^2+\delta_c||\Psi_{1cs}(k)||^2\\&+\delta_c||\Psi_{2cs}(k)||^2.\end{aligned}$$

Combining terms, the first difference is given by

$$\begin{aligned}\Delta V \leq &-(1-3k_{vmax}^2-3\lambda_{mmax}-3\lambda_{mmax}^2-k_{1c}^2k_{vmax}^2-k_{1c}^2\lambda_{mmax}-k_{1c}^2\lambda_{mmax}^2)||r(k)||^2\\&-[(1-k_{1c}^2)-\frac{\delta_c}{k_{1c}^2}]||\tilde{w}^T(k)\varphi||^2-[(1-k_{1c}^2)-\frac{k_{1c}^2}{\delta_c}]||\tilde{\lambda}_m(k)||^2\\&||\frac{r(k)}{(r^T(k)r(k)+c_c)}||^2.\end{aligned}\tag{5.61}$$

Hence $\Delta V\leq 0$ provided the gains are taken as $k_{1c}\leq min(a_c,b_c,c_c)$ and $\delta_c<\frac{1}{4}$, $k_{vmax}\leq\sqrt{\frac{1-3\lambda_{mmax}-3\lambda_{mmax}^2-k_{1c}^2\lambda_{mmax}-k_{1c}^2\lambda_{mmax}^2}{3+k_{1c}^2}}$, where $a_c=\sqrt{\frac{1+\sqrt{1-4\delta_c}}{2}}$,

$b_c = \sqrt{\frac{1-\sqrt{1-4\delta_c}}{2}}$ and $c_c = \sqrt{\frac{\delta_c}{1+\delta_c}}$. Thus $\Delta V \le 0$ provided the gains are selected as defined above. Hence $r(k)$, $\tilde{w}(k)$ and $\tilde{\lambda}_m(k)$ are bounded, provided if $r(k_0)$, $\tilde{w}(k_0)$ and $\tilde{\lambda}_m(k_0)$ are bounded in the compact set S. Additionally by using [17] and summing on both sides and taking $k \to \infty$, it can be found that the tracking error $||r(k)||$ approaches zero as $k \to \infty$. Hence $r(k)$ converges asymptotically. □

In deriving the above theorem, $\hat{\lambda}_m(k)$ of the robust control term was assumed to be a tunable parameter. But in an ideal case, it would be good enough to assume the parameter to be a constant [12]. As a consequence, the filter tracking in (5.50) is modified as

$$r(k+1) = k_v r(k) + \Psi_{1c}(k) + \frac{\lambda_m r(k)}{r^T(k)r(k)+c_c} + \varepsilon(k) \tag{5.62}$$

where $v(k) = \frac{\lambda_m r(k)}{r^T(k)r(k)+c_c}$. Next we present the following corollary.

Corollary 5.2. *Let $x_{nd}(k)$ be the desired trajectory, and the initial conditions be bounded in a compact set S. Consider bounded uncertainties and the control law defined in (5.46). Let the neural network weight update law be given from (5.54) as*

$$\hat{w}(k+1) = \hat{w}(k) + \alpha_c \varphi(k) r^T(k+1) \tag{5.63}$$

where $\alpha_c > 0$ is the learning rate. Then, the filter tracking error $r(k)$ and the neural network weight estimates $\hat{w}(k)$ are locally asymptotically stable.

Proof. Let the Lyapunov function candidate as

$$V = r^T(k)r(k) + \frac{1}{\alpha_c} tr[\tilde{w}^T(k)\tilde{w}(k)]. \tag{5.64}$$

Performing similar mathematical steps as in Theorem 5.2, the final step of the proof would be given by

$$\begin{aligned} \Delta V \le &-(1-3k_{vmax}^2 - 6\lambda_{mmax}^2 - 4\alpha_c\varphi_{max}^2 k_{vmax}^2 - 8\alpha_c\varphi_{max}^2\lambda_{mmax}^2)||r(k)||^2 \\ &-(1-4\alpha_c\varphi_{max}^2)||\tilde{w}^T(k)\varphi(k)||^2. \end{aligned} \tag{5.65}$$

Hence from (5.65), $\Delta V < 0$ provided the gains are taken as $\alpha_c \le 1/4\varphi_{max}^2$, $k_{vmax} < 1/2$, and $\lambda_{mmax} = \sqrt{\frac{1-3k_{vmax}^2 - 4\alpha_c\varphi_{max}^2 k_{vmax}^2}{(6+8\alpha_c\varphi_{max}^2)}}$.

Thus the filtered tracking error $r(k)$ and the neural network estimation error $\tilde{w}(k)$ are locally asymptotically stable. □

Next, we extend the above results for a three layer neural network controller. Though the stability analysis is difficult, a multilayer neural network will render a more accurate function approximation.

Remark 5.2. The addition of the robust control input along with the Assumptions 5.1 and 5.2 render asymptotic stability both for estimation error and tracking error provided the gains are selected appropriately. The selection of the gains will be different for the case of estimation or system identification and control. This selection is obtained from the Lyapunov proof.

5.3.3.2 Three Layer Neural Network Controller

In this section, a three layer neural network is considered. By using (5.4), the neural network output of a nonlinear function in (5.42) could be written as

$$\hat{h}(x) = \hat{W}_3^T(k)\hat{\varphi}_3(\hat{W}_2^T(k)\hat{\varphi}_2(\hat{W}_1^T(k)\hat{\varphi}_1(x(k)))) \tag{5.66}$$

where the actual weights of the neural networks are defined in the early part of this chapter. Define the weight estimation errors as $\tilde{W}_1(k) = W_1 - \hat{W}_1(k)$, $\tilde{W}_2(k) = W_2 - \hat{W}_2(k)$, and $\tilde{W}_3(k) = W_3 - \hat{W}_3(k)$.

Next the following fact can be stated. The activation functions are bounded by known positive values so that $||\hat{\varphi}_1(k)|| \leq \varphi_{1max}$, $||\hat{\varphi}_2(k)|| \leq \varphi_{2max}$ and $||\hat{\varphi}_3(k)|| \leq \varphi_{3max}$. Define activation function vector error as
$\tilde{\varphi}_1(k) = \varphi_1 - \hat{\varphi}_1(k)$, $\tilde{\varphi}_2(k) = \varphi_2 - \hat{\varphi}_2(k)$ and $\tilde{\varphi}_3(k) = \varphi_3 - \hat{\varphi}_3(k)$.
Then by using (5.66) in the control input (5.46), the control input can be expressed as

$$\begin{aligned} u(k) = x_{nd}(k+1) - \hat{W}_3^T(k)\hat{\varphi}_3(\hat{W}_2^T(k)\hat{\varphi}_2(\hat{W}_1^T(k)\hat{\varphi}_1(x(k)))) + k_v r(k) + \\ v(k) - \lambda_{c_1} e_n(k) - \ldots\ldots - \lambda_{c_{n-1}} e_2(k). \end{aligned} \tag{5.67}$$

Then, the closed-loop filtered error dynamics becomes

$$r(k+1) = k_v r(k) + v(k) + \Psi_{1m}(k) + d(k) + \varepsilon_1(k) + W_3^T\tilde{\varphi}_3(x(k)) \tag{5.68}$$

where $\Psi_{1m}(k) = \tilde{W}_3^T(k)\tilde{\varphi}_3(x(k))$. Select the robust term as $v(k) = \frac{\hat{\lambda}_p(k)r(k)}{r^T(k)r(k)+c_m}$. Apply the robust control term in (5.68), (5.68) could be rewritten as
$r(k+1) = k_v r(k) + \frac{\hat{\lambda}_p(k)r(k)}{r^T(k)r(k)+c_m} + \Psi_{1m}(k) + d(k) + \varepsilon_1(k) + W_3^T\tilde{\varphi}_3(x(k))$.
Next by adding and subtracting $\frac{\lambda_p^* r(k)}{r^T(k)r(k)+c_m}$ in the above equation, the filtered error dynamics can be written as

$$r(k+1) = k_v r(k) - \Psi_{2m}(k) + \Psi_{1m}(k) + \varepsilon(k) + \frac{k_2\lambda_p^* r(k)}{r^T(k)r(k)+C_m} \tag{5.69}$$

where $\Psi_{2m}(k) = \frac{\tilde{\lambda}_p(k)r(k)}{r^T(k)r(k)+c_m}$, $\varepsilon(k) = d(k) + \varepsilon_1(k) + W_3^T\tilde{\varphi}_3(x(k))$.

The following theorem guarantees asymptotic stability of the closed-loop system using the proposed control law in (5.67).

Theorem 5.3. *Let $x_{nd}(k)$ be the desired trajectory and the initial conditions be bounded in a compact set S. Consider bounded uncertainties in the system along with the control law proposed in (5.66), are acting on the nonlinear system. Let the* *input and hidden layer neural network weight tuning be provided by*

$$\hat{w}_1(k+1) = \hat{w}_1(k) - \alpha_1 \hat{\varphi}_1(k)[\hat{y}_1(k) + B_1 k_v r(k)]^T \tag{5.70}$$

$$\hat{w}_2(k+1) = \hat{w}_2(k) - \alpha_2 \hat{\varphi}_2(k)[\hat{y}_2(k) + B_2 k_v r(k)]^T \tag{5.71}$$

where $\hat{y}_i(k) = \hat{w}_i^T(k)\hat{\varphi}_i(k)$ *and* $||B_i|| \leq K_i$, $i = 1,2$. *Take the weight update law for the third layer as*

$$\hat{w}_3(k+1) = \hat{w}_3(k) + \alpha_3 \hat{\varphi}_3(k) r^T(k+1) \tag{5.72}$$

where $\alpha_i > 0$, $\forall i = 1,2,3$, *denotes the learning rate or adaptation gains. Let the parameter of the robust control term be tuned online via*

$$\hat{\lambda}_p(k+1) = \hat{\lambda}_p(k) - \gamma_p \frac{r^T(k+1)r(k)}{r^T(k)r(k) + c_m} \tag{5.73}$$

where $\gamma_p > 0$ *is the learning rate, and* $c_m > 0$ *is a constant. Then, the filter tracking error* $r(k)$ *is locally asymptotically stable, the neural network weight estimates* $\hat{w}_1(k), \hat{w}_2(k)$ *and* $\hat{w}_3(k)$, *and the parameter estimate* $\hat{\lambda}_p(k)$ *are bounded.*

The robust control term used for the three layer neural network controllers is similar to the single layer neural network controller.

Proof. Consider a Lyapunov candidate as

$$\begin{aligned} V =& r^T(k)r(k) + \frac{1}{\alpha_1} tr[\tilde{w}_1^T(k)\tilde{w}_1(k)] + \frac{1}{\alpha_2} tr[\tilde{w}_2^T(k)\tilde{w}_2(k)] + \frac{1}{\alpha_3} tr[\tilde{w}_3^T(k)\tilde{w}_3(k)] \\ &+ \frac{1}{\gamma_p}\tilde{\lambda}_p^2(k). \end{aligned}$$

The first difference is given by

$$\begin{aligned} \Delta V =& r^T(k+1)r(k+1) - r^T(k)r(k) + \frac{1}{\alpha_1} tr[\tilde{w}_1^T(k+1)\tilde{w}_1(k+1) - \tilde{w}_1^T(k)\tilde{w}_1(k)] \\ &+ \frac{1}{\alpha_2} tr[\tilde{w}_2^T(k+1)\tilde{w}_2(k+1) - \tilde{w}_2^T(k)\tilde{w}_2(k)] \\ &+ \frac{1}{\alpha_3} tr[\tilde{w}_3^T(k+1)\tilde{w}_3(k+1) - \tilde{w}_3^T(k)\tilde{w}_3(k)] \\ &+ \frac{1}{\gamma_p}[\tilde{\lambda}_p^2(k+1) - \tilde{\lambda}_p^2(k)]. \end{aligned} \tag{5.74}$$

Substituting (5.69) to (5.73) in (5.74), collecting terms together and completing square yields

$$
\begin{aligned}
\Delta V \le\; & 3r^T(k)k_v^T k_v r(k) + 3\varepsilon^T(k)\varepsilon(k) + 3\frac{\lambda_p^{*2} r^T(k)r(k)}{(r^T(k)r(k)+c_m)^2} \\
& - \Psi_{1m}^T(k)\Psi_{1m}(k) - \Psi_{2m}^T(k)\Psi_{2m}(k) + 2\Psi_{1m}^T(k)\Psi_{2m}(k) - r^T(k)r(k) \\
& + 5\alpha_3\hat{\varphi}_3^T\hat{\varphi}_3 r^T(k)k_v^T k_v r(k) + 5\alpha_3\hat{\varphi}_3^T\hat{\varphi}_3\Psi_{1m}^T(k)\Psi_{1m}(k) + 5\alpha_3\hat{\varphi}_3^T\hat{\varphi}_3\Psi_{2m}^T(k)\Psi_{2m}(k) \\
& + 5\alpha_3\hat{\varphi}_3^T\hat{\varphi}_3\varepsilon^T(k)\varepsilon(k) + 5\alpha_3\hat{\varphi}_3^T\hat{\varphi}_3\frac{\lambda_p^{*2} r^T(k)r(k)}{(r^T(k)r(k)+c_m)^2} + 5\gamma_p r^T(k)k_v^T k_v r(k) \\
& + 5\gamma_p\Psi_{1m}^T(k)\Psi_{1m}(k) + 5\gamma_p\Psi_{2m}^T(k)\Psi_{2m}(k) + 5\frac{\gamma_p\lambda_p^{*2} r^T(k)r(k)}{(r^T(k)r(k)+c_m)^2} + 5\gamma_p\varepsilon^T(k)\varepsilon(k) \\
& + \frac{W_{1max}^2||\hat{\varphi}_1(k)||^2}{(2-\alpha_1||\hat{\varphi}_1(k)||^2)} + \frac{W_{2max}^2||\hat{\varphi}_2(k)||^2}{(2-\alpha_2||\hat{\varphi}_2(k)||^2)} + \frac{k_{vmax}^2||r(k)||^2\sum_{i=1}^2 K_i^2}{(2-\alpha_i\hat{\varphi}_i^T(k)\hat{\varphi}_i(k))} \\
& - (2-\alpha_1\hat{\varphi}_1^T(k)\hat{\varphi}_1(k))||\hat{W}_1^T(k) - \frac{(1-\alpha_1\hat{\varphi}_1^T(k)\hat{\varphi}_1(k))}{(2-\alpha_1\hat{\varphi}_1^T(k)\hat{\varphi}_1(k))}(W_1^T\hat{\varphi}_1(k) + B_1 k_v r(k))|| \\
& - (2-\alpha_2\hat{\varphi}_2^T(k)\hat{\varphi}_2(k))||\hat{W}_2^T(k) - \frac{(1-\alpha_2\hat{\varphi}_2^T(k)\hat{\varphi}_2(k))}{(2-\alpha_2\hat{\varphi}_2^T(k)\hat{\varphi}_2(k))}(W_2^T\hat{\varphi}_2(k) + B_2 k_v r(k))|| \\
& + \frac{2k_{vmax}||r(k)||\sum_{i=1}^2 K_i\varphi_{imax}W_{imax}}{(2-\alpha_i\hat{\varphi}_i^T(k)\hat{\varphi}_i(k))}.
\end{aligned}
\tag{5.75}
$$

Apply the Cauchy-Schwarz inequality ($2ab \le a^2 + b^2$) to the last term and using the following assumption, the proof is completed. □

Assumption 5.3 *Using Assumption 5.2, the term* (ε)*, and the activation function vectors of the neural network are assumed to be bounded above [13, 20] as*

$$
\begin{aligned}
& \frac{\sum_{i=1}^2 W_{imax}^2(2-\alpha_i||\hat{\varphi}_i(k)||^2 + K_i^2)||\hat{\varphi}_i(k)||^2}{(2-\alpha_i||\hat{\varphi}_i(k)||^2)^2} \\
& +(5\gamma_p + 5\alpha_3\hat{\varphi}_3^T\hat{\varphi}_3 + 3)\varepsilon^T(k)\varepsilon(k) \le \lambda_p^* r^T(k)r(k).
\end{aligned}
$$

Remark 5.3. Although this may appear to be a restrictive assumption, but this is better than assuming these terms are upper bounded by a constant. Using the definition $k_{2p} = \frac{\delta_p}{k_{1p}}$, where $\delta_p > 0$. Taking $\gamma_p = \alpha_3\varphi_{3max}^2$, $\alpha_3 \le \frac{k_{1p}^2}{10\varphi_{3max}^2}$, $\Psi_{1ms}(k) = \frac{\tilde{w}^T(k)\varphi(x,u)}{k_{1p}}$, and $\Psi_{2ms}(k) = \frac{\tilde{\lambda}_p(k)r(k)}{k_{2p}(r^T(k)r(k)+c_m)}$, additionally, by using Assumption 5.2, taking the norm, (5.75) becomes

$$\Delta V \le -(1 - 3k_{vmax}^2 - 3\lambda_{pmax}^2 - \lambda_{pmax} - k_{1p}^2 k_{vmax}^2 - k_{1p}^2\lambda_{pmax}^2 - k_{vmax}^2\beta^2)||r(k)||^2$$
$$- [(1 - k_{1p}^2) - \frac{\delta_p}{k_{1p}^2}]||\tilde{w}_3^T(k)\varphi_3||^2 - [(1 - k_{1p}^2) - \frac{k_{1p}^2}{\delta_p}]||\tilde{\lambda}_p(k)||^2||$$
$$\frac{r(k)}{(r^T(k)r(k) + c_m)}||^2$$
$$- (2 - \alpha_1\hat{\varphi}_1^T(k)\hat{\varphi}_1(k))||\hat{W}_1^T(k) - \frac{(1 - \alpha_1\hat{\varphi}_1^T(k)\hat{\varphi}_1(k))}{(2 - \alpha_1\hat{\varphi}_1^T(k)\hat{\varphi}_1(k))}(W_1^T\hat{\varphi}_1(k) + B_1 k_v r(k))||$$
$$- (2 - \alpha_2\hat{\varphi}_2^T(k)\hat{\varphi}_2(k))||\hat{W}_2^T(k) - \frac{(1 - \alpha_2\hat{\varphi}_2^T(k)\hat{\varphi}_2(k))}{(2 - \alpha_2\hat{\varphi}_2^T(k)\hat{\varphi}_2(k))}(W_2^T\hat{\varphi}_2(k) + B_2 k_v r(k))|| \tag{5.76}$$

where $\beta^2 = \frac{\Sigma_{i=1}^2(2-\alpha_i||\hat{\varphi}_i(k)||^2 K_i^2+1)}{(2-\alpha_i||\hat{\varphi}_i(k)||^2)^2}$. Hence, from (5.76), $\Delta V \le 0$ if the gains are taken as $a_M = \sqrt{\frac{1+\sqrt{1-4\delta_p}}{2}}$, $b_M = \sqrt{\frac{1-\sqrt{1-4\delta_p}}{2}}$ and $c_M = \sqrt{\frac{\delta_p}{1+\delta_p}}$. Then $k_{1p} \le min(a_M, b_M, c_M)$ and $\delta_p \le 1/4$. Finally, we have $k_{vmax} \le \sqrt{\frac{1-3\lambda_{pmax}^2-\lambda_{pmax}-k_{1p}^2\lambda_{pmax}^2}{3+k_{1p}^2+\beta^2}}$.

Hence the filter tracking error $r(k)$ and the neural network weight estimates $\hat{w}_1(k)$, $\hat{w}_2(k)$ and $\hat{w}_3(k)$, and the parameter estimate $\hat{\lambda}_p(k)$ are bounded. Additionally, summing both sides and taking limits by using [17], it can be seen that the tracking error $||r(k)||$ approaches zero as $k \to \infty$. Hence $r(k)$ converges asymptotically. In the next section, some simulation results are explained.

5.4 Simulation Results

Consider the following nonlinear discrete-time system [17]

$$x_1(k+1) = x_1(k)x_2(k),$$
$$x_2(k+1) = x_1^2(k) - sin(x_2(k)) + u(k) + d(k). \tag{5.77}$$

The objective is to track a sinusoidal input of magnitude one unit for a total time interval of 30 s. The value of $k_v = -0.0001$ and a sampling interval of 10 ms was considered. A single layer neural network controller with 12 sigmoid activation functions was used, and the weights of the network and the robust control term were tuned using (5.54) and (5.55). Taking $\delta_c = 0.248$, one can determine $\alpha_c = \gamma_c = 0.0199$, and $c_c = 0.013$. The initial neural network weights were chosen to be zero and initial parameter value $\hat{\lambda}_m(k_0)$ is randomly chosen. Finally, the initial conditions of the system were chosen to be $[0.001, 0.001]$. Since $n = 2$, we take $\lambda_{c_1} = 0.01$.

Figure 5.8 shows the response of three controllers: the neural network controller without the robust control term, the neural network controller with the robust control term (referred to here as asymptotic controller), and finally, by removing the neural network along with the robust control term which becomes a proportional controller, i.e., $u(k) = k_v r(k)$. From the simulation, it is very obvious that the asymptotic controller yields a highly satisfactory performance except at the time instant when peaks are observed when there is a change in sign. Additionally, the contribution of the neural network is also observed between the neural network controller and a simple proportional controller without the neural network.

In Figure 5.9, the representative neural network weights of the asymptotic controller are given, where they appear to be bounded. Thus these simulation results validate the performance of the proposed neural network controller.

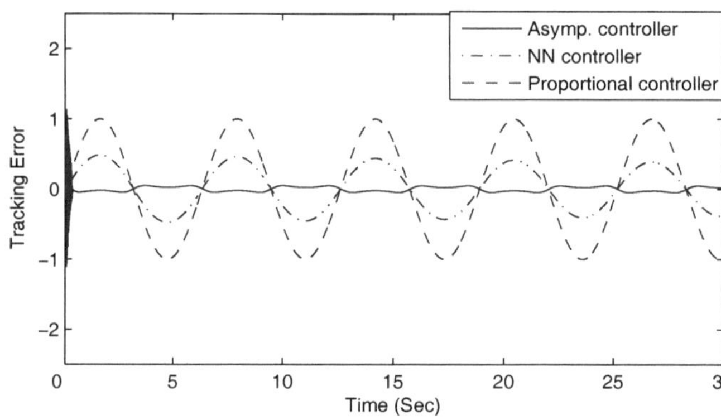

Fig. 5.8 Performance of the controllers

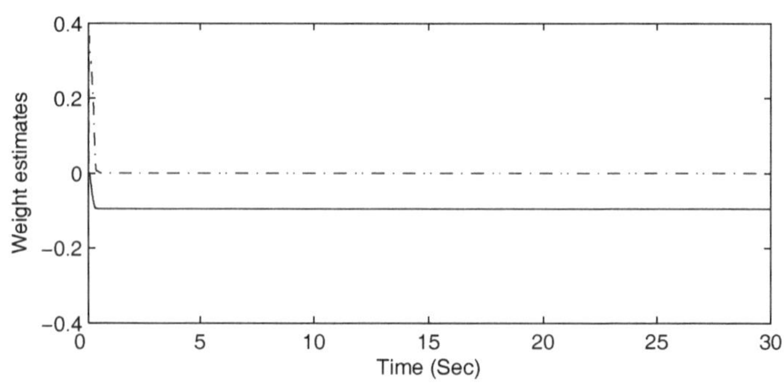

Fig. 5.9 Representative weight estimates of the asymptotic controller

5.5 Conclusions

In this chapter, a new estimation and control scheme for a nonlinear discrete-time system with guarantees of asymptotic stability was presented. The proposed estimation scheme was used to identify unknown nonlinear discrete-time systems that were widely used in the literature. Later, a suite of neural network controllers were developed for nonlinear discrete-time systems. By using the robust control term and based on certain mild assumptions on the neural network approximation error and the system uncertainties and in the presence of bounded disturbances, the asymptotic stability of the estimation and the control scheme were presented. No *a priori* learning phase for the neural network is needed. Simulation results are included to verify theoretical conjectures.

References

1. K. J. Åatröm and P. Eykhoff, System identification-A Survey, *Automatica*, vol. 7, pp. 123-162, 1971.
2. L. Ljung, Asymptotic behavior of the extended Kalman filter as a parameter estimator for linear systems, *IEEE Trans. on Automatic Control*, vol. Ac-24, no.1, pp. 36-50, 1979.
3. T. Soderstorm, L.Ljung, and I. Gustavsson, A theoretical analysis of recursive identification algorithms, *Automatica*, vol. 14, pp. 231-244, 1978.
4. G. Box, G. M. Jenkins, and G. C. Reinsel, *Time series analysis: forecasting and control*, Prentice Hall, 3rd edition, NJ, USA, 1994.
5. C. H. Luh and J.Y. Duh, Analysis of nonlinear system using NARMA models, *Structural Engineering*, vol. 13, no. 1, pp. 11-21, 2006.
6. K. S. Narendra and K. Parthasarathy, Identification and control of dynamical systems using neural networks, *IEEE Trans. on Neural Networks*, vol. 1, no. 1, pp. 4-27, 1990.
7. S. Jagannathan, *Neural network control of nonlinear discrete-time systems*, CRC Press, NY, USA, 2006.
8. E. B. Kosmatopoulous, M. M. Polycarpou, M. A. Christodoulou, and P. A. Ioannou, High-order neural network structures for identification of dynamical systems, *IEEE Trans. on Neural Networks*, vol. 6, no. 2, pp. 422-431, 1995.
9. H. K. Khalil, *Nonlinear Systems*, Prentice Hall, 3rd cd., NJ, USA, 2002.
10. A. R. Barron, Universal approximation bounds for superposition of a sigmoidal function, *IEEE Trans. on Information Theory*, vol. 39, no. 3, pp. 930-945, 1993.
11. S. S. Ge, C. C. Hang, and T. Zhang, Adaptive neural network control of nonlinear systems by state and output feedback, *IEEE Trans. on System, Man, and Cybernetics- Part B*, vol. 29, no. 6, pp. 818-828, 1999.
12. T. Hayakawa, W. M. Haddad, and N. Hovakimyan, Neural network adaptive control for a class of nonlinear uncertain dynamical systems with asymptotic stability guarantees, *IEEE Trans. on Neural Networks*, vol. 19, no. 1, pp. 80-89, 2008.
13. C. M. Kwan, D. M. Dawson, and F. L. Lewis, Robust adaptive control of robots using neural network: global tracking stability, *Proc. of the 34th Conference on Decision and Control*, New Orleans, LA, pp. 1846-1851, 1995.
14. S. Jagannathan and F. L. Lewis, Discrete-Time neural net controller for a class of nonlinear dynamical systems, *IEEE Trans. on Automatic Control*, vol. 41, no. 11, pp. 1693-1699, 1996.
15. F. L. Lewis, A. Yesildirek, and K. Liu, Multilayer neural-net robot controller with guaranteed tracking performance, *IEEE Trans. on Neural Networks*, vol. 7, pp. 387-398, 1996.

16. S. Jagannathan and F. L. Lewis, Multilayer discrete-time neural-net controller with guaranteed performance, *IEEE Trans. on Neural Networks*, vol. 7, no.1, pp. 107-130, 1996.
17. P. M. Patre, W. MacKunis, K. Kaiser, and W. E. Dixon, Asymptotic tracking for uncertain dynamics systems via a multilayer NN feed forward and RISE feedback control structure, *Proc. of the 2007 American Control Conference*, NewYork City, NY, USA, pp. 5989-5994, 2007.
18. B. Xian, D. M. Dawson, M. S. de Queiroz, and J. Chen, A continuous asymptotic tracking control strategy for uncertain nonlinear systems, *IEEE Trans. on Automatic Control*, vol. 49, no. 7, pp. 1206-1211, 2004.
19. S. Huang, K. K. Tan, and T. H. Lee, Decentralized control design for large-scale systems with strong interconnections using neural networks, *IEEE Trans. on Automatic Control*, vol. 48, no. 5, pp. 805-810, 2003.
20. J. Gertler, Survey of model-based failure detection and isolation in complex plants, *IEEE Control Systems Magazine*, vol. 8, pp. 3-11, 1988.
21. B. T. Thumati and S. Jagannathan, An online approximator-based fault detection framework for nonlinear discrete-time systems, *Proc. of the 46th IEEE Conference on Decision and Control (CDC)*, New Orleans, LA, USA, pp. 2608-2613, 2007.
22. S. Jagannathan and F. L. Lewis, Identification of nonlinear dynamical systems using multilayered neural networks, *Automatica*, vol. 32, no. 11, pp. 1707-1712, 1996.

Chapter 6
Neural Networks Based Probability Density Function Control for Stochastic Systems

Xubin Sun, Jin Liang Ding, Tianyou Chai and Hong Wang

Abstract This chapter presents the recent development of neural network based output probability density function (PDF) shaping for stochastic distribution systems, where the purpose of controller design is to select proper feedback control laws so that the probability density function of the system output can be made to follow a target distribution shape. To start with, a survey on the stochastic distribution control (SDC) is given. This is then followed by the description of several neural networks approximations to the output PDFs. To illustrate the use of neural networks in the control design, an example of grinding process control using SDC theory is included here.

6.1 Stochastic Distribution Control

SDC systems are widely seen in practical processes, where the aim of the controller design is to realize the shape control of the distributions of certain random variables in the process. Once the PDFs of these variables are used to describe their distri-

Xubin Sun
Institute of Automation, Chinese Academy of Sciences, Beijing, P.R. China, e-mail: xubin.sun@ia.ac.cn

Jinliang Ding
Key Laboratory of Integrated Automation of Process Industry, Northeastern University Shenyang, P.R. China, e-mail: tychai@mail.neu.edu.cn

Tianyou Chai
Key Laboratory of Integrated Automation of Process Industry, Northeastern University Shenyang, P.R. China, e-mail: tychai@mail.neu.edu.cn

Hong Wang
The University of Manchester, Manchester, United Kingdom, e-mail: hong.wang@manchester.ac.uk

butions, the control task is to obtain control signals so that the output PFDs of the system are made to follow their target PDFs.

Indeed, stochastic systems are widely seen in control engineering practice. This is because almost all the control systems are subjected to random signals such as those originated from the system parameter variations and sensor noises, etc. As such, one of the important issues in the controller design for practical systems is to minimize the randomness in the closed-loop system. This has therefore motivated studies on minimum variance control whose purpose is to ensure minimum contained uncertainties in the controlled system outputs or the tracking errors. Indeed, even nowadays most of the stochastic control design has been focused on the control of the mean and the variance of stochastic systems. In general these developments are largely based upon the assumptions that the system variables are of Gaussian types. Such assumptions, albeit strict to many practical systems, allow control engineers to make use of the well-established stochastic theory to deal with the required controller design and closed-loop system analysis. Indeed, the rather rich literature of the treatment of Gaussian systems have always been regarded as a natural basis for the design and analysis of stochastic control systems. The control of the whole shape of the output PDF for complex stochastic systems has been studied in response to the increased demand from many practical systems ([11-12, 15]), where for this type of system, the actual controlled output is the shape of its output PDFs and the inputs are only related to time. This means that the controlled PDF is a function of both the time and the space and can therefore be generally expressed as a partial differential equation (PDE). This PDE model can be formulated from the general expression of many population balance equations such as the widely used particulate system model [1], where the aim of the controller design is to ensure that the shape of the output PDF can follow a target distribution shape. In this context, one can see that the general solution framework of the above PDE can be used. However, in practice it is generally difficult to obtain such a PDE, and even if such a PDE is available, the controller design would be still difficult to perform.

This type of control is termed as the SDC, which is a new research area where simple and implementable solutions for industrial processes are required. In comparison with the traditional stochastic control theory where only output mean and variances are of concern, SDC can offer a much better solution and is of course not restricted to Gaussian input cases. Indeed, this is a challenging problem and such systems are seen in many material processing industries. The problem of controlling the output PDF is long standing. Solutions to such a challenging problem have been sought by theorists and engineers in the control community. In theory, the first paper was published in *Automatica* in 1996 by Karney (see survey in [11]), where the process is represented by a PDF and the control is also expressed in a PDF format. As such, the purpose of controller design is to obtain the PDF of the controller so that the closed-loop PDF will follow its target PDF. However, as the control input to the process is always crispy (i.e., one input value at any time instant), the realization of the PDFs of the controller appears to be difficult. In this regard, the first practically implementable control strategy was developed by the the first author of this paper at UMIST in 1996 [11]. Since then, fast developments have been seen and at present there are around 15 research groups in the world actively seeking solutions to controller design and their applications. Special sessions and special issues have been seen in various international control conferences and refereed journals. In 2000, a book (which is unique to its subject) by the first author of this chapter was published by Springer London, summarizing the earlier developments in this new research area.

In general, the methods developed so-far can be classified into the following three groups:

1. Output PDF control using neural networks.
2. Output PDF control using system input-output models.
3. Output PDF control using Ito differential equations.

6.1.1 Neural Networks in Stochastic Distribution Control

One of the key issues in the development of SDC has been to approximate measurable output PDF of the concerned system using neural networks. Once a proper neural network has been selected to realize such an approximation, the output PDF control can be realized by the use of the dynamical models between the neural network weights and the control inputs. Although there are many control methods for the stochastic system, we only present the application of neural networks used in SDC. The features of function approximation of neural networks to output PDFs will be described in the stochastic distribution system for modeling and control purposes.

It is well known that the output variable of a stochastic system cannot be fully represented by its mean and variance especially for non-Gaussian stochastic systems, therefore it is necessary to use the PDF of the output as the system output. However, the PDF dynamic model of such a system is difficult to realize unless a partial differential equation is used in line with the well-known Ito stochastic differential equation [12].

Among several neural networks, B-spline networks are the most popular ones used in SDC. In this context, there are four kinds of B-spline functions, including: linear B-spline function, rational B-spline function, square root B-spline approximation, and rational square root B-spline approximation [12, 14]. Except for these B-spline approximations, radial basis functions (RBFs) have also been adopted recently in system modeling and control [1, 11].

6.1.1.1 B-spline Neural Networks

There are many approaches to realize the required output PDF approximation, such as the Gaussian approximation method, a reduced sufficient statistics algorithm, the Wavelet approximation method, and Monte Carlo sampling. Some of the above methods can be directly used in a PDF approximation in a SDC system, but an important reason to use B-spline basis functions here is their localization feature where any changes in parts of B-spline do not influence other parts of B-spline basis functions.

B-spline neural networks are examples of associative memory networks. The input space is defined over an n-dimensional lattice with basis functions defined for each cell. Nonzero basis function can be located by calculation and training is localized.

The input axis is divided into intervals by a set of values called knots. Knots serve as "address locations" for the univariate B-spline functions. The knots can be placed in any manner desired by the designer. Each input axis can also have different numbers of knots placed in different ways. The knot vector contains r interior knots and k exterior knots. Taking the one dimension for example, the interior knots should satisfy:

$$y^{\min} < \lambda_1 \leq \lambda_2 \leq \cdots \leq \lambda_r < y^{\max} \tag{6.1}$$

where $y^{\min}$ and $y^{\max}$ are the minimum and maximum of the input axis. The exterior knots locate at two sides of input space which satisfy:

$$\lambda_{-(m-1)} < \cdots < \lambda_0 = y^{\min} \tag{6.2}$$

$$y^{\max} = \lambda_{r+1} < \cdots < \lambda_{r+m} \tag{6.3}$$

where m is the number of exterior knots. The performance of the B-spline intensely correlates with its order which not only determines the differentiable order, but also effective region of the basis functions. For a kth order univariate B-spline, it spans on the k lattice and is k-order differentiable. For each lattice, there are k nonzero basis functions projecting on it. With the order increasing, B-spline basis functions become more and more smooth. Univariate B-spline can be formulated in an iterative way:

$$\begin{aligned} B_{k,j}(x) &= \frac{x-\lambda_j}{\lambda_{j+k-1}-\lambda_j} B_{k-1,j}(x) + \frac{\lambda_{j+k}-x}{\lambda_{j+k}-\lambda_{j+1}} B_{k-1,j+1}(x) \\ B_{1,j}(x) &= \begin{cases} 1 & x \in [\lambda_{j-1}, \lambda_j) \\ 0 & x \notin [\lambda_{j-1}, \lambda_j). \end{cases} \end{aligned} \tag{6.4}$$

In Figure 6.1, the first B-spline basis function of order 1, 2, and 3 are demonstrated separately.

6.1.1.2 Linear B-spline Modeling

Define the PDF $\gamma(x, u(t))$ is the continuous function of $u(t)$ and x. $u(t)$ is uniform bounded for $t \in [0, \infty)$. For arbitrary $u(t)$ and all $x \in [a, b]$, PDF $\gamma(x, u(t))$ is the continuous function of x. Based on the B-spline neural network theory, we know that there exists a B-spline network that can guarantee the following inequality hold:

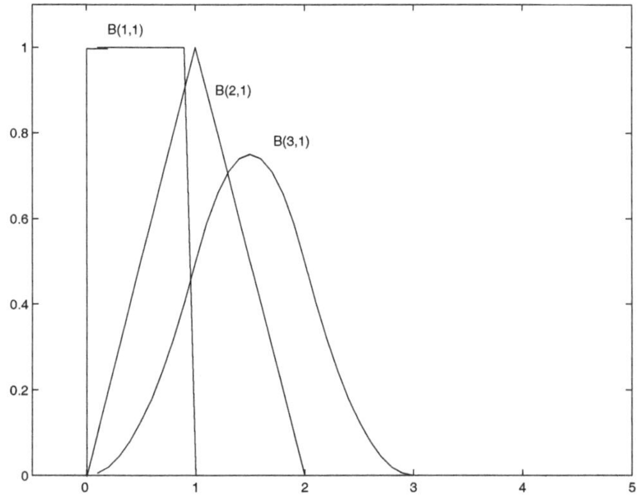

Fig. 6.1 B-spline basis functions

$$\left| \gamma(x,u(t) - \sum_{i=1}^{n} v_i B_i(x) \right| \leq \delta \tag{6.5}$$

where δ is a prescribed arbitrarily small positive number; $B_i(x), (i = 1, 2, \cdots, n)$ are basis functions defined on $[a,b]$; $v_i, (i = 1, 2, \cdots, n)$ are the corresponding weights of the basis functions, n is the number of basis functions.

When the basis functions are fixed, then the weights of the basis functions are determined by the input $u(t)$, that is to say the weights are the functions of input $u(t)$, then (6.5) can be further expressed as the following:

$$\gamma(x,u(t)) = \sum_{i=1}^{n} v_i(u(t))B_i(x) + e \tag{6.6}$$

where e is the approximation error, satisfying $|e| < \delta$. The change of the PDF $\gamma(x,u(t))$ can be viewed being aroused by the weights v_i as showed in Figure 6.2. To be convenient, the approximation error e is always neglected. The system dynamics can be expressed by the dynamic function of weights as the following:

$$\dot{V}(t) = AV(t) + Bu(t) \tag{6.7}$$

where

$$V(t) = [v_1(t), v_2(t), \cdots, v_{n-1}(t)]^T$$

$$A \in R^{(n-1)\times(n-1)}$$

$$B \in R^{(n-1)\times 1}.$$

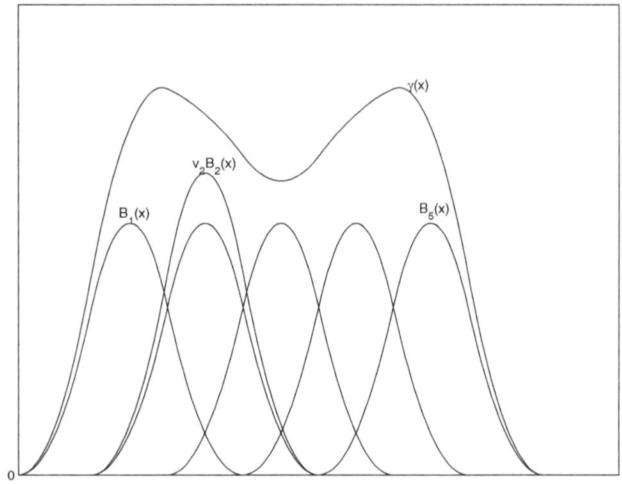

Fig. 6.2 Function approximation demonstration by linear B-spline

The constraint that the integration of the PDF equals 1 should be satisfied, that is:

$$\int_a^b \gamma(x,u(t))dx = 1 \tag{6.8}$$

This means that only $n-1$ weights among the n weights are independent. Then, define

$$\begin{array}{l} b_i = \int_a^b B_i(x)dy \\ b^T = \begin{bmatrix} b_1\ b_2 \cdots\ b_{n-1} \end{bmatrix} \in R^{1\times(n-1)} \\ C_1(x) = \begin{bmatrix} B_1(x)\ B_2(x) \cdots\ B_{n-1}(x) \end{bmatrix} \in R^{1\times(n-1)} \\ L(x) = b_n^{-1}B_n(x) \in R^{1\times 1} \\ C_0(x) = C_1(x) - \frac{B_n(x)}{b_n}b^T \in R^{1\times(n-1)}. \end{array} \tag{6.9}$$

Then the system model can be further formulated by:

$$\begin{cases} \dot{V} = AV + Bu \\ \gamma(x,u) = C_0(x)V + L(x). \end{cases} \tag{6.10}$$

The relation of the linear B-spline weights are linear and only $n-1$ are independent, so it is relatively easy to design the controller using the developed algorithm. But the PDF nonnegative constraint cannot be guaranteed sometime in weights training. So we need to develop a new approximation way to guarantee the approximated PDF nonnegative.

6.1.1.3 Square Root B-spline Model

Square root B-spline can satisfy the requirement to keep the PDF nonnegative, because the curve it approximated is the square root of the PDF, and not the PDF itself. Then the dynamic part of (6.10) stays unchanged, while the output equation becomes a nonlinear form:

$$\begin{cases} \dot{V} = AV + Bu \\ \sqrt{\gamma(x,u)} = C_1(x)V + v_n B_n(x) \end{cases} \tag{6.11}$$

where all the parameters have the same meaning as defined above. Obviously, this model is also a degree-reduction model, and the dynamic equation is $n-1$ order.

In the output equation of the square root model, the relation between weights becomes nonlinear.

In the linear B-spline model or square root B-spline model, there are only $n-1$ weights, which are independent and used in system modeling, when the total number of weights are n. This is determined by the natural constraint that the integral of PDF equals to one. In fact, we expect all the weights are independent of each other. To meet this requirement, the following rational B-spline is put forward.

6.1.1.4 Rational B-spline Model

It will make the controller design more complex that only $n-1$ weights are independent among total n weights, especially for the square root B-spline model. Therefore, the rational B-spline model is given by:

$$\begin{cases} \dot{V} = AV + Bu \\ \gamma(x,u) = \frac{\sum_{i=1}^{n} v_i B_i(x)}{\sum_{i=1}^{n} v_i b_i} \end{cases} \tag{6.12}$$

where all parameters are defined as before. Obviously, the PDF in this model naturally satisfies the integration constraint. As the linear B-spline model, rational B-spline model approximates the PDF itself, which makes the controller design relatively easier than the square root B-spline model. But the rational B-spline model has the same disadvantage that the PDF nonnegative cannot be guaranteed in some cases.

6.1.1.5 Rational Square Root B-spline Model

As the rational B-spline model and square root model have complementary advantages, the combination of these two models will present a new model, called the rational square root B-spline model.

$$\begin{cases} \dot{V} = AV + Bu \\ \sqrt{\gamma(x,u)} = \frac{C(x)V}{\sqrt{V^T EV}} \end{cases} \tag{6.13}$$

where

$$\begin{array}{l} C(x) = \left[B_1(x) \; B_2(x) \; \cdots \; B_n(x) \right] \\ E = \int_a^b C^T(x)C(x)dx \\ V = \left[v_1 \; v_2 \; \cdots \; v_n \right]^T. \end{array} \tag{6.14}$$

In (6.13), the natural constraints, the PDF integration and nonnegative constraints, are all satisfied. We should note here that the weights are not real weights in rational and rational square root B-spline models, so we can call them pseudo weights. The real weights for these two models are given by:

$$v_i \Big/ \sum_{j=1}^n v_j \int_a^b B_j(x)dx$$

$$v_i \Big/ \sqrt{\sum_{i,j=1}^n v_i v_j \int_a^b B_i(x)B_j(x)dx.}$$

These four B-spline models are related to each other and can transform among each other under some given conditions. The relationship between linear B-spline weights are simply linear, and a linear B-spline model will change to a rational B-spline model if order-reduction is applied to it. If order-reduction is applied to the square root B-spline model or rational B-spline model is used to approximate the square root of the PDF, then we will obtain a rational square root B-spline model, which inherits robustness property from the square root B-spline model, also with relative low order and no weights constraints. Such advantages provide more convenience for the controller design. Based on the relation of the four models, a relation map is showed in Figure 6.3.

6.1.1.6 RBF Networks

The basis functions discussed above need to be adjusted to get a better function approximation result in some cases such as iterative learning control for stochastic distribution systems [1, 11]. It is also possible to try other neural networks to find a better way to adjust parameters of the basis functions in an iterative manner. Indeed, the RBF network is an option at curve-fitting in high-dimensional space.

RBF network is a class of forward network based on the theory of function approximation. The RBF network learning process is equivalent to find a best-fit plane of the training data in a multi-dimension space. As multi-layer perceptrons, the construction of RBF networks involves three layers: input layer, hidden layer and output layer. The hidden layer adopts nonlinear transformation from input space to the hidden space and it is always of high dimensionality. The dimension of the hidden layer is directly related to the capacity of the network to approximate a smooth

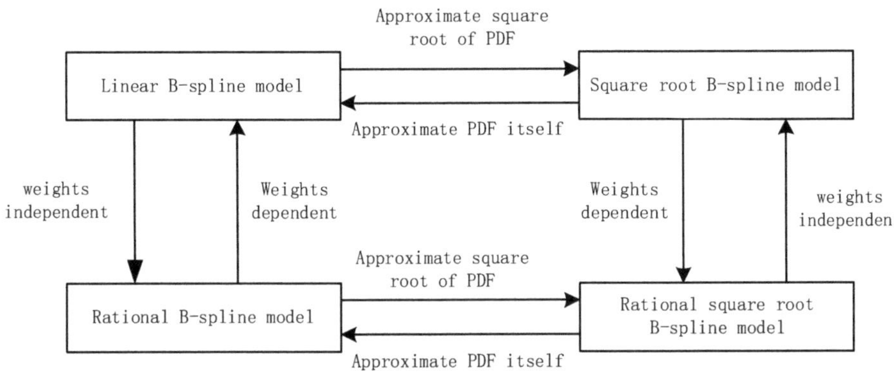

Fig. 6.3 Relation between four kinds of B-spline models

input-output mapping. Of course, the higher the dimension of the hidden layer, the more accurate the approximation will be.

The transformation function of each neuron in RBF network constructs a basis function of the approximation plate, and this is why we call it RBF network. RBF network is a regional approximation network, where there are only few neurons are used to determine the network output for the a certain regional area in the input space.

The nonlinear transformation functions of the hidden layer, that are also viewed as the basis functions, are always selected in Gaussian type:

$$B_i(\mathbf{x}) = \Psi_i(\|\mathbf{x} - \mathbf{c}_i\| / \sigma_i) = e^{-\|\mathbf{x}-\mathbf{c}_i\|^2 / \sigma_i^2} \tag{6.15}$$

where $\mathbf{x}$ is the input vector, $\mathbf{c_i}$ is the center vector of the ith basis function, and σ_i is parameter that determines the width of the basis function around the center. Gaussian type RBF has the following advantages:

- it is arbitrary degree derivative;
- its structure is simple and theory analysis is easy to apply.

6.2 Control Input Design for Output PDF Shaping

Once the structure of the network is selected, the control of the PDF shape can be regarded as the control of the weights and biases of the considered network. As such, many existing control methods can be directly used to formulate the required control laws for the weights control as modeled by [10] to [13]. In particular, when the dynamics are linear and the output PDF is approximated by a B-spline neural network in [10], a compact solution can be generated which minimizes the following performance index

$$J = \int_a^b (\gamma(y,u) - g(y))^2 dy + u^T R u \tag{6.16}$$

where $g(y)$ is the targeted PDF to be tracked, $u(t)$ is the control input and R is a weighting matrix that applies constraints for the input. The integration is calculated with respect to a definition domain represented by $[a,b]$ which defines the output PDFs and the first term represents the functional distance between and $g(y)$.

Of course, other performance functions can also be used to represent the distance between the actual output PDF and the target PDF. The minimization of these performance functions have generally produced numerical methods for the implementation of the closed-loop control. Once the control algorithms are obtained, closed-loop stability can be performed using standard techniques such as linear matrix inequalities, or simply a straightforward application of Lyapunov stability theory ([12]). In the next section, we will present a case study so as to illustrate how the SDC can be applied.

6.3 Introduction of the Grinding Process Control

The simulated ball mill grinding circuit in this section consists of a ball mill, a pump sump and two identical hydrocyclones. The control objective is to guarantee the product particle size index, and at the same time to maximize the mill grinding capability. The object can also be expressed as making the output particle size distribution to follow the given distribution. This is a SDC problem and the output PDF control method can be used in this part in terms of system modeling and control by taking the fresh ore as the input and by setting the fresh water proportional to the fresh ore feed. The flow rates of the sump water and pump are controlled by a feedforward controller to keep the sump concentration, sump water level and hydrocyclone feed inside their required intervals. The control results of the simulated grinding circuit are presented in this chapter.

In ore dressing processes, grinding circuit is an energy consuming process, and also an important process determining the final product quality. The grinding circuit is a complex system characterized by highly coupled, multiple inputs and frequently changing working conditions. Therefore it is still a challenging topic on how to guarantee the product quality in the varying working conditions including the rigidity and particle size distribution of the fresh ore. At present, research into grinding circuit focuses on two kind of models, namely the simulated model and the real model. Regarding the real model, Duarte *et al.* [3] designed a sub-optimal control scheme for the Section C of the CODELCO-Andina concentrator plant in Chile. Yianatos *et al.* [13] realized grinding capacity enhancement by solid concentration control of hydrocyclone underflow, which was tested in EI teniente Division, Codelco in Chile. As to the simulated model, Richard Lestage, *et al.* [8] realized real-time optimization using steady-state linear programming supervisory control. Radhakrishnan [9] applied a supervisory control to a ball mill grinding circuit using the model presented by Herbst *et al.* [6] and similar models appeared also in the PHD thesis of Ming Tie [10] and papers of Austin [2] and King [7]. Although these papers have presented some information about particle size distribution, they still use the traditional method to realize the grinding circuit control without using the particle size distribution (PSD) information provided by the system model as a possible feedback signal. As such, it is necessary to explore possibilities of applying particle distribution measure to constitute direct feedback, where the techniques developed for the SDC can be used.

The mill model used in this chapter is based on the model developed by Yianatos [13], where the breakage function and selection are used to evolve the dynamic process of particle size breakage. Since particle size distribution (PSD) can be easily obtained in such a grinding model, the tradition product index (such as 70%, -200 mesh) can be substituted by the PSD. This would provide much detailed information about the particle size. With the development of the online PSD measuring instruments and the stochastic distribution theory, it is necessary and possible to design the controller including the information of PSD. The output PDF control method developed by Wang [12] can therefore be used in the grinding circuit by taking the PSD, or so-called PDF, as the system output. The control objective is to make the output PSD follow the target PSD as closely as possible. Firstly, for this purpose, the PSD is approximated by a set of basis functions, where a set of weights will be related to the output PSD. Secondly, the dynamic relationship of the second order nature between the fresh ore and the output PSD is modeled. Finally, an optimal control algorithm is designed for this system. The simulation results are presented in the end of this chapter, and these results show that the output PDF method is effective in the modeling and control of the grinding circuit.

6.4 Model Presentation

6.4.1 Grinding Circuit

There are three main production phases in ore processing plant including shaft furnace roasting process, the grinding process and the magne in ore-dressing process. Ball mill grinding circuit studied in this chapter is based on part of the JiuQuan Ore-Dressing Plant in China. The grinding circuit consists of a ball mill, a sump, a sump pump and two identical hydrocyclones (Figure 6.4). In this plant, the fresh ore is fed into the mill by a vibratory feeder, and the fresh water is fed into the mill to adjust the slurry concentration. Another part of feeding is the recycle slurry from the underflow of the hydrocyclones.

The overflow discharging slurry from the ball mill will flow into the pump sump, and some water is added into the sump to dilute the discharging slurry.

The slurry in the sump is injected into the hydrocyclone by a pump, whose rate should be controlled to guarantee the pressure and sump slurry height. The hydrocyclone separates the ore particles into two parts with the coarser ones underflowing as recycle particles and finer ones overflowing as accepted particles.

The variable flow process is showed in Figure 6.5, where u_1 is the fresh ore, u_2 is the fresh water, u_3 is the flow rate of the sump water, u_4 is the pump rate, ρ is the density of the ore, R is the rigidity of the ore, Q, C, M are the slurry volume, slurry concentration and PSD respectively, the subscripts A, B, C, D, E stand for the mill feed, the mill discharge, the hydrocyclone feed, the hydrocyclone overflow and the recycle. The devices specifications in the grinding circuit are listed in Table 6.1.

Table 6.1 Devices specifications

Ball Mill			Hydrocyclone				Sump	
Diameter	Length	Effective Volume	Diameter	Diameter (Inlet)	Finder Diameter	Spigot Diameter	Base Area	Height
3.2 m	3.5 m	25.3 m^3	350 mm	80 mm	100 mm	50 mm	1 m^2	3.5 m

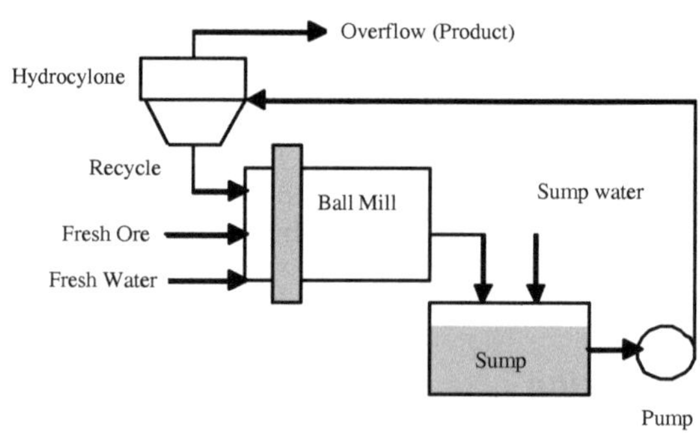

Fig. 6.4 Ball mill grinding circuit scheme

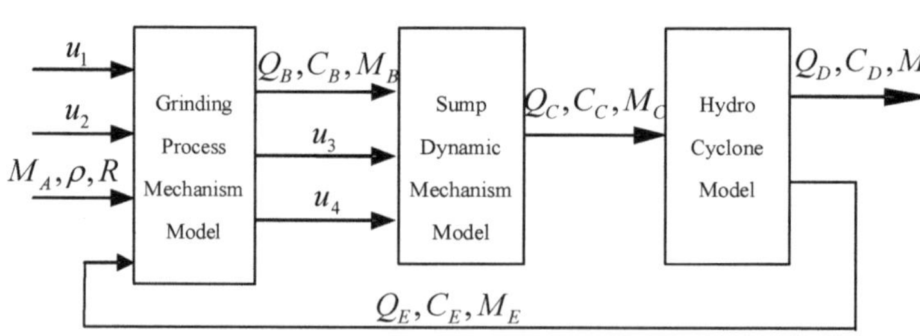

Fig. 6.5 Material variable flow map

6.4.2 Grinding Process Dynamic Model

6.4.2.1 Ball Mill Dynamic Model

Ball mill is an important grinding device in the ore-dressing process. Certain steel balls will be added into the mill, which will collide with the ore particles and also with themselves caused by the mill rotation. The modeling approach, which is different from the traditional modeling method, based on discretized particle size distribution, can present the evolvement of the particles. As size reduction in the mill is a complicated process and the feed ore in the mill will break into finer particles repeatedly, dynamic equations will be formulated based on parameters named as the selection function and the breakage function.

The selection function s_i is defined a percentage rate at which particles break out of the ith interval. Obviously, the selection function is determined by some state variables such as mill power, mill load and feed ore properties such as rigidity and particle size distribution. To formulate the selection function, a parameter, energy specific selection function, is defined for the top size interval, which is a constant determined by the material properties and size distribution. This parameter will be obtained by experiment. The selection function of the 1th (top) interval will be formulated by the energy specific selection function as:

$$s_1 = \frac{s_1^E P}{HC_B}. \tag{6.17}$$

The ith selection function is related with the top selection function s_1 as:

$$s_i = s_1 \exp(\zeta_1 In\sqrt{\frac{d_i d_{i+1}}{d_1 d_2}} + \zeta_2 \left(In\sqrt{\frac{d_i d_{i+1}}{d_1 d_2}}\right)^2) \tag{6.18}$$

where ζ_1 , ζ_2 are constants and d_i d_{i+1} d_1 d_2 are the particle diameters corresponding to size intervals i, $i+1$,1 and 2 respectively.

Breakage function b_{ij} is a mass fraction of broken products from the primary breakage of material in the size interval indexed by j which appears in the size interval indexed by i, so the cumulative breakage function can be defined by the following equations:

$$\begin{aligned} B_{i,j} &= \sum_{k=i}^{N} b_{k,j} \\ B_{i,j} &= \alpha_1 (\tfrac{d_i}{d_{j+1}})^{\alpha_2} + (1-\alpha_1)(\tfrac{d_i}{d_{j+1}})^{\alpha_3} \end{aligned} \tag{6.19}$$

where α_1 ,α_2, α_3 are constants for a particular ore, N is the total particle size intervals, d_i and d_{j+1} are particle diameters corresponding to the ith and $j+1$th size interval. The breakage function is only related to the solid fraction of the mill load. The relationship can be formulated between the variable α_1 and the solid fraction C_2.

$$\alpha_1 = \phi_1 + \phi_2 C_2^{\tau} \tag{6.20}$$

where ϕ_1, ϕ_2 , τ are constants.

When all the selection functions and breakage functions are determined, the dynamic evolving equation of the ith particle interval in the ball mill can be formulated as:

$$\begin{cases} \dot{S_1} = u_1 M_{A1} + Q_E C_E \rho M_{E1} - Q_B C_B \rho M_{B1} - r_1 S_1 \\ \dot{S_2} = u_1 M_{A2} + Q_E C_E \rho M_{E2} - Q_B C_B \rho M_{B2} - r_2 S_2 + \sum\limits_{k<2} b_{2,k} r_k \\ \vdots \\ \dot{S}_{N-1} = u_1 M_{A(N-1)} + Q_E C_E \rho M_{E(N-1)} - Q_B C_B \rho M_{B(N-1)} \\ \quad - r_{N-1} S_{N-1} + \sum\limits_{k<N-1} b_{N-1,k} r_k \\ \dot{S_N} = u_1 M_{AN} + Q_E C_E \rho M_{EN} - Q_B C_B \rho M_{BN} + \sum\limits_{k<N} b_{N,k} r_k \end{cases} \tag{6.21}$$

where $S_i (i = 1, \ldots, N)$ is the particle mass of the ith interval in the mill, r_i is the breakage function of the ith interval.

6.4.2.2 Hydrocyclone Dynamic Model

The sump can be seen as an integrator to realize a material balance and particle balance of different size intervals. The adjusted separation efficiency of the hydrocyclone is formulated based on the Rosin-Rammeler equation:

$$E_i^{'} = 1 - \exp(-0.693(\frac{\overline{d_i}}{d_{50}})^z) \tag{6.22}$$

$$z = \frac{\alpha_4}{Q_C^{0.15}} \exp(-1.6 \frac{Q_E}{Q_C}) \tag{6.23}$$

where α_4 is a constant, d_{50} is the cut size, and $\overline{d_i}$ is the particle geometric average diameter of the ith interval. The cut size d_{50} can be defined as:

$$In(d_{50}) = \alpha_5 + \alpha_6 \log Q_C + \alpha_7 C_E - 0.5 \log(\rho_s - 1) \tag{6.24}$$

where α_5, α_6, α_7 are all constants, ρ_s is the slurry specific gravity. The flow fraction of hydrocyclone underflow to the feed is defined as:

$$R_f = \alpha_8 + \alpha_9 \frac{Q_E(1 - C_E)}{Q_C(1 - C_C)} \tag{6.25}$$

where α_8 , α_9 are both constants. The separation efficiency for the ith particle interval is given by:

$$E_i = E_i^{'}(1 - R_f) + R_f. \tag{6.26}$$

6.5 System Modeling and Control of Grinding Process

The control main object of the grinding circuit is to guarantee that the overflow particles of the hydrocylone meet a requirement, such as the particles less that 200 mesh should be more than 70% while maximizing the mill capacity, which will be substituted by a new index that makes the output PSD follow a given PSD as closely as possible while maximizing the mill capacity.

6.5.1 Control System Structure

As the PSD can be formulated, which can be obtained by measuring instruments, the output PSD can be seen as the output of the whole grinding circuit. The particles in the ball mill are stochastically distributed, and the discrete PSD can be seen as the probability distribution function of the particles in the mill. So the recently developed output PDF control method can be used in the modeling and control of the ball mill grinding circuit system.

The traditional product particle index (such as 70%, -200mesh) can be expressed by the integral of the PSD function (Figure 6.6), so the PSD product index (a given PSD) is a more complex and exact product index.

There are four manipulated variables in this grinding circuit: u_1 is between (0, 100) ton/hour with a corresponding feeder frequency between (0, 50)Hz; u_2 is between (0, 40) ton/hour with a corresponding valve position between (0, 100%); u_3 is between (0, 200) ton/hour with a corresponding valve position between (0, 100%); u_4 is between (0, 100) ton/hour with corresponding valve position between (0, 50%).

Based on the analysis above, the fresh ore will be used to control the output PSD, while u_2 is set to be proportional with u_1, the pump rate and the sump water feed are

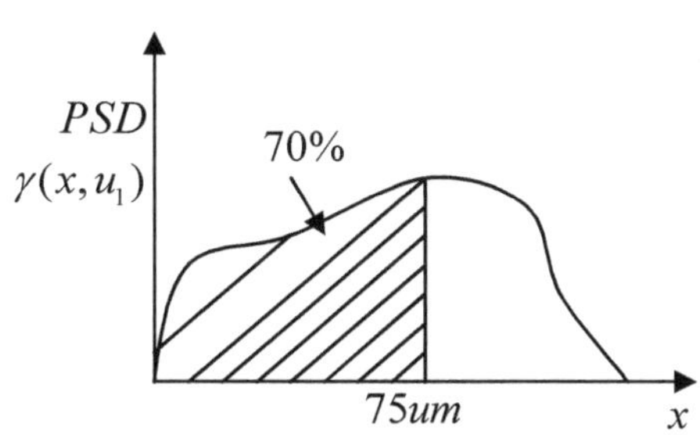

Fig. 6.6 The relationship between PSD and traditional product index

realized by feedforward control to keep the sump concentration, the hydrocyclone feed and the sump height in a required interval. As such, the grinding circuit can be seen as a single-input single-output output PDF control system, whose input is the fresh ore feed u_1 and the output is the PSD of the hydrocyclone overflow; the control system structure is shown in Figure 6.7.

Denote $g(x)$ as the target PSD, $\gamma(x,u_1)$ as the system output PSD, and K as the proportional coefficient between u_1 and u_2. The PDF controller is the main controller to make the system output PSD follow the target PSD as closely as possible by tuning the input u_1. Controller 2 is a feedforward controller to make the slurry volume concentration, sump water level and the cyclone feed in a required interval, which are [30% 60%], (0 3.5) meter and [30 90] ton/hour respectively. Define the sump water level as L_s, and then the control rule of controller 2 is as follows:

- when $0 < L_s < 0.5, u_3 = 200\,t/h, u_4 = 30\,t/h$
- when $0.5 \leq L_s < 1, u_4 = 60L_s\ t/h, u_3\ is\ determined\ by$ $\frac{L_sC_s+Q_BC_B-Q_CC_C}{u_3+L_s-Q_c+Q_B} = 0.2L_s + 0.1$
- when $1 \leq L_s < 2.5, u_3 = 200\,t/h, u_4 = 45\,t/h$
- when $1 \leq L_s < 2.5, u_3 = 200\,t/h, u_4 = 45\,t/h$
- when $2.5 \leq L_s < 3, u_3 0\,t/h, u_4 = 60L_s - 90\,t/h$
- when $3 \leq L_s < 3.5, u_3 = 0\,t/h, u_4 = 90\,t/h$

6.5.2 PDF Modeling Using Basis Function

The particle size is divided into 15 intervals as shown in Figure 6.8, where the PSD histogram and corresponding curve are in a one to one correspondence. So the PSD function is defined as $\gamma(x,u_1)$, where x is a vector with 15 particle size intervals, and which can be approximated by RBF Gaussian basis functions.

$$\gamma(x,u_1(k)) = \sum_{i=1}^{n} v_i(u(k))B_i(x) \tag{6.27}$$

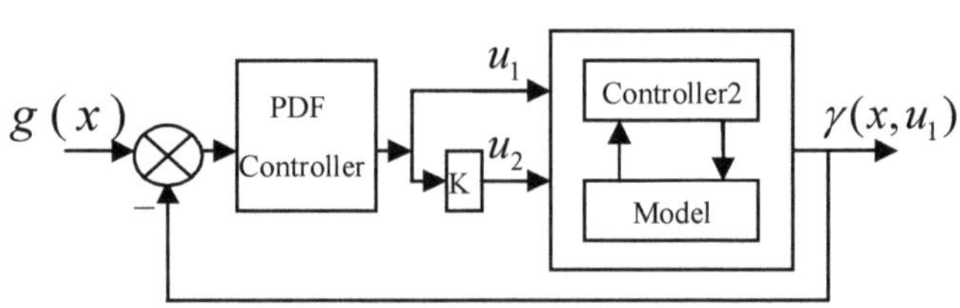

Fig. 6.7 Control system structure

where $v_i(i = 1,\dots,n)$ is the weights of the basis functions and $B_i(x)(i = 1,\dots,n)$ is the basis functions defined as follows:

$$B_i(x) = \mathrm{e}^{-(x-\beta_i)^2/\sigma_i^2} \tag{6.28}$$

where $\beta_i, \sigma_i(i = 1,\dots,n)$ are the mean and variance respectively of the ith basis function. These parameters are listed in Table 6.2.

Table 6.2 Parameters of the basis functions

i	1	2	3	4	5	6	7	8	9	10	11	12	13	14	15
β_i	0.15	0.20	0.25	0.30	0.35	0.40	0.45	0.50	0.55	0.60	0.65	0.70	0.75	0.8	0.85
σ_i	0.08	0.08	0.08	0.08	0.08	0.08	0.08	0.08	0.08	0.08	0.08	0.08	0.08	0.08	0.08

Obviously, the PSD, a PDF function, should satisfy the constraint:

$$\int_a^b \gamma(x,u_1)dx = \sum_{i=1}^{n} v_i(u_1)b_i(x) = 1 \tag{6.29}$$

where $b_i = \int_a^b B_i(x)dx$.

Therefore, there only $n-1$ weights are independent among the n weights, so the PSD function can be expressed by the $n-1$ independent weights.

$$\begin{aligned} \gamma(x,u_1) &= \sum_{i=1}^{n-1} v_i(u_1)B_i(x) + v_n(u_1)B_n(x) \\ &= \sum_{i=1}^{n-1} v_iB_i(x) + \frac{1-\sum_{i=1}^{n-1} v_ib_i(x)}{b_n(x)}B_n(x) \\ &= \mathbf{C}(x)V(u_1) + L(x) \end{aligned} \tag{6.30}$$

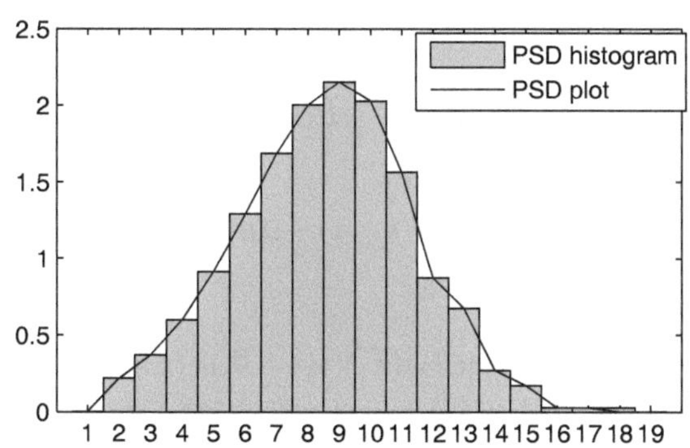

Fig. 6.8 PSD histogram and plot demonstration

where

$$C^T(x) = \begin{bmatrix} B_1(x) - \frac{B_n(x)}{b_n(x)} b_1(x) \\ B_2(x) - \frac{B_n(x)}{b_n(x)} b_2(x) \\ \vdots \\ B_{n-1}(x) - \frac{B_n(x)}{b_n(x)} b_{n-1}(x) \end{bmatrix} \in R^{(n-1)\times 1}, \tag{6.31}$$

$$V(u_1(k)) = \left[v_1(u_1(k)) \; v_2(u_1(k)) \; \cdots \; v_{n-1}(u_1(k)) \right]^T \in R^{(n-1)\times 1}, \tag{6.32}$$

$$L(x) = (\int_a^b B_n(x)dx)^{-1} B_n(x) \in R^{1\times 1}. \tag{6.33}$$

6.5.3 System Modeling and Control

Based on the system dynamic response analysis of the grinding circuit, a first-order dynamic model is used as follows:

$$\begin{aligned} V(k+1) &= \overline{A_1}V(k) + \overline{B}_1 u_1(k) + \overline{C} \\ \gamma(x, u_1(k)) &= C(x)V(u_1) + L(x) \\ &= \sum_{i=1}^{n-1} C_i(x) v_i(u_1(k)) + L(x) \end{aligned} \tag{6.34}$$

where $\overline{A_1} \in R^{(n-1)\times(n-1)}$ are diagonal matrixes, and $\overline{B_1}, \overline{C} \in R^{(n-1)\times 1}$ are column vectors.

The control sequence will be obtained by minimizing the following performance function:

$$J_{k-1}(u_1) = \int_a^b [\gamma(x, u_1(k)) - g(x)]^2 dx \tag{6.35}$$

where $g(x)$ is the target PSD function. Substitute the system model (6.34) into (6.35):

$$J_k(u_1) = \int_a^b [C(x)(\overline{A_1}V(k) + \overline{B_1}u_1(k) + \overline{C}) - g(x)]^2 dx. \tag{6.36}$$

Define

$$[u_1(k) = u_1(k-1) + \Delta u_1(k)], \tag{6.37}$$

then (6.36) can be further expressed as:

$$\begin{aligned} J_k(u_1) = \textstyle\int_a^b [C(x)(\overline{A_1}V(k) + \overline{B}_1 u_1(k-1) \\ + \overline{B_1}\Delta u_1(k)) + L(x) - g(x)]^2 dx. \end{aligned} \tag{6.38}$$

By minimizing the performance function, the input at time k can be obtained:

$$\Delta u_1(k) = -\frac{F}{\int_a^b C(x)\overline{B_1 B_1}^T C^T(x)dx} \tag{6.39}$$

where

$$\begin{aligned} F = \int_a^b C(x)\overline{B_1}[C(x)\overline{A_1}V(k) + C(x)\overline{B_1}u_1(k-1) \\ + C(x)\overline{C} + L(x) - g(x)]dx. \end{aligned} \tag{6.40}$$

Finally, the input at time k is defined as:

$$u_1(k) = u_1(k-1) - \alpha\Delta u_1(k) \tag{6.41}$$

where α is the step length.

6.6 System Simulation and Results

6.6.1 Simulation Parameters

The particle size of the grinding circuit is discretized into 15 intervals, and the boundaries of these intervals are as the following:

Table 6.3 Parameters of the basis functions

No.	1	2	3	4	5	6	7	8	
Particle size	0	0.0191	0.0234	0.0287	0.0351	0.0430	0.0527	0.0645	0.07
No.	9	10	11	12	13	14	15		
Particle size	0.0968	0.1185	0.1452	0.1778	0.2177	0.2667	0.5000		

So the first interval is $[0, 0.0191]$ and the other 14 intervals can be easily obtained using these values by turn. Obviously, the interval lengths are different, but they are mapped into the axis with the same interval length without changing the PSD essentially.

The initial variable state of the mill and sump is:

Table 6.4 The initial state of the mill and sump

Slurry	Mill	Mass	Water	sump
8.3397 m^3	81.1%	7 ton	0.5063 m	50%

With the proportional coefficient $K = 0.08$, the PSD of fresh ore is:

Table 6.5 Particle size distribution of the fresh ore

No.	1	2	3	4	5	6	7	8	9	10	11	12	13	14	15
%	0	0	0	0	0	0.3	0.7	1	1	2	2	3	5	15	31

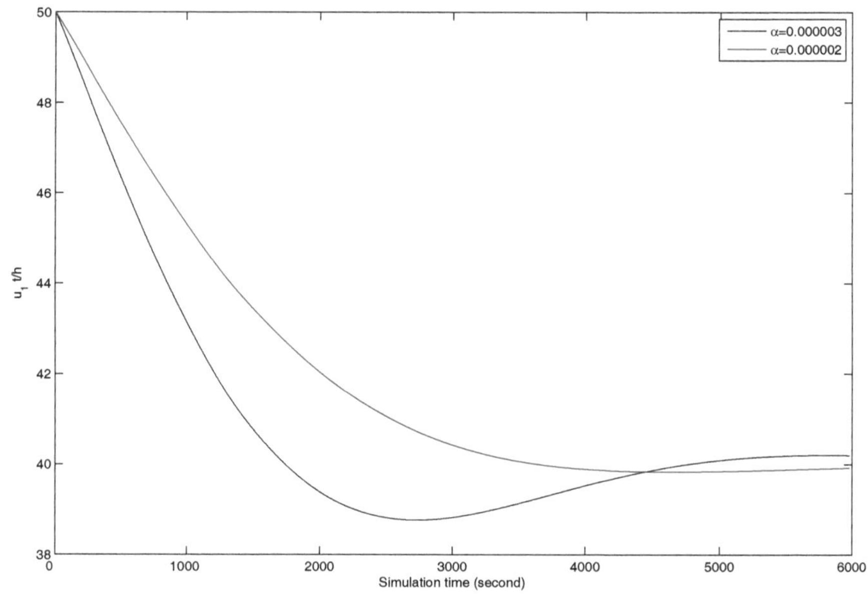

Fig. 6.9 The control sequence of fresh ore feed

6.6.2 Simulation Results

The following simulation results are obtained based on the model and control algorithm stated above with the given initial conditions. Figure 6.9 shows the control

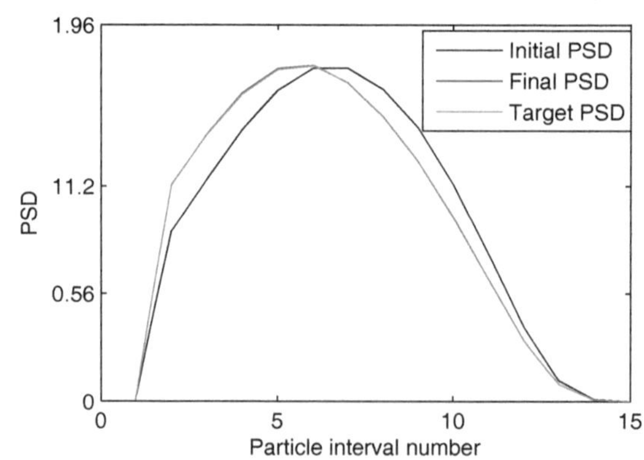

Fig. 6.10 The initial, final and target PSD illustration

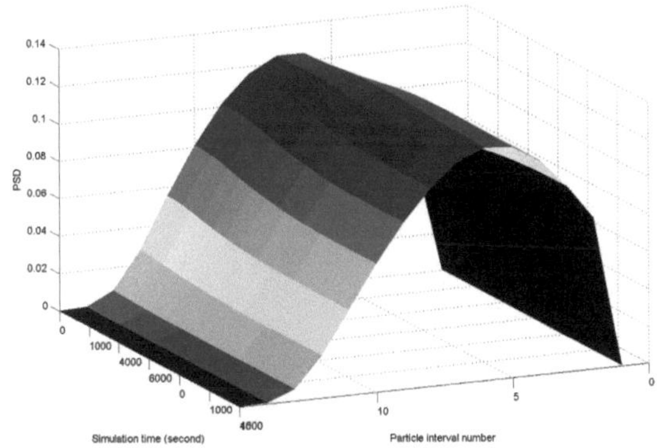

Fig. 6.11 System output PSD 3D mesh

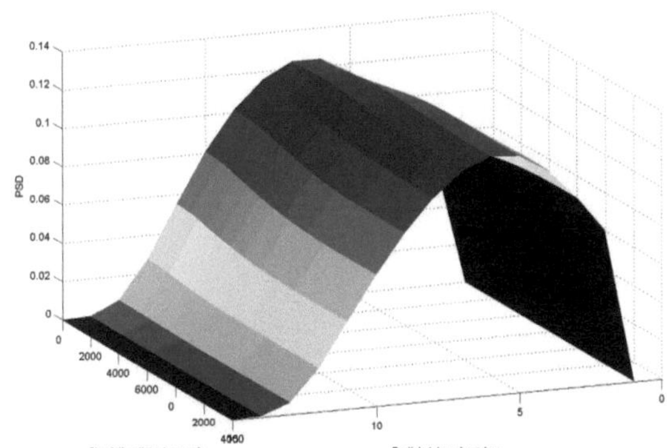

Fig. 6.12 Modeled PSD 3D mesh

sequence with two different step lengths. It can be seen that bigger step length will fasten the control sequence to the target state. All the next figures are all based on the simulation when $\alpha = 0.000002$. Figure 6.10 illustrates the initial PSD, final PSD and the target PSD, and the obtained final PSD well follows the target one. The product index of the initial PSD using the traditional method is 76.72% with less mill capacity, while the product index of the final PSD is 72.63% with a bigger capacity. Figures 6.11 and 6.12 present the evolvement of system output PSD and modeled PSD by 3D mesh. Figure 6.13 gives the control sequence of the sump water feed. Figure 6.14 gives the control sequence of each hydrocyclone feed. Figure 6.15 gives the sump concentration response. Figure 6.16 gives the sump water level response.

6.7 Conclusions

Different from traditional methods for grinding circuits, the output PDF modeling and control method used in this chapter takes PSD as its output, which can provide

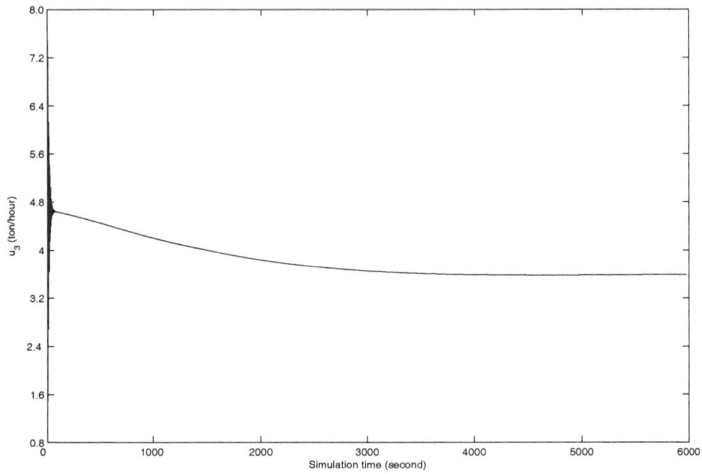

Fig. 6.13 The control sequence of sump water feed

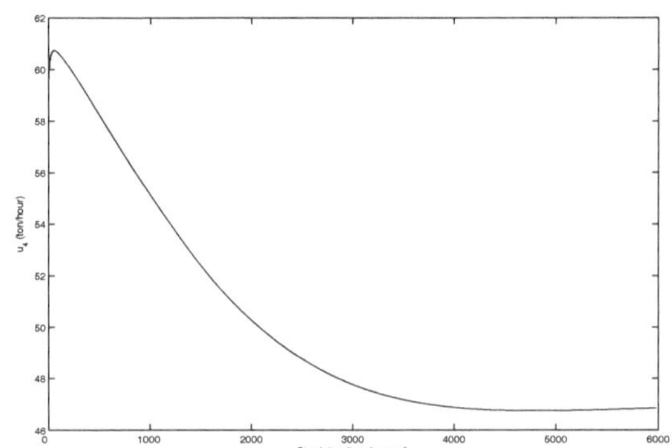

Fig. 6.14 The control sequence of each hydrocyclone feed

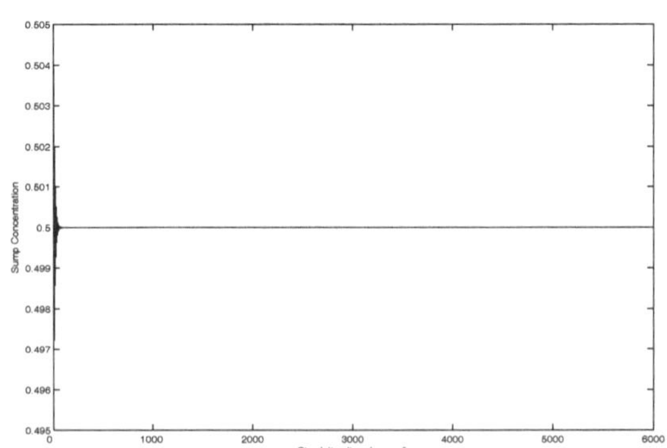

Fig. 6.15 Sump concentration response

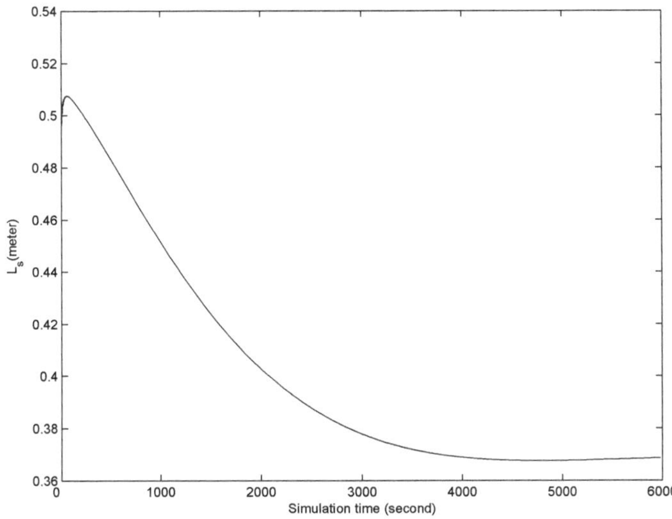

Fig. 6.16 Sump water level response

more information of the particle size. From the results obtained in this chapter, it can be said that the output PDF method is a promising method in processes such as ore dressing.

Acknowledgements This chapter is sponsored by the National Natural Science Foundation of P.R. China (60534010, 60828007), the 111 Project (B08015) and the Sci-Tech Program of Ministry of Education (308007).

References

1. Afshar P, Yue H and Wang H (2007) Robust iterative learning control of output PDF in non-Gaussian stochastic sytems using Youla parametrization. American Control Conference 2007: 576–581.
2. Austin LG (1999) A discussion of equations for the analysis of batch grinding data. Powder Technology 106:71–77.
3. Duarte M, *et al.* (1998) Grinding operation optimization of the codelco andina concentrator plant. Minerals Engineering 11(12):1119–1142.
4. Gonzalez G, *et al.*(1986) A dynamic compensation for static particle size distribution estimators. ISA Trans 25(1):47–51.
5. Haykin S (1999) Neural networks: a comprehensive foundation. Prentice Hall, New Jersey, USA.
6. Herbst JA, Kinnenberg DJ, Rajamani K (1997) Estimill-a program for grinding simulation and parameter estimation with linear medels. Met. Eng. Dept., Univ of Utah, Salt Lake City, Utah, USA.
7. King RP (2001) Modeling and simulation of mineral processing systems. Butterworth-Heinemann Press, Oxford.

8. Lestage R, Pomerleau A, Hodouin D (2002) Constrained real-time optimization of a grinding circuit using seteady-state linear programming supervisory control. Powder Technology 124:253–263.
9. Radhakrishnan VR (1999) Model based supervisory control of a ball mill grinding circuit. Journal of Process Control. 9:195–211.
10. Tie M (2006) Research on hybrid intelligent modeling and applications for several metallurgical industry processes with integrated complexities. PhD thesis, Northeastern University, Shenyang, P.R.China.
11. Wang AP, Afsha P and Wang H (2008) Complex stochastic systems modelling and control via iterative machine learning. Neurocomputing 71:2685-2692.
12. Wang H (2000) Bounded dynamic stochastic distributions: Modelling and Control. Springer-Verlag, London, UK.
13. Yianotos JB, Lisboa MA, Baeza DR (2000) Grinding capacity enhancement by solid concentration control of hydrocyclone underflow. Minerals Engineering 15:317–323.
14. Zhou JL (2005) PDF control and its application in filtering. PhD thesis, Institute of Automation Chinese Academy of Sciences, Beijing, P.R.China.
15. Zhou JL, Yue H and Wang H (2005) Shaping of output PDF based on the rational square-root B-spline model. ACTA Automatica Sinica 31(3):343-351.

Chapter 7
Hybrid Differential Neural Network Identifier for Partially Uncertain Hybrid Systems

Alejandro García, Isaac Chairez and Alexander Poznyak

Abstract This chapter presents a hybrid differential neural network (DNN) -identifier has demonstrated excellent results even in the presence of perturbations. Convergence analysis is realized considering the practical stability of identification error for a general class of hybrid systems. As can be seen in the numerical examples this algorithm could be easily implemented. In this sense the *artificial modeling* strategy of the continuous subsystems could be used in the automatic control implementation.

7.1 Introduction

Hybrid dynamic systems (HDS) arise whenever one mixes logical decision-making with a continuous-time process [16]. The continuous/discrete-time subsystems are represented by a set of differential/difference equations whereas the logical/decision making subsystem (supervisor) could be represented by many ways [7], for instance: Petri nets, fuzzy logic decision system, static neural networks, and commonly by automats [2], [10]. Some physical systems that can be considered as examples of hybrid systems are: computer disk drives, transmission and stepper motors, constrained robotic systems, intelligent vehicle/highway systems, sampled-data systems, discrete event systems, etc. [4]. Most of the literature regarding hybrid systems analysis commonly assumes the existence of mathematical models for each contin-

---

Alejandro García
Departamento de Control Automatico, CINVESTAV-IPN, A.P. 14-740, Av.IPN 2508, México D.F., 07360, México, e-mail: agarcia@ctrl.cinvestav.mx

Isaac Chairez
Bioelectronic Section, UPIBI-IPN, México, e-mail: apoznyak@ctrl.cinvestav.mx

Alexander Poznyak
Departamento de Control Automatico, CINVESTAV-IPN, A.P. 14-740, Av.IPN 2508, México D.F., 07360, México, e-mail: apoznyak@ctrl.cinvestav.mx

uous subsystem as well as the knowledge of the switching law. However, systems identification strategies for those cases when only partial or even null knowledge about subsystems structure is available, are still a challenge for the control community. In this sense, it is possible to find some approaches mainly focused on linear subsystems identification, and less commonly, in nonlinear systems identification. Regarding nonlinear approaches, it is known that if a mathematical model of a nonlinear continuous-time process (non-hybrid) is incomplete or partially known, it is possible to take advantage of the function approximation capacity of an artificial neural network (NN), which may straightforwardly substitute unknown system uncertainties by NN which are defined by a specific mathematical model with a number of free-parameters (*weights*) to be adjusted [5], [13]. By this reason, some proposals to incorporate NN in HDS identification have been developed, but considering the NN identifier as a global model (in average sense) for HDS [1], [9], [12], [6], losing the fundamental nature of the HDS and frequently without a formal proof of its convergence.

In the present chapter a hybrid identifier (HI) for a class of uncertain HDS is presented. The structure of HI is based on DNN approach, the feedback properties of the applied DNN's, avoiding many problems related to the global extremum search converting the learning process to an adequate feedback design [13], [8]. This hybrid the DNN identifier tackles the case when mathematical models of continuous-time subsystems (of HDS) are unknown and only some data collections are available. We also assume that the logical/decision making subsystem is well established and known. Hybrid DNN identifier consists in a pool of DNN with the same elements as the number of subsystems of the HDS and their selection is determined by a logical/decision making system with the same structure as HDS, but evaluated with the identified information. The main difference of the approach depicted here against other algorithms based on NNs is the convergence proof by the technique developed by Xuping Xu and Guisheng Zhai [15] of practical stability for HDS based on Lyapunov-like analysis; in our case, it is possible to obtain the DNN adaptive law to adjust the free parameters.

The chapter organization is as follows: the first section introduces the basic concepts and nomenclature about HDS and the structure of partially uncertain HDS is presented. The second section corresponds to the hybrid DNN Identifier description and the practical stability method presentation, a main theorem about convergence is also included in this section. In the third section two numerical examples are developed: the first one corresponds to a restricted mechanical system and the second, to the classical hybrid system of two interconnected tanks. In both cases the good behavior of the proposed hybrid DNN identifier is shown. The last section corresponds to the conclusion and perspectives. Finally, an appendix containing the extended convergence proof is added.

7.2 Hybrid System

7.2.1 Representation

Many representations for finite hybrid systems have been developed during the last decade (however, there is not standard). In this chapter a hybrid system is considered as in [15], which gives the description as:

$$\begin{gathered}\dot{x}(t) = f_i(x,t) + \xi(t) \\ x \in \mathbb{R}^n,\ f_i : \mathbb{R}^n \times \mathbb{R} \to \mathbb{R}^n,\ i \in I := \{1 \dots M\}.\end{gathered} \tag{7.1}$$

For the system (7.1) the active subsystem at each instant is specified by switching sequences. Given $x(t_0)$, a switching sequence is of the form

$$\varpi = ((t_0, i_0), (t_1, i_1), \dots, (t_k, i_k), \dots, (t_M, i_M))$$

where

$$(t_0 \le t_1 \le \dots \le t_k \le \dots \le t_M, i_k \in I),$$

and specifies that subsystem i is active within $[t_k, t_{k+1})$. All of them are nonzero sequences. It is also assumed that the hybrid system has a discontinuous state jump governed by

$$x(t) := g_{i,j}\left(x\left(t_k^-\right), t\right), \tag{7.2}$$

when it switches from subsystem i_{k-1} to $i_k(x\left(t_k^-\right), t_k)$ at time t_k. Each function $g_{i,j}$ $(i, j \in I, i \neq j)$ characterizes the jump from subsystem i to j.

A behavior study of the hybrid system (7.1)-(7.2), usually specifies some infinite time interval $T = [t_0, \infty)$, or a finite time interval $T = [t_0, t_f]$. In this study it will be also considered that the state vector $x(t)$ always belongs to the interior of the union of regions $Z_i \subset \Re^n$ even in the presence of noise, i.e.:

$$x(t) \in int\left(\bigcup_{i=1}^{M} Z_i\right) \quad \forall t \in T. \tag{7.3}$$

Switching sequences are generated by a switching law defined as follows [15].

Definition 7.1. Given a time interval T, a switching law S over T is a mapping $S : \Re^n \to \Sigma_T$ which specifies a switching sequence $\varpi \varepsilon \Sigma_T$ for any initial state $x(t_0)$. Here $\Sigma_T \triangleq \{$switching sequence Σ over $T\}$ and is a nonZeno switching sequence, (i.e., Σ_T is a sequence that switch a finite number of times in any finite time interval).

S over a given T is often determined by some rules or algorithms, which describe how to generate a switching sequence for a given $x(t_0)$, rather than mathematical equations.

Definition 7.2. A hybrid system without state jump at switching instants is called a switched system, and can be denoted as well as the system defined in (7.1).

It will be also assumed that for every $x(t_0)$ and Σ_T the hybrid or switched system possesses a unique solution over T.

7.2.2 Uncertain Hybrid Systems

If each $i-$*component* of the right hand side of (7.1) is unknown or uncertain, it can be approximated by a set of nonlinear functions $\bar{f}_i(x(t),u(t)\mid W_i(t))$ where $\bar{f}_i \in \Re^n$ defines the approximative mapping depending on the time-varying parameters $W_i(t)$. These parameters should be adjusted by a concrete *adaptation law* (in our case, that adaptive algorithm is derived by a practical stability analysis based on Lyapunov theory). According to the DNN approach [13], a nonlinear function $\bar{f}_i(x(t),u(t)\mid W_i(t))$ may be decomposed into two parts: the first one approximates the linear dynamics part by a *Hurwitz* fixed matrix $A_i \in \Re^{n\times n}$ (selected by the designer) and a nonlinear part is approximated by variable time parameters $W_i(t)$ with *sigmoid* multipliers, that is:

$$\bar{f}_i(x(t),u(t)\mid W_i(t)) := A_i x(t) + W_{1i}(t)\sigma_i(x(t)) + W_{2i}(t)\varphi_i(x(t))u(t) \tag{7.4}$$

where $A_i \in \Re^{n\times n}$, $W_{(1,2)i}(t) \in \Re^{n\times p}, \sigma(\cdot) \in \Re^{p\times 1}, \varphi(\cdot) \in \Re^{p\times r}$ and $u(t) \in \Re^{r\times 1}$ is an external bounded input but not a control, i.e. $U^{adm} := \{u(t) : \|u(t)\| \le \Upsilon_u < \infty\}$, this approximation considers the general case which involves the frequent situation in hybrid systems where $u(t)=0$. The activation vector-functions $\sigma(\cdot)$ and $\varphi(\cdot)$ are usually selected as a function with sigmoid-type components, i.e.,

$$\sigma_r(x(t)) := a_r\left(1+b_r\exp\left(-\sum_{s=1}^{n} c_s x_s(t)\right)\right)^{-1}, \tag{7.5}$$

and

$$\varphi_{s,s}(x(t)) := a_{r,s}\left(1+b_{r,s}\exp\left(-\sum_{e=1}^{n} c_e x_e(t)\right)\right)^{-1}.$$

It is easy to see that the activation functions satisfy the following sector conditions

$$\left\|\sigma(x(t)) - \sigma(x'(t))\right\|^2_{\Lambda_\sigma} \le L_\sigma \left\|x(t) - x'(t)\right\|^2_{\Lambda'_\sigma} \tag{7.6}$$

$$\left\|\varphi(x(t)) - \varphi(x'(t))\right\|^2_{\Lambda_\varphi} \le L_\varphi \left\|x(t) - x'(t)\right\|^2_{\Lambda'_\varphi} \tag{7.7}$$

and are globally bounded on $\Re^n$. In (7.4), the constant parameters A_i as well as the time-varying parameters $W_i(t)$ should be properly adjusted to guarantee a good state approximation. Notice that for any fixed matrices $W_{1i}(t) = \hat{W}_{1i}$, $W_{2i}(t) = \hat{W}_{2i}$ the dynamics (7.1) could always be represented as

$$\dot{x}(t) = A_i x(t) + \hat{W}_{1i}\sigma(x(t)) + \hat{W}_{2i}\varphi_i(x(t))u(t) + \tilde{f}_i(t) + \xi(t) \tag{7.8}$$

where $\tilde{f}_i(t) := f_i(x(t)) - \bar{f}_i(x(t) \mid \hat{W}_i)$ is referred to as the modeling error vector-field and the vector ξ_t represents the whole noises affecting the state. This approximation remains valid inside each region $Z_i \subset \Re^n$. In view of the corresponding boundedness property and the condition (7.3), the following upper bound for the unmodeled dynamics $\tilde{f}_i(t)$ takes place:

$$\begin{gathered} \left\|\tilde{f}_i(t)\right\|^2_{\Lambda_f} \leq \tilde{f}_0 \; \forall i \in [1,M] \\ \tilde{f}_0 > 0; \; \Lambda_f > 0, \; \Lambda_f = \Lambda_f^{\top}. \end{gathered} \tag{7.9}$$

7.3 Hybrid DNN Identifier

Let us introduce the following hybrid DNN identifier:

$$f_i(x,\hat{x},t) := A_i\hat{x}(t) + W_{1i}(t)\sigma_i(\hat{x}(t)) + W_{2i}(t)\varphi_i(\hat{x}(t))\,u(t) \tag{7.10}$$

working inside the region $Z_i \subset \Re^n$. Here, the weights matrices $W_{1i}(t)$ and $W_{2i}(t)$ supply the adaptive behavior and accurate representation of the uncertain nonlinear system for this class of identifier if they are adjusted in an adequate manner. We suggest the following nonlinear weight *updating* law derived from the practical stability analysis described in the appendix:

$$\dot{W}_{1i}(t) = -\alpha_{Q_i}\tilde{W}_{1i}(t) - \tilde{k}_{1i}2k_{1i}P_i\Delta_t\sigma_i^{\top}(\hat{x}(t)) \tag{7.11}$$

$$\dot{W}_{2i}(t) = -\alpha_{Q_i}\tilde{W}_{2i}(t) - \tilde{k}_{2i}P_i\Delta_t u^{\top}(t)\varphi_i^{\top}(\hat{x}(t)) \tag{7.12}$$

where $\alpha_{Q_i} = 2^{-1}\lambda\min(P_i^{-\frac{1}{2}}Q_iP_i^{-\frac{1}{2}})$, $Q_i, P_i \in \Re^{n\times n}, Q_i, P_i > 0; \tilde{W}_{ji}(t) := W_{ji}(t) - \hat{W}_{ji},; \tilde{k}_{ji} \in \Re,\ k_{ji} > 0, j = 1,2\ \Delta(t) := \hat{x}(t) - x(t)$ with P_i positive define solution of the following set of Riccati algebraic equations:

$$\begin{gathered} P_iA_i + A_i^{\top}P_i + P_iR_iP_i + \tilde{Q}_i = 0 \\ R_i = \hat{W}_{1i}\Lambda_{1i}\hat{W}_{1i}^T + \Lambda_{4i},\ i = 1,..,M \\ \tilde{Q}_i = L_{\sigma}\Lambda_{1i}^{-1} + L_{\varphi}\Upsilon_u\Lambda_{5i}^{-1} + Q \end{gathered} \tag{7.13}$$

where Λ_{1i}, Λ_{4i} and Λ_{5i} are positive definite matrices ($\Lambda_{1i}, \Lambda_{4i}, \Lambda_{5i} \in \Re^{n\times n}$). To improve the behavior of these adaptive laws, the matrix $\hat{W}_{1i}$ can be *provided* by one of the, so-called, training algorithms. Some approaches are presented in [3] and [14].

7.3.1 Practical Stability of Hybrid Systems

The following definitions introduce the main concept of practical stability of hybrid systems [15]:

Definition 7.3. (ε-Pr*actical Stability Over T*) Assume that a time interval T and a switching law S over T are given. Given $\varepsilon > 0$, hybrid system (7.1)-(7.2) is said to be ε-practically stable over T under S if there exists a $\delta > 0$ (δ depends on and the interval T) such that $x(t)\varepsilon B[0,\varepsilon], \forall t \varepsilon T$, whenever $x(t_0)\varepsilon B[0,\delta]$.

7.3.1.1 A Direct Method for Practical Stability

This method was suggested by Xuping Xu and Guisheng Zhai in [15] using practical Lyapunov-like functions. Similar to the Lyapunov stability theory for hybrid systems, it applied the aforementioned direct method for the ε-practical stability of a hybrid system using practical Lyapunov-like functions.

Definition 7.4. ($\varepsilon-$*Practical Lyapunov-like Functions*) Given a time interval T and a switching law S over a continuously differentiable real-valued function $V(x,t)$ satisfying $V(0,t) = 0, \forall t \varepsilon T$ is said to be $\varepsilon-$*practical Lyapunov-like function over T under S* if there exist a Lebesgue integrable function $\phi(x,t)$, positive constants μ and γ, such that for any trajectory $x(t)$ generated by S that starts from $x(t_0)$ and its corresponding switching sequence $\sigma = ((t_0,i_0),(t_1,i_1),\ldots,(t_k,i_k),\ldots)$, the following conditions hold:

1. $\dot{V}(x_t,t) \leq \phi(x_t,t)$, *a.e.* $t \in T$.
2. $V(g_{i_{k-1},i_k(x(t_k^-),t_k)},t_k) \leq \mu V(x_{t_k^-},t_k)$ at any switching instant t_k.
3. $\int_{t_0}^{t} \mu^{N(\tau,t)}\phi(x_\tau,\tau)d\tau < \inf_{x \notin B[0,\varepsilon]} V(x,t) - \mu^{N(t_0,t)}\gamma, \ \forall t \in T$.

In 1. $\dot{V}(x_t,t)$ denotes the derivative of $V(x,t)$ along $x(t)$, i.e., $\dot{V}(x(t),t) = V_x(x(t),t) \cdot f_i(x(t),t) + V_t(x(t),t)$ where i_t is the index of the active subsystem at time instant t. In 3. $N(a,b)$ denotes the number of switching during the interval (a,b).

Theorem 7.1. *[15] Given a time interval T and a switching law S over T, hybrid system (7.1)-(7.2) is ε-practically stable over T under S if there exists an $\varepsilon-$practical Lyapunov-like function $V(x,t)$ over T under S.*

7.3.2 Stability of Identification Error

Hereafter, it is assumed that:

1. The class of the functions $f_i : \Re^n \to \Re^n$ is Lipschitz continuous in $x \in X$, that is, for all $x, x' \in X$ there exist constants $L_{1,2}$ such that

$$\begin{aligned} \|f(x,u) - f(y,v)\| &\leq L_1 \|x-y\| + L_2 \|u-v\| \\ \|f(0,0,t)\|^2 &\leq C_1 \end{aligned} \tag{7.14}$$

$x,y \in \Re^n$; $u,v \in \Re^m$; $0 < L_1, L_2 < \infty$.

2. The noise $\xi(t)$ in the system (7.1) are uniformly (on t) bounded such as

$$\|\xi_i(t)\|^2_{\Lambda_\xi} \leq \Upsilon_\xi, \ \forall i \in [1, M] \tag{7.15}$$

where Λ_ξ is a known *normalizing* non-negative definite matrix which permits to operate with vectors having components of different physical nature.

3. Without loss of generality let us assume that the switching sequence $g_{i,j}\left(x\left(t_k^-\right), t\right)$ is defined as follows

$$g_{i,j}\left(x\left(t_k^-\right), t\right) = \Theta_{i,j} x\left(t_k^-\right) \tag{7.16}$$

with $\left\|\Theta_{i,j}\right\| \leq 1$.

The following theorem describes the performance of the identification algorithm based on DNN methodology:

Theorem 7.2. *(Identification Error Stability) Under Assumptions 7.14) and (7.15), assuming that (7.6), (7.7) and (7.9) are valid, considering the DNN identifier (7.10) with the adaptive laws (7.11-7.12), and if there exist matrices* $\Lambda_{ri} = \Lambda_{ri}^{\mathsf{T}} > 0$, $\Lambda_{r,i} \in \Re^{n\times n}$, $r = 1\ldots5$, $Q \in \Re^{n\times n}$ *such that the set (7.13) of the Riccati equations has a solution, then the hybrid system* $\Delta_t := \hat{x}(t) - x(t)$ *is* ε*-practically stable, with* ε *holding:*

$$\frac{\psi M T}{1 + \alpha_Q T} < \varepsilon$$

where:

$$\psi = M\left(\Upsilon_\xi + \tilde{f}_0\right) > 0.$$

(See proof in appendix).

Corollary 7.1. *The* $\varepsilon-$*practical stability, attained in the identification process based on DNN, implies the following upper bounds for the state and the weights trajectories:*

$$\begin{array}{c} \|\Delta(t)\| \leq M\sqrt{\varepsilon_1 \lambda_{\min}(P)} \\ \operatorname{tr}\left\{\tilde{W}_{1i}^{\mathsf{T}}(t)\,\tilde{W}_{1i}(t)\right\} \leq 2k_{1i} M \varepsilon_2 \\ \operatorname{tr}\left\{\tilde{W}_{2i}^{\mathsf{T}}(t)\,\tilde{W}_{2i}(t)\right\} \leq 2k_{2i} M \varepsilon_3 \end{array}$$

where ε_1, ε_2 *and* ε_3 *are such that*

$$\varepsilon_1 + \varepsilon_2 + \varepsilon_3 \leq \varepsilon.$$

Corollary 7.2. *If* T *is finite, the best attainable upperbound* ε *is*

$$\varepsilon := \frac{M\psi T}{1 + \alpha_Q T},$$

meanwhile if $T \to \infty$ *then* $\varepsilon \to \dfrac{M\psi}{\alpha_Q}$. *So, the restrictions given in previous theorem are not sufficient conditions to get the asymptotic convergence of the identification process. Nevertheless, if external perturbations and error approximation vanish* ($\Upsilon_\xi = \tilde{f}_0 = 0$) *then the asymptotic convergence is reached too.*

7.4 Examples

In this section numerical simulation for two implementation examples are presented: a spring-mass system and the two interconected tanks system. In both cases $u(t) = 0$, and also they belong to the particular subclass of HDS called switched systems (see Definition 7.2).

7.4.1 Spring-mass System

7.4.1.1 System Description

The mechanical system modeled in [4] is used to test the hybrid DNN identifier. Figure 7.1 depicts the system. A block of mass m supported by two springs of stiffness $k/2$ at both ends, is placed on a lever which can rotate about an axis at its midpoint. There is a viscous friction of viscosity c between the mass and the lever. Two proximity sensors, denoted by SR and SL, are located at a distance ls from the midpoint of the lever at the right and left sides to detect the block while moving on the lever. The lever rotates with an angular velocity ω, which is controlled by an actuator. x is

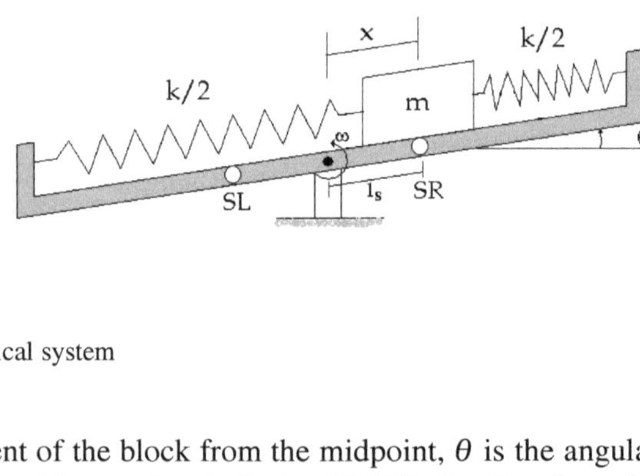

Fig. 7.1 Mechanical system

the displacement of the block from the midpoint, θ is the angular displacement of the lever measured in counter clockwise (CCW) direction from the horizontal and g is the gravitational acceleration. θ is assumed to be limited between $-\frac{\pi}{2}$ and $\frac{\pi}{2}$. The lever either rotates in clockwise (CW) or CCW direction.

The switching sequence S over T can be expressed as follows: after the block passes over the sensor SR, the lever rotates with a constant angular velocity of $\omega = p$, whereas after it passes over SL, the lever rotates with $\omega = -p$. That is, the lever keeps rotating in the same direction until the block passes over the other sensor. Defining $x_1(t) := x(t)$; $x_2(t) = \dot{x}_1(t)$; $x_3(t) = \theta$, then continuous state dynamics for each particular state is modeled by:

$$f_1(x,t) = \begin{bmatrix} x_2 \\ -\frac{k}{m}x_1 - \frac{c}{m}x_2 - g\sin(x_3) \\ -p \end{bmatrix} \tag{7.17}$$

and

$$f_2(x,t) = \begin{bmatrix} x_2 \\ -\frac{k}{m}x_1 - \frac{c}{m}x_2 - g\sin(x_3) \\ p \end{bmatrix}. \tag{7.18}$$

7.4.1.2 Hybrid DNN Identifier

As in the hybrid system (7.17)-(7.18) $u(t) = 0$, a hybrid DNN identifier structure is proposed as:

$$f_1(x,\hat{x},t) = A_1 x(t) + W_{11}(t)\sigma_i(x(t))$$
$$f_2(x,\hat{x},t) = A_2 x(t) + W_{12}(t)\sigma_i(x(t))$$

As both subsystems are very similar, we have proposed similar parameters as follows:

$$A_1 = \begin{bmatrix} -2.6 & 0 & 0 \\ 0 & -1.64 & 0 \\ 0 & 0 & -2.2 \end{bmatrix} A_2 = \begin{bmatrix} -2.0 & 0 & 0 \\ 0 & -1.4 & 0 \\ 0 & 0 & -2.8 \end{bmatrix}$$

and the implemented adaptive law is defined as (7.11) with:

$$\hat{W}_{1i} = \begin{bmatrix} 0.09 & 0.064 & 0.08 \\ 0.11 & 0.11 & 0.2 \\ 0.05 & 0.05 & 0.06 \end{bmatrix} \quad i = 1,2.$$

Figure 7.2 depicts the block diagram for hybrid system and the hybrid DNN identifier. It is worth noticing that the existence of noise $\xi(t)$ in the available information is being considered. $\xi(t)$ is a pseudowhite *white noise* signal with amplitude=0.05.

7.4.1.3 Simulation Results

Figure 7.3 depicts the identification results for the state x_1 taking into account the noise presence. It is possible to verify the transition moments between subsystems and also the time interval in which each DNN is working.

Figures 7.4 and 7.5 depict the identification results for x_2 and x_3.

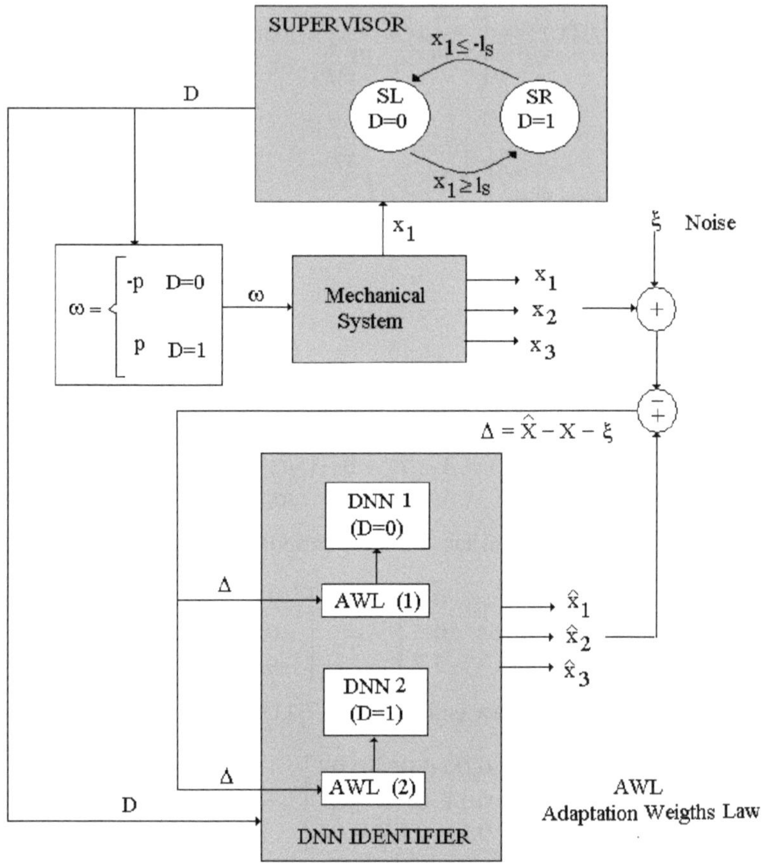

Fig. 7.2 Block diagram of spring-mass hybrid system and DNN identifier

7.4.2 Two Interconnected Tanks System

7.4.2.1 System Structure

In this section the model regarded with a hydraulic system composed of two tanks is employed to test the hybrid DNN identifier. Figure 7.6 depicts its structure. A similar model has been presented by [11]. The systems consists of two cylindrical tanks L_1 and L_2. Tank L_1 has an input on-off pump Z and output electrical valve V_2, in a similar way tank L_2 has an input valve V_1 and output valve V_2. Valves V_1 and V_2 are identical on-off electrical valves. Tanks are interconnected by a pipe using the valve V_2. The system can be represented as a HDS with two continuos variables, h_1 and h_2, representing the height of the fluid in the tanks and three binary inputs Z,V_1

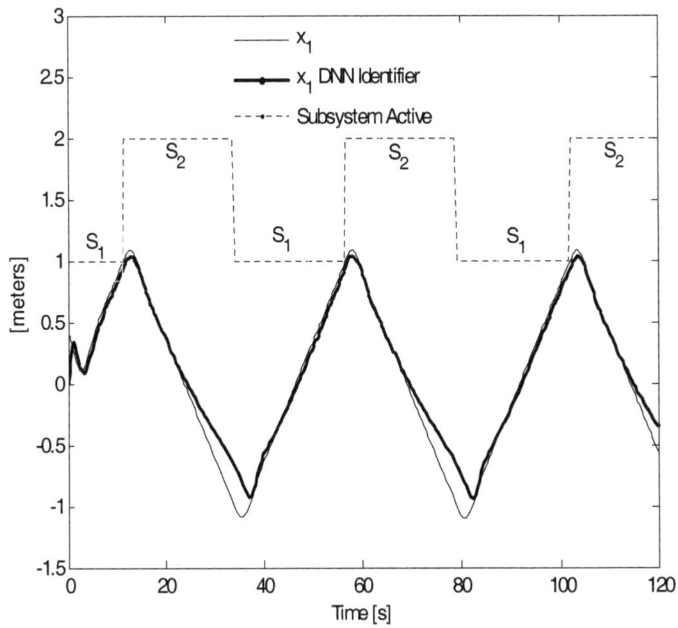

Fig. 7.3 x_1 identification and the activation time for each subsystem and DNN

and V_2. An important assumption is that the dynamics of the pump and valves are very fast. The mathematical model of each subsystem is represented by:

$$f_i(x,t) = \begin{bmatrix} \dot{h}_{1,i}(x,t) \\ \dot{h}_{2,i}(x,t) \end{bmatrix} = \frac{1}{S} \begin{bmatrix} Q_{Z,i} - Q_{V_{1i}} \\ Q_{V_{1i}} - Q_{V_{2i}} \end{bmatrix} \qquad i = 1,2...8 \tag{7.19}$$

where:

$$\begin{aligned} Q_{Z,i} &= DZ && Z \varepsilon \{0,1\} \\ Q_{V_{1i}} &= A \operatorname{sign}(h_1 - h_2) \sqrt{2g\,|h_1 - h_2|} V_1 && V_1 \varepsilon \{0,1\} \\ Q_{V_{2i}} &= A\sqrt{2gh_2} V_2 && V_2 \varepsilon \{0,1\}. \end{aligned} \tag{7.20}$$

So, there exist eight subsystems according to the combination of binary inputs Z, V_1 and V_2 for this hybrid system:

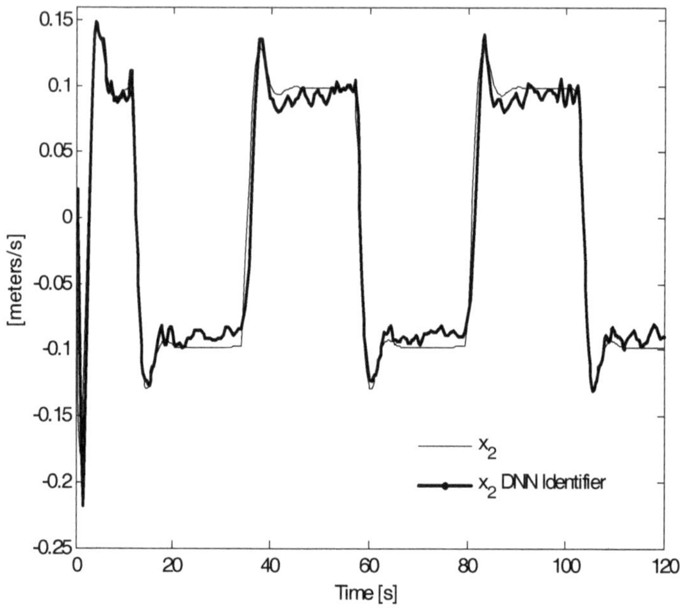

Fig. 7.4 Identification of x_2

Subsystem	V_1	V_2	Z
S_1	0	0	0
S_2	0	0	1
S_3	0	1	0
S_4	0	1	1
S_5	1	0	0
S_6	1	0	1
S_7	1	1	0
S_8	1	1	1

7.4.2.2 Identifier Structure

A hybrid DNN identifier structure for the hybrid dynamic system (7.19)-(7.20) is proposed as:

$$f_i(x,\hat{x},t) = A_i x(t) + W_{1i}(t)\sigma_i(x(t)) \qquad i = 1\ldots 8.$$

Parameters A_i and $\hat{W}_{1i}$ are selected for each $f_i(x,\hat{x},t)$ as was shown in the previous example. The adaptive law implemented is defined by (7.11). Figure 7.7 depicts the

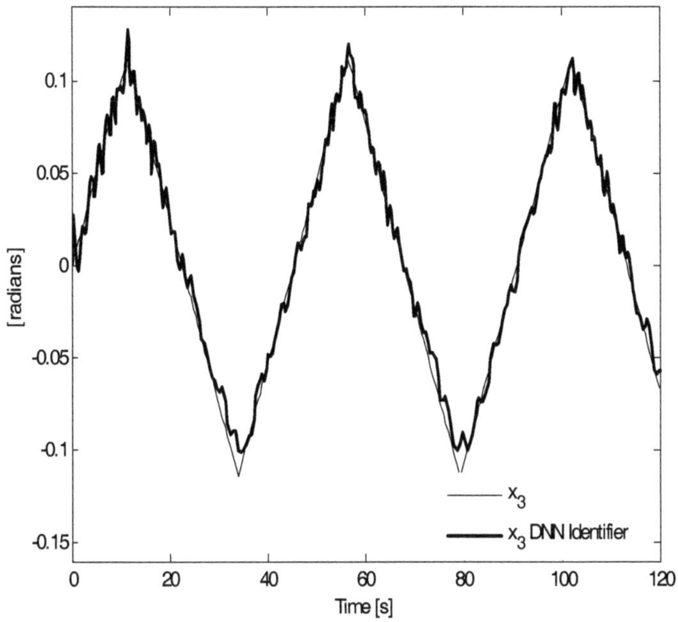

Fig. 7.5 Identification of x_3

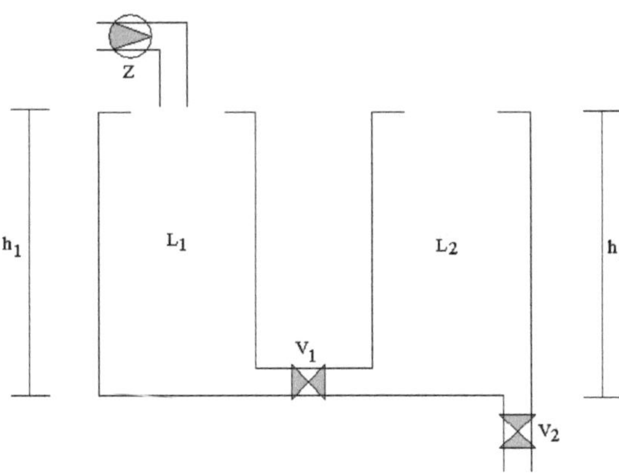

Fig. 7.6 Two tanks interconnected

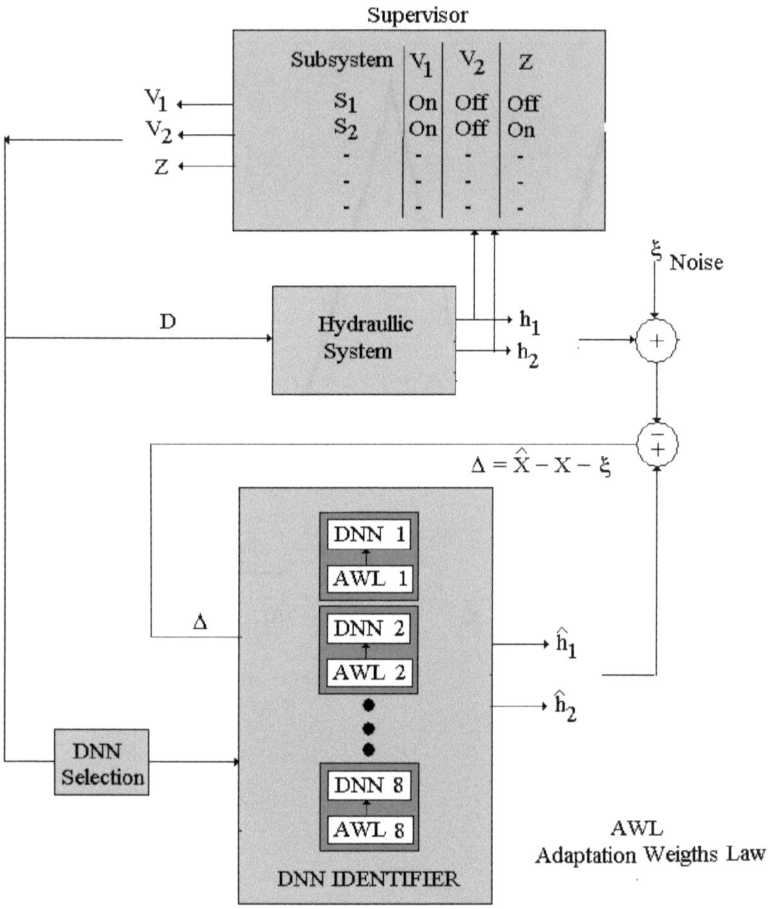

Fig. 7.7 Block diagram of hybrid system and DNN identifier

block diagram for the hybrid system and the hybrid DNN identifier; continuous state is affected by noise $\xi(t)$.

7.4.2.3 Simulation Results

In order to carry out the simulation we have considered the situation when V_2 is available for a human operator (this condition was represented by an aleatory sequence in the simulation) and some logical control is in charge to maintain the levels of h_1 and h_2 in the next intervals:

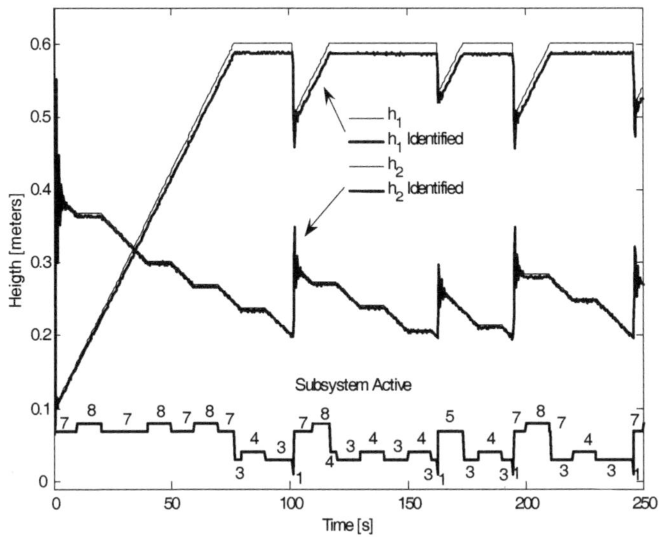

Fig. 7.8 Identification for h_1 and h_2, and subsystems active

$$0.6m \leq h_1 \leq 0.8m$$
$$0.2m \leq h_2 \leq 0.5m$$

employing Z and V_1. Parameters were fixed at: $S = 1.54m^2$, $A = 30m^2$, $.D = 0.1m^3/s$, $g = 9.21m/s^2$. Figure 7.8 depicts the identification results for h_1 and h_2 taking into account the presence of noise and the aleatory switching sequence of *on* and *off* state for V_2, even more of a verification of subsystems and the time interval in which each DNN is working. As it can be seen, not all subsystems are presented for the current simulation condition. A good correspondence between the hybrid DNN identifier and the hybrid system is obtained.

7.5 Conclusions

The suggested hybrid DNN identifier has demonstrated excellent results even in the presence of perturbations. Convergence analysis was realized considering the practical stability of an identification error for a general class of hybrid systems. As can be seen in the numerical examples this algorithm could be easily implemented. In this sense the *artificial modeling* strategy of the continuous subsystems could be used in the automatic control implementation.

Appendix

To prove Theorem 7.2, let us define the identification error as $\Delta(t) = \hat{x}(t) - x(t)$. The dynamics of $\Delta_i(t)\ t \in T$ is governed by the following ordinary differential equation:

$$\begin{array}{c} \dot{\Delta}_i(t) := A_i\hat{x}(t) + W_{i1}(t)\sigma_i(\hat{x}(t)) \\ +W_{2i}(t)\varphi_i(\hat{x}(t))u(t) - A_ix(t) \\ -\hat{W}_{1i}\sigma_i(x(t)) - \hat{W}_{2i}\varphi_i(x(t))u(t) - \tilde{f}_i(t) - \xi_{it} = \\ A_i\Delta(t) + \left[W_{i1}(t) - \hat{W}_{1i}\right]\sigma_i(\hat{x}(t)) + \\ \hat{W}_{1i}\left[\sigma_i(\hat{x}) - \sigma_i(x)\right] - \tilde{f}_i(t) - \xi_{it} + \\ \left[W_{2i}(t) - \hat{W}_{2i}\right]\varphi_i(x(t))u(t) + \\ \hat{W}_{2i}\left[\varphi_i(\hat{x}(t)) - \varphi_i(x(t))\right]u(t). \end{array} \tag{7.21}$$

Defining: $\tilde{W}_{1i}(t) := W_{i1}(t) - \hat{W}_{1i}$, $\tilde{W}_{2i}(t) := W_{2i}(t) - \hat{W}_{2i}$, $\tilde{\sigma}_i(x(t),\hat{x}(t)) := \sigma_i(\hat{x}) - \sigma_i(x)$, $\tilde{\varphi}(x(t),\hat{x}(t)) := \varphi_i(\hat{x}(t)) - \varphi_i(x(t))$ we obtain:

$$\begin{array}{c} \dot{\Delta}_i(t) = A_i\Delta(t) + \tilde{W}_{1i}(t)\sigma_i(\hat{x}(t)) + \\ \hat{W}_{1i}\tilde{\sigma}_i(x(t),\hat{x}(t)) + \tilde{W}_{2i}(t)\sigma_i(\hat{x}(t))\varphi_i(\hat{x}(t))u(t) + \\ \hat{W}_{2i}\tilde{\varphi}(x(t),\hat{x}(t))u(t) - \tilde{f}_i(t) - \xi_{it}. \end{array} \tag{7.22}$$

Considering the following Lyapunov-like (energetic) function as

$$\begin{array}{c} V(t) = \sum\limits_{i=1}^{M} \|\Delta_i(t)\|_{P_i}^2 + 2^{-1}\sum\limits_{i=1}^{M} k_{1i}^{-1}\operatorname{tr}\left\{\tilde{W}_{1i}(t)^\intercal \tilde{W}_{1i}(t)\right\} + \\ 2^{-1}\sum\limits_{i=1}^{M} k_{2i}^{-1}\operatorname{tr}\left\{\left[\tilde{W}_{2i}(t)\right]^\intercal \tilde{W}_{2i}(t)\right\}, \end{array}$$

which first time derivative is expressed by:

$$\begin{array}{c} \dot{V}(t) = 2\sum\limits_{i=1}^{M} \Delta_i^\intercal(t)P_i\dot{\Delta}_i(t) + \sum\limits_{i=1}^{M} \operatorname{tr}\left\{k_{1i}^{-1}\tilde{W}_{1i}(t)^\intercal \dot{W}_{1i}(t)\right\} + \\ \sum\limits_{i=1}^{M} \operatorname{tr}\left\{k_{2i}^{-1}\tilde{W}_{2i}(t)^\intercal \dot{W}_{2i}(t)\right\}. \end{array}$$

Taking into account the dynamics of the system (7.22), we have

$$\begin{aligned}
\dot{V}_t = \sum_{i=1}^{M} 2\Delta_i^{\top}(t) P_i A_i \Delta_i(t) + \\
2\sum_{i=1}^{M} \Delta_i^{\top}(t) P_i \hat{W}_{1i} \tilde{\sigma}_i(x(t), \hat{x}(t)) + 2\sum_{i=1}^{M} \Delta_i^{\top}(t) P_i \tilde{W}_{1i}(t) \sigma_i(\hat{x}(t)) + \\
2\sum_{i=1}^{M} \Delta_i^{\top}(t) P_i \tilde{W}_{2i}(t) \sigma_i(\hat{x}(t)) \varphi_i(\hat{x}(t)) u(t) + \\
2\sum_{i=1}^{M} \Delta_i^{\top}(t) P_i \hat{W}_{2i} \tilde{\varphi}(x(t), \hat{x}(t)) u(t) - \\
2\sum_{i=1}^{M} \left(\Delta_i^{\top}(t) P_i \tilde{f}_i(t) - 2\Delta_i^{\top}(t) P_i \xi_i(t)\right) + \\
\sum_{i=1}^{M} \mathrm{tr}\left\{k_{1i}^{-1} \tilde{W}_{1i}(t)^{\top} \dot{W}_{1i}(t)\right\} + \sum_{i=1}^{M} \mathrm{tr}\left\{k_{2i}^{-1} \tilde{W}_{2i}(t)^{\top} \dot{W}_{2i}(t)\right\}.
\end{aligned} \tag{7.23}$$

Notice that the first term is equivalent to (7.23):

$$2\Delta_i^{\top}(t) P_i A_i \Delta_i(t) = \Delta_i^{\top}(t)\left(P_i A_i + A_i^{\top} P_i\right)\Delta_i(t).$$

For estimating each next terms in expression (7.23) we will use the next inequality:

$$XY^{\top} + YX^{\top} \leq X\Lambda X^{\top} + Y\Lambda^{-1}Y^{\top} \tag{7.24}$$

valid for any $X, Y \in R^{r\times s}$ and any $0 < \Lambda = \Lambda^{\top} \in R^{s\times s}$ [13]. Considering (7.24) and (7.14) and (7.15). We get:

1.
$$\begin{aligned}
2\Delta_i^{\top}(t) P_i \hat{W}_{1i} \tilde{\sigma}_i(x(t), \hat{x}(t)) \leq \\
\Delta_i^{\top}(t) P_i \hat{W}_{1i} \Lambda_1 \hat{W}_{1i}^{\top} P_i \Delta_i(t) + \\
L_{\sigma} \Delta_i^{\top}(t) \Lambda_{1i}^{-1} \Delta_i(t).
\end{aligned}$$

2.
$$2\Delta_i^{\top}(t) P_i \tilde{f}_i(t) \leq \Delta_i^{\top}(t) P_i \Lambda_{3i} P_i \Delta_i(t) + \tilde{f}_0.$$

3.
$$2\Delta_i^{\top}(t) P_i \xi_i(t) \leq \Delta_i^{\top}(t) P_i \Lambda_{4i} P_i \Delta_i(t) + \Upsilon_{\xi}.$$

4.
$$\begin{aligned}
2\Delta_i^{\top}(t) P_i \hat{W}_{2i} \tilde{\varphi}(x(t), \hat{x}(t)) u(t) \leq \\
\Delta_i^{\top}(t) P_i \hat{W}_{2i} \Lambda_{5i} \hat{W}_{2i}^{\top} P_i \Delta_i(t) + \\
L_{\varphi} \Upsilon_u \Delta_i^{\top}(t) \Lambda_{5i}^{-1} \Delta_i(t).
\end{aligned}$$

Bringing together results given in 1.-4., we can rewrite the derivative $\dot{V}_t$ as:

$$\dot{V}_t \leq -\alpha_Q \sum_{i=1}^{M} \left[\left\| \Delta_i^{\intercal}(t) \right\|_{P_i}^2 + 2^{-1} k_{1i}^{-1} \operatorname{tr}\left\{ \tilde{W}_{1i}(t)^{\intercal} \tilde{W}_{1i}(t) \right\} + 2^{-1} k_{2i}^{-1} \operatorname{tr}\left\{ \tilde{W}_{2i}(t)^{\intercal} \tilde{W}_{2i}(t) \right\} \right] + \sum_{i=1}^{M} \Delta_i^{\intercal}(t) \left(P_i A_i + A_i^{\intercal} P_i + P_i R_i P_i + \tilde{Q}_i \right) \Delta_i(t) + \sum_{i=1}^{M} 2\Delta_i^{\intercal}(t) P_i \tilde{W}_{1i}(t) \sigma_i(\hat{x}_t) + 2\Delta_i^{\intercal}(t) P_i \tilde{W}_{i,2}(t) \varphi_i(\hat{x}(t)) u(t) + \sum_{i=1}^{M} \operatorname{tr}\left\{ k_{1i}^{-1} \tilde{W}_{1i}(t)^{\intercal} \dot{W}_{1i}(t) \right\} + \sum_{i=1}^{M} \operatorname{tr}\left\{ k_{2i}^{-1} \tilde{W}_{2i}(t)^{\intercal} \dot{W}_{2i}(t) \right\} + \sum_{i=1}^{M} \left(\Upsilon_{\xi} + \tilde{f}_0 \right) + \alpha_{Q_i} \sum_{i=1}^{M} \left[\tilde{k}_{1i} \operatorname{tr}\left\{ \tilde{W}_{1i}(t)^{\intercal} \tilde{W}_{1i}(t) \right\} + \tilde{k}_{2i} \operatorname{tr}\left\{ \tilde{W}_{2i}(t)^{\intercal} \tilde{W}_{2i}(t) \right\} \right]$$

with

$$\begin{array}{c} R_i = \hat{W}_{1i} \Lambda_{1i} \left[\hat{W}_{1i} \right]^T + \Lambda_{4i} \\ \tilde{Q}_i = L_{\sigma} \Lambda_{1i}^{-1} + L_{\varphi} \Upsilon_u \Lambda_{5i}^{-1} + Q \\ \alpha_{Q_i} = 2^{-1} \lambda \min(P_i^{-\frac{1}{2}} Q P_i^{-\frac{1}{2}}) > 0 \\ \tilde{k}_{ji} = 2^{-1} k_{j,i}^{-1}, \\ i = 1..M, j = 1,2. \end{array} \tag{7.25}$$

Since the Riccati equation admits the same positive definite solution P and considering the adapting laws (7.11) and (7.12) we obtain:

$$\dot{V}_t \leq -\alpha_Q V_t + \psi$$

where $\alpha_Q := \min_{i=1:M} (\alpha_{Q_i})$. The first condition of Definition 7.4 is held, with:

$$\phi(\Delta_t, t) = -\alpha_Q V_t + \psi$$

where:

$$\psi = M \left(\Upsilon_{\xi} + \tilde{f}_0 \right) > 0.$$

So, for any set of $\left(\Delta_i(t), \tilde{W}_{1i}(t), \tilde{W}_{2i}(t) \right)$ such that

$$\begin{array}{c} \left\| \Delta_i(t) \right\|_{P_i}^2 + 2^{-1} k_{1i}^{-1} \operatorname{tr}\left\{ \tilde{W}_{1i}(t)^{\intercal} \tilde{W}_{1i}(t) \right\} \\ +2^{-1} k_{2i}^{-1} \operatorname{tr}\left\{ \tilde{W}_{2i}(t)^{\intercal} \tilde{W}_{2i}(t) \right\} \in [0, \varepsilon] \end{array}$$

holds (in fact, this is the condition where the practical stability is attained), then $\dot{V}_t \leq -\alpha_Q \varepsilon + \psi_t$. The last condition implies simultaneously that

$$\begin{array}{c} \left\| \Delta_i(t) \right\| \leq \sqrt{\varepsilon_1 \lambda_{\min}(P_i)} \\ \operatorname{tr}\left\{ \tilde{W}_{1i}^{\intercal}(t) \tilde{W}_{1i}(t) \right\} \leq 2k_{1i} \varepsilon_2 \\ \operatorname{tr}\left\{ \tilde{W}_{2i}^{\intercal}(t) \tilde{W}_{2i}(t) \right\} \leq 2k_{2i} \varepsilon_3 \end{array}$$

where ε_1, ε_2 and ε_3 are such that

$$\varepsilon_1 + \varepsilon_2 + \varepsilon_3 \leq \varepsilon.$$

The second condition of Definition 7.4 is automatically attained with $\mu = 1$. Finally, according the last condition given in Definition 7.4, the DNN identifier is $\varepsilon-$practically stable over T if there exists γ such that $0 < \gamma < \varepsilon$ and

$$\int_{t_0}^{t} (-\alpha_Q \varepsilon + M\psi) d\tau < \inf_{x \notin B[0,\varepsilon]} V(x,t) - \gamma, \ \forall t \varepsilon T.$$

So, one has

$$\int_{t_0}^{t} (-\alpha_Q \varepsilon + M\psi) d\tau < \varepsilon - \gamma < \varepsilon.$$

In particular if $T = [t_0, t]$ then, practical stability is assured if

$$\frac{\psi MT}{1 + \alpha_Q T} < \varepsilon.$$

The proof is complete. □

References

1. Back A, Chen T-P (1997) Approximation of hybrid systems by neural networks. In Proc. Int Conf on Neural Information Processing, Springer-Verlag 326–329.
2. Branicky M (1998) Multiple Lyapunov functions and other analysis tools for swiched and hybrid systems. IEEE Trans. on Autom. Control 43:475–482.
3. Chairez I, Poznyak A, Poznyak T (2006) New sliding-mode learning law for dynamic neural network observer. IEEE Trans. on Circuits and Syst. II 53:1338-1342.
4. Dogruel M, Arif M (2005) Hybrid state approach for modelling electrical and mechanical systems. Mathematical and Computer Modelling 41:759–771.
5. Haykin S (1994) Neural Networks. A comprehensive foundation. IEEE Press, New York.
6. Holderbaum W (2005) Application of neural network to hybrid systems with binary inputs. IEEE Trans. on Neural Networks 18(4):1254–1261.
7. Henzinger T, Ho P-H, Wong-Toi H (1998) Algorithmic analysis of nonlinear hybrid systems. IEEE Trans. on Autom. Control 43:540–554.
8. Lewis F, Yesildirek A, Liu K (1996) Multilayer neural-net robot controller with guaranteed tracking performance. IEEE Trans. on Neural Networks 7(2):1–11.
9. Li P, Cao J (2007) Global stability in switched recurrent neural networks with time-varing delay via nonlinear measure. Nonlinear Dyn. 49:295–305.
10. Li X, Yu W, Peres S (2002) Adaptive fuzzy petri nets for supervisory hybrid systems modeling. In 15th Triennial World Congress, Barcelona Spain.
11. Messai N, Riera B, Zaytoon J (2008) Identification of a class of hybrid dynamic systems with feed-forward neural networks: About the validity of the global model. Nonlinear Analysis: Hybrid Systems 2:773–785.
12. Narendra K, Balakrishnan J, Ciliz M (1995) Adaptation and learning using multiple models, switching, and tuning. IEEE Control Syst. Mag. 15:37–51.
13. Poznyak A, Sanchez E, Yu W (2001) Differential neural networks for robust nonlinear control (identification, state estimation and trajectory tracking, World Scientific.
14. Stepanyan V, Hovakimyan N (2007) Robust adaptive observer design for uncertain systems with bounded disturbances. IEEE Trans. on Neural Networks 18(5):1392–1403.

15. Xu X, Zhai G (2005) Practical stability and stabilization of hybrid and switched systems. IEEE Trans. on Autom. Control 50:1897–1903.
16. Ye H, Michel A, Hou L (1998) Stability theory for hybrid dynamical systems. IEEE Trans. on Autom. Control 43:461–474.

Chapter 8
Real-time Motion Planning of Kinematically Redundant Manipulators Using Recurrent Neural Networks

Jun Wang, Xiaolin Hu and Bo Zhang

Abstract With the wide deployment of kinematically redundant manipulators in complex working environments, obstacle avoidance emerges as an important issue to be addressed in robot motion planning. In this chapter, the inverse kinematic control of redundant manipulators for obstacle avoidance task is formulated as a convex quadratic programming (QP) problem subject to equality and inequality constraints with time-varying parameters. Compared with our previous formulation, the new scheme is more favorable in the sense that it can yield better solutions for the control problem. To solve this time-varying QP problem in real time, a recently proposed recurrent neural network, called an improved dual neural network, is adopted, which has lower structural complexity compared with existing neural networks for solving this particular problem. Moreover, different from previous work in this line where the nearest points to the links on obstacles are often assumed to be known or given, we consider the case of obstacles with convex hull and formulate another time-varying QP problem to compute the critical points on the manipulator. Since this problem is not strictly convex, an existing recurrent neural network, called a general projection neural network, is applied for solving it. The effectiveness of the proposed approaches is demonstrated by simulation results based on the Mitsubishi PA10-7C manipulator.

Jun Wang
Department of Mechanical and Automation Engineering, The Chinese University of Hong Kong, Shatin, N.T., Hong Kong, e-mail: jwang@mae.cuhk.edu.hk

Xiaolin Hu
State Key Laboratory of Intelligent Technology and Systems, Tsinghua National Laboratory for Information Science and Technology (TNList), and Department of Computer Science and Technology, Tsinghua University, Beijing 100084, P.R. China, e-mail: xiaolin.hu@gmail.com

Bo Zhang
State Key Laboratory of Intelligent Technology and Systems, Tsinghua National Laboratory for Information Science and Technology (TNList), and Department of Computer Science and Technology, Tsinghua University, Beijing 100084, P.R. China, e-mail: xiaolin.hu@gmail.com

8.1 Introduction

As the automation industry develops, robot manipulators are required to work in more and more complex and dynamic environments. An important issue to be considered is how to effectively avoid static or moving objects in the workspace. Kinematically redundant manipulators are those having more degrees of freedom than required to perform given end-effector moving tasks. Being dexterous and flexible, they have been used for avoiding obstacles, for singularity, and for optimizing various performance criteria in addition to tracking desired end-effector trajectories. Of those versatile applications, obstacle avoidance is extremely important for successful motion control when there exist obstacles in the workspace.

To accomplish the obstacle avoidance task, it is often demanded to determine the *critical points* on manipulator links that have minimum (or near minimum) distances to obstacles in real time. However, in some recent studies [16, 33] the critical points considered were assumed to be available without any additional effort. This assumption oversimplified reality. Indeed, even if the obstacles are static in the workspace, the critical points usually change their positions on the surfaces when the manipulator is moving. In sensor-based control, obstacles can be determined by the synthetic information of sensor fusion technology, e.g., utilizing vision, ultrasonic and infrared sensors [21, 23]; while in model-based control, they can be derived via online distance minimization between the manipulators links and obstacles [4, 19, 23]. In the chapter, the obstacles are assumed to be 3D convex polyhedron and a convex time-varying QP problem is formulated to determine the critical points on the manipulator without assuming the prior information of the critical points.

Obtaining critical points on the manipulator is just a prerequisite. The next step is to plan the manipulator desired velocity and direct it away from the obstacles. Many strategies about the velocity control have been reported. A popular branch of methods for real-time obstacle avoidance is to apply the pseudoinverse formulation for obtaining a general solution at velocity level, which contains a minimum L_2-norm solution and a homogeneous solution (e.g., [5, 10, 18, 20, 22, 26]). The homogeneous solution is selected such that a secondary goal of obstacle avoidance is achieved while accomplishing the primary goal of trajectory tracking. From the mathematical viewpoint, pseudoinverse-based methods inherently treat the inverse kinematics of redundant manipulators as optimization problems. In contrast, another branch of methods formulate the inverse kinematic control problem into explicit optimization problems [2–4, 7, 11, 27, 32, 33]. This treatment has an advantage in including practical considerations such as joint limits, joint velocity limits and obstacle avoidance into problems [3, 4, 11, 32, 33]. In particular, [33] proposed a QP formulation for obstacle avoidance. In this chapter, we will present an improved version of this QP formulation.

It is seen that the entire control scheme proposed in this chapter is comprised of solving two time-varying QP problems, which are called the *critical point problem* and the *joint velocity problem*, respectively, for convenience.

The bottleneck of the formulations based on optimization techniques is that they entail intensive computation during the manipulator movement. To solve such a time-varying optimization problem, conventional optimization techniques are usually incompetent, especially in dynamic and/or uncertain environments. Parallel and distributed approaches such as recurrent neural networks to real-time obstacle avoidance are deemed desirable. In the past twenty years, recurrent neural networks for optimization have been widely explored since the pioneering work of Tank and Hopfield [12, 25]. Reported results of numerous investigations have shown many desirable features (e.g., see [8, 29] and references therein). In particular, many neural networks have been developed to solve the inverse kinematics problems focusing on achieving pseudoinverse solutions [6], minimum torque [27, 28], minimum kinetic energy [32], optimal grasping force [1, 31], and so on. In this chapter, two existing recurrent neural networks will be adopted for solving two formulated QP problems. It will be shown that the two neural networks are superior to most other existing counterparts in the literature for solving the particular problems.

The remainder of the chapter is organized as follows. In Section 8.2, the obstacle avoidance problem is formulated as two time-varying QP problems with some preliminary results presented in [16]. In Section 8.3, two neural networks are proposed for solving the QP problems. Section 8.4 reports the simulation results and Section 8.5 concludes the chapter.

8.2 Problem Formulation

8.2.1 Critical Point Problem

The obstacle avoidance task is to identify in real time the critical points on the manipulator that are closest to the obstacles, then to determine desired velocities for the joints and to keep the manipulator away from the obstacles. At any time, there is a point O on every obstacle, defined as the *obstacle point*, and a point C on every link, defined as the *critical point*, such that $|OC|$ is the minimum distance between the particular obstacle-link pair (see Figure 8.1). If the obstacle point O is known or given at anytime, it is easy to locate the critical point C on the link by considering two possible cases as shown in Figure 8.1. This is the case considered in [33] and [16]. Actually, it can be said equivalently that in those two studies, the whole obstacle was represented by an abstract point without any geometric shape. As discussed in the introduction, this assumption oversimplified real situations.

We now discuss in a model-based environment where obstacles are 3D bodies, how to obtain the representative points O on these objects and the corresponding critical points C on a manipulator. The problem is formulated as a quadratic optimization problem:

$$\begin{aligned}&\text{minimize } \tfrac{1}{2}\|(x,y,z)^T-(\tilde{x},\tilde{y},\tilde{z})^T\|_2^2\\&\text{subject to } (x,y,z)^T\in\mathscr{U},(\tilde{x},\tilde{y},\tilde{z})^T\in\mathscr{V}\end{aligned}$$

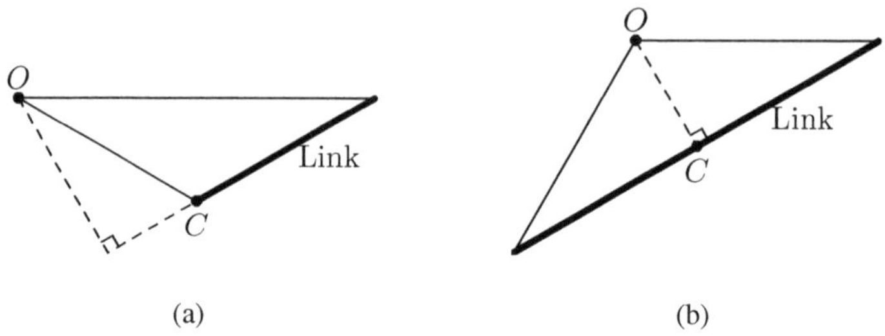

Fig. 8.1 Critical point location (a) case 1, (b) case 2

where $\mathscr{U}$ and $\mathscr{V}$ stand for the subsets of $\Re^3$ occupied by the obstacle and the manipulator, respectively. For every link and every obstacle we should seek the optimum $(x,y,z)^T$ and $(\tilde{x},\tilde{y},\tilde{z})^T$ in the above problem, which represent the position of O on the obstacle and the position of C on the link, respectively. For simplicity, throughout the chapter the links are approximated by line segments and the detailed shape information (such as radius, and thickness) is ignored. (But in principle, more realistic configurations of links can be considered.) Then for every link we have

$$\mathscr{V} = \left\{ (\tilde{x},\tilde{y},\tilde{z})^T \in \Re^3 \,\middle|\, \begin{matrix} G\cdot(\tilde{x},\tilde{y},\tilde{z})^T = d, \\ \underline{\eta} \le (\tilde{x},\tilde{y},\tilde{z})^T \le \overline{\eta} \end{matrix} \right\},$$

where

$$G = \begin{pmatrix} \bar{y}_2 - \bar{y}_1 & \bar{x}_1 - \bar{x}_2 & 0 \\ \bar{z}_2 - \bar{z}_1 & 0 & \bar{x}_1 - \bar{x}_2 \end{pmatrix},$$
$$d = \begin{pmatrix} (\bar{y}_2 - \bar{y}_1)\bar{x}_1 - (\bar{x}_2 - \bar{x}_1)\bar{y}_1 \\ (\bar{z}_2 - \bar{z}_1)\bar{x}_1 - (\bar{x}_2 - \bar{x}_1)\bar{z}_1 \end{pmatrix}.$$

In the above description, $(\bar{x}_1,\bar{y}_1,\bar{z}_1)^T$ and $(\bar{x}_2,\bar{y}_2,\bar{z}_2)^T$ stand for the position vectors of the two terminals of the link, and $\underline{\eta},\overline{\eta} \in \Re^3$ stand for respectively the lower and upper bounds of $(\tilde{x},\tilde{y},\tilde{z})^T$. If $\mathscr{U}$ is a convex polyhedron, without loss of generality, it can be written as

$$\mathscr{U} = \{(x,y,z)^T \in \Re^3 | A\cdot(x,y,z)^T \le e\}$$

where $A \in \Re^{q\times 3}$ and $e \in \Re^q$ are constants. In this case, the optimization problem becomes a (convex) QP problem. By introducing a new vector $w = (x,y,z,\tilde{x},\tilde{y},\tilde{z})^T$, the problem can be put into the following compact form,

$$\begin{array}{l} \text{minimize } \frac{1}{2}w^TQw \\ \text{subject to } Kw \in \Pi, w \in \mathscr{W} \end{array} \tag{8.1}$$

where

$$Q = \begin{pmatrix} I & -I \\ -I & I \end{pmatrix}, K = \begin{pmatrix} A & 0 \\ 0 & G \end{pmatrix},$$
$$\Pi = \left\{ v \in \Re^{q+2} \middle| \begin{pmatrix} -\infty \\ d \end{pmatrix} \leq v \leq \begin{pmatrix} e \\ d \end{pmatrix} \right\},$$
$$\mathscr{W} = \left\{ w \in \Re^{6} \middle| \begin{pmatrix} -\infty \\ \underline{\eta} \end{pmatrix} \leq w \leq \begin{pmatrix} \infty \\ \overline{\eta} \end{pmatrix} \right\}.$$

By solving (8.1), and the obstacle point O, critical point C, together with the distance between them can be obtained simultaneously. Note that (8.1) is a time-varying optimization problem as the the parameters $K, \Pi, \mathscr{W}$ change with time while the manipulator or the obstacle itself is moving.

The problem (8.1) is established for every link of the manipulator versus every obstacle, not for all links versus all obstacles. In this sense, it is a local strategy, instead of a global one. Otherwise, it is impossible to formulate a convex optimization problem. However, as the number of links is very limited, a global solution can be easily obtained by using simple comparison or winner-takes-all operation.

8.2.2 Joint Velocity Problem: An Inequality-constrained Formulation

Taking obstacle avoidance into consideration, the forward kinematics of a serial-link manipulator can be described by the following augmented forward kinematics equation:

$$r_e(t) = f_e(\theta(t)), \quad r_c(t) = f_c(\theta(t)) \tag{8.2}$$

where $r_e \in \Re^m$ is an m-dimensional position and orientation vector of the end-effector, and $r_c \in \Re^3$ is the position vector of a critical point, $\theta(t)$ is the joint vector of the manipulator, and f_e and f_c are nonlinear functions of the manipulator with respect to respectively the end-effector and the critical point. If there exists more than one critical point, multiple equations, similar to the second one, are present.

The manipulator path planning problem (also called the inverse kinematics problem or kinematic control problem) is to find the joint variable $\theta(t)$ for any given $r_e(t)$ and $r_c(t)$ through the inverse mapping of (8.2). Unfortunately, it is usually impossible to find an analytic solution due to the nonlinearity of $f_e(\cdot)$ and $f_c(\cdot)$. The inverse kinematics problem is thus usually solved at the velocity level with the relation

$$J_e(\theta)\dot{\theta} = \dot{r}_e, \quad J_c(\theta)\dot{\theta} = \dot{r}_c, \tag{8.3}$$

where $J_e(\theta) = \partial f_e(\theta)/\partial\theta$ and $J_c(\theta) = \partial f_c(\theta)/\partial\theta$ are the Jacobian matrices, and $\dot{r}_e$ is the desired velocity of the end-effector, $\dot{r}_c$ is the desired velocity of the critical point which should be properly selected to effectively direct the link away from the obstacle. However, in practice it is often difficult to determine the suitable magnitude of escape velocity $\dot{r}_c$. Even worse, if there are p critical points, we should have $3p$ equalities of $J_c(\theta)\dot{\theta} = \dot{r}_c$; then, if $3p > n$, these equalities will be overdetermined. Because of these considerations, in [33], $J_c(\theta)\dot{\theta} = \dot{r}_c$ in (8.3) is replaced by

$$J_N(\theta)\dot{\theta} \leqslant 0, \tag{8.4}$$

where $J_N(\theta) = -\mathrm{sgn}(\overrightarrow{OC}) \diamond J_c(\theta)$. The vector matrix multiplication operator $\diamond$ is defined as $u \diamond V = [u_1V_1, u_2V_2, \cdots, u_pV_p]^T$, where u is a column vector and V_i denotes the ith row of a matrix V.

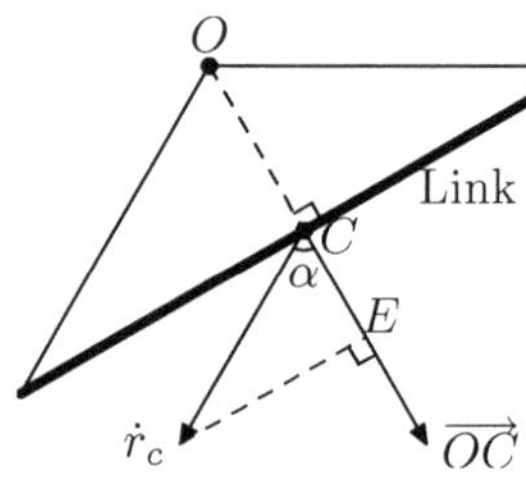

Fig. 8.2 Escape direction

Then the motion planning and obstacle avoidance problem of a kinematically redundant manipulator was formulated in a time-varying local optimization framework as:

$$\begin{aligned} &\text{minimize } c(\dot{\theta}(t)) \\ &\text{subject to } J_e(\theta(t))\dot{\theta}(t) = \dot{r}_e(t) \\ &\qquad\qquad J_N(\theta(t))\dot{\theta}(t) \leqslant 0 \\ &\qquad\qquad l \leqslant \dot{\theta}(t) \leqslant h \end{aligned} \tag{8.5}$$

where l and h are respectively lower and upper bounds of the joint velocity vector, and $c(\cdot)$ is an appropriate convex cost function, for example, $\|\dot{\theta}(t)\|_1$, $\|\dot{\theta}(t)\|_2^2$, $\|\dot{\theta}(t)\|_\infty$, selected according to the task needs [18], [3], [5]. To avoid singularity configurations or drift in repeated operations, a linear term of $\dot{\theta}$ can be added. The first constraint corresponds to the primary goal of tracking a given end-effector motion, and the second constraint is to achieve the secondary goal of obstacle avoidance, which is included whenever the shortest distance between the obstacle and the manipulator is within a prescribed safety margin. When there exists more than one obstacle, multiple constraints, similar to the second constraint in (8.5) should be included. By putting the motion requirement of the critical points as a constraint instead of a part of the objective function (e.g., [11]), the solution obtained from (8.5) is ensured to drive the manipulator to avoid obstacles compulsively, if only there exists a feasible solution to the optimization problem.

8.2.3 Joint Velocity Problem: An Improved Formulation

Changing the equality $J_c(\theta)\dot{\theta} = \dot{r}_c$ to (8.4) does bring some advantages to the problem formulation. But there exists room for further improvement. Note that the geometric interpretation of (8.4) is that each component of $\dot{r}_c$ should take the same sign as corresponding component of $\overrightarrow{OC}$. In other words, the permissible directions of $\dot{r}_c$ constitute $1/8$ of the entire space $\Re^3$ in that $\dot{r}_c \in \Re^3$ (see Figure 8.2). An alternative inequality is introduced as follows

$$\overrightarrow{OC}^T J_c(\theta)\dot{\theta} \geqslant 0. \tag{8.6}$$

The physical interpretation of this inequality is obvious from Figure 8.2: $\dot{r}_c$ should lie in the same side with $\overrightarrow{OC}$ with respect to the link. Then, it expands the permissible region of $\dot{r}_c$ from $1/8$ to $1/2$ of $\Re^3$ space [16]. Simulation studies in Section 8.4 will validate the superiority of the new scheme over that in [33].

In order not to introduce too much restriction on the feasible space and to consequently obtain a better optimum, the inequality constraint (8.6) should be applied only when $|\overrightarrow{OC}|$ is less than some threshold. However, such an idea suffers from discontinuity problem as explained in [33]. To conquer this difficulty, a time varying parameter is introduced

$$b_0(t) = s(|\overrightarrow{OC}|)\max(0, -\overrightarrow{OC}^T J_c(\theta)\dot{\theta})$$

where $s(d)$ is a distance-based smoothing function defined as

$$s(d) = \begin{cases} 1, & \text{if } d \geqslant d_2 \\ \sin^2\left(\frac{\pi}{2}\frac{d-d_1}{d_2-d_1}\right), & \text{if } d_1 < d < d_2 \\ 0, & \text{if } d \leqslant d_1 \end{cases} \tag{8.7}$$

with d_1, d_2 standing for the inner safety threshold and outer safety threshold, respectively ($d_2 > d_1$), provided by users. Then (8.6) is replaced with

$$-\overrightarrow{OC}^T J_c(\theta)\dot{\theta} \leqslant b_0(t). \tag{8.8}$$

For an detailed discussion of this smoothing technique, see [33].

In view that $\overrightarrow{OC} = (x_c - x_o,\ y_c - y_o,\ z_c - z_o)^T$, (8.8) can be rewritten as

$$-(x_c - x_o,\ y_c - y_o,\ z_c - z_o)J_c(\theta)\dot{\theta} \leqslant b_0(t).$$

Suppose there are p critical points. Let $b = (b_0, \cdots, b_0)^T \in \Re^p$ and

$$L(\theta) = \begin{pmatrix} (x_{o_1} - x_{c_1},\ y_{o_1} - y_{c_1},\ z_{o_1} - z_{c_1})J_{c_1}(\theta) \\ \vdots \\ (x_{o_p} - x_{c_p},\ y_{o_p} - y_{c_p},\ z_{o_p} - z_{c_p})J_{c_p}(\theta) \end{pmatrix},$$

where the subscripts o_i and c_i constitute an obstacle-point-critical-point pair. For example, if there are two obstacles and three serial links, then $p = 6$ because each obstacle corresponds to a critical point on each link. By these definitions, the above inequality becomes

$$L(\theta(t))\dot{\theta}(t) \leqslant b(t). \tag{8.9}$$

Substituting the second constraint in (8.5) by (8.9) yields the following formulation if the L_2-norm cost function is adopted,

$$\begin{aligned} \text{minimize } & \frac{1}{2}\|\dot{\theta}(t)\|_2^2 \\ \text{subject to } & J_e(\theta(t))\dot{\theta}(t) = \dot{r}_e(t), \\ & L(\theta(t))\dot{\theta}(t) \leqslant b(t), \\ & l \leqslant \dot{\theta}(t) \leqslant h. \end{aligned} \tag{8.10}$$

Note from (8.4) that each critical point results in three inequalities because $J_N(\theta) \in \Re^{3\times n}$. If there are p critical points, then we have in total $3p$ inequalities; while in (8.10) we have only p inequalities. It is evident that more constraints in the problem imply a more complex problem, as a consequence, a more complex designed neural network for solving it is needed. This can be regarded as another advantage of the new QP formulation (8.10).

Like problem (8.1), problem (8.10) is also a time-varying optimization problem. The entire kinematic control process is shown in Figure 8.3, similar to most optimization based kinematic controls in [2–4, 7, 11, 27, 33]. The desired end-effector velocity $\dot{r}_e$ and the obstacle information are fed into the optimizer as its inputs. In sensor-based control, the critical point is detected with respect to the current pose of the manipulator; in model-based control, it is determined by solving problem (8.1). The time-varying parameters of problems (8.1) and (8.10) are determined by the pose of the manipulator and the position of the obstacle. The optimal joint rate $\dot{\theta}$ that could make the manipulator avoid obstacles is generated as the output of the "Optimizer" block. By taking integration of the joint velocities with the known initial values, we can get the joint motions that the manipulator should follow.

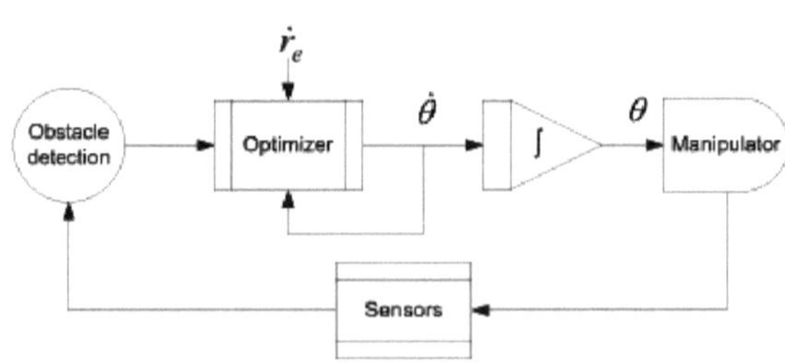

Fig. 8.3 Kinematic control process

8.3 Neural Network Models

The bottleneck of the proposed method in the last section for kinematic control of manipulators is the intensive computation for solving the QP problems (8.1) and (8.10) (see the "Optimizer" blocks and "Obstacle detection" in Figure 8.3). If conventional numerical algorithms are used, two processing units with very high performance are required. We suggest here two neural networks for solving (8.1) and (8.10), respectively, which can be realized by hardware systems connected by dense interconnections of simple analog computational elements. Because of the analog nature of the circuit system, recurrent neural networks have great potential in handling on-line optimization problems [12,25].

8.3.1 Model Selection

Before presenting the neural networks, an activation function is introduced first. Suppose $\mathscr{X}$ is a box set, i.e., $\mathscr{X} = \{x \in \Re^n | \underline{x}_i \le x_i \le \overline{x}_i, \forall i = 1,...,n\}$, then $P_{\mathscr{X}}(x) = (P_{\mathscr{X}_i}(x_i), \cdots, P_{\mathscr{X}_i}(x_n))^T$ with

$$P_{\mathscr{X}_i}(x_i) = \begin{cases} \underline{x}_i, \ x_i < \underline{x}_i, \\ x_i, \ \underline{x}_i \le x_i \le \overline{x}_i, \\ \overline{x}_i, \ x_i > \overline{x}_i. \end{cases} \tag{8.11}$$

Note $\overline{x}_i$ can be $+\infty$ and $\underline{x}_i$ can be $-\infty$. In particular, if $\underline{x} = 0$ and $\overline{x} = \infty$, the operator becomes $P_{\Re^n_+}(x)$, where $\Re^n_+$ denotes the nonnegative quadrant of $\Re^n$. To simplify the notation, in this case it is written as x^+. The definition can be simplified as $x^+ = (x_1^+, \cdots, x_n^+)^T$ with $x_i^+ = \max(x_i, 0)$ as well.

Consider solving problem (8.1). Because Q is positive semidefinite only and many neural networks such as those in [17,24,28,33] can not be applied to solve the problem (8.1). But according to [13], this problem can be formulated as an equivalent generalized linear variational inequality, and as a consequence, it can be solved by a general projection neural network (GPNN) which as studied in [9,14,30]. When specialized to solve (8.1) the dynamic equation of the GPNN is shown as follows (cf. (17) in [13]):

$$\varepsilon \frac{du}{dt} = (M+N)^T \{-Nu + P_{\mathscr{W} \times \Pi}((N-M)u)\}, \tag{8.12}$$

where $\varepsilon > 0$ and $u = (w^T, v^T)^T$ is the state vector and

$$M = \begin{pmatrix} Q & -K^T \\ 0 & I \end{pmatrix}, N = \begin{pmatrix} I & 0 \\ K & 0 \end{pmatrix}.$$

The output vector of the neural network is $w(t)$, which is part of the state vector $u(t)$. The architecture of the neural network can be drawn similarly as in [9] or [30], which is omitted here for brevity. Since Q is positive semidefinite, according to [13], the neural network is globally asymptotically convergent to an optimal solution of (8.1).

For solving (8.10) there exist many recurrent neural networks and some of them will be presented in the next subsection. By observing the special form of the problem, in the chapter we propose to use the improved dual neural network (IDNN) which was developed recently and presented in [15]. For solving this problem, the IDNN is tailored as

- state equation

$$\varepsilon \frac{d}{dt}\begin{pmatrix}\rho \\ \varsigma\end{pmatrix} = -\begin{pmatrix}\rho - (\rho + L(\theta)v - b)^{+} \\ Je(\theta)v - \dot{r}_e\end{pmatrix}, \tag{8.13a}$$

- output equation

$$v = P_{\Omega}(-L(\theta)^T \rho + Je(\theta)^T \varsigma), \tag{8.13b}$$

where $\varepsilon > 0$ is a constant scaling factor, $\Omega = [l, h]$ is a box set, $(\rho^T, \varsigma^T)^T$ is the state vector and v is the output vector as the estimate of $\dot{\theta}$. According to [15], the neural network is globally asymptotically convergent to the optimal solution of (8.10).

The architectures of the above GPNN and IDNN can be drawn similarly as in [9, 15, 30], which are omitted here for brevity.

8.3.2 Model Comparisons

We now show the superiority of the selected two recurrent neural networks by comparing them with several existing recurrent neural networks for motion control of kinematically redundant manipulators. The main comparison criteria are their abilities to solve the two problems formulated in preceding subsections (i.e., (8.1) and (8.10)) and their model complexities. Usually, any neural network of this kind comprises some amplifiers (for realizing nonlinear activation functions), integrators (usually using electric capacitors, whose number equals the number of states of the governing state equation), adders, multipliers[1], and a large number of interconnections. Among them, the number of amplifiers, integrators and interconnections are crucial in determining the structural complexity of the neural network. In order to facilitate the hardware realization and to ensure robust and precise solutions, these crucial elements should be as few as possible.

Clearly, the IDNN (8.13) has n activation functions $P_{\Omega}(\cdot)$ and p activation functions $(\cdot)^{+}$, which entail a total of $n + p$ amplifiers in a circuit realization. In addition, its state vector $(\rho^T, \varsigma^T)^T$ is of dimension $\Re^{p+m}$, and correspondingly, it entails $p + m$ integrators. It is meaningless to count the exact number of connections within a neural network without actually implementing it in an analog circuit because different design techniques will result in a different number of connections.

[1] The number of multipliers actually dominates the number of connections among a large-scale network.

Anyway, the magnitude of the number should be the same. If we omit the constants and first order terms about n,m,p, the order of the connections among this IDNN is $O[n(p+m)]$, which can be determined by counting the number of multiplications in (8.13). In the same way, we can conclude that implementation of the GPNN (8.12) entails $q+8$ amplifiers, as well as $q+8$ integrators, and $O[(q+8)^3]$ connections.

In [24], the obstacle avoidance was formulated as a set of equality constraints and the joint velocity limits are not considered. The formulated QP problem is

$$\begin{aligned} &\text{minimize } \frac{1}{2}\|\dot{\theta}\|_2^2, \\ &\text{subject to } J_e(\theta)\dot{\theta} = \dot{r}_e, \\ &\qquad\qquad\quad J_c(\theta)\dot{\theta} = \dot{r}_c. \end{aligned}$$

A Lagrangian neural network with its dynamic equation defined as follows was proposed for solving it:

$$\varepsilon \frac{d}{dt}\begin{pmatrix} v \\ u \end{pmatrix} = \begin{pmatrix} -v - J^T u \\ Jv - \dot{r} \end{pmatrix}, \tag{8.14}$$

where ε is a positive scaling constant, $J = [{J_e}^T, {J_c}^T]^T$ and $\dot{r} = [\dot{r}_e^T, \dot{r}_c^T]^T$. The output of the neural network, denoted by v in above equation, is the joint velocity vector $\dot{\theta}$ of the manipulators. Since this neural network can handle equality constraints only, it solves neither problem (8.10) nor problem (8.1).

In contrast, the primal-dual neural network (PDNN) proposed in [28] can solve QP problems with not only equality constraints but also bound constraints, which is an improved version of its predecessor in [32]. Problem (8.10) considered in this chapter can be converted to such a QP problem by introducing slack variables:

$$\begin{aligned} &\text{minimize } \frac{1}{2}\|x\|_2^2 \\ &\text{subject to } Dx = a, \quad \underline{\zeta} \leqslant x \leqslant \overline{\zeta}, \end{aligned}$$

where

$$\begin{aligned} x &= \begin{pmatrix} \dot{\theta} \\ \mu \end{pmatrix}, \underline{\zeta} = \begin{pmatrix} l \\ 0 \end{pmatrix}, \overline{\zeta} = \begin{pmatrix} h \\ \infty \end{pmatrix}, \\ D(\theta) &= \begin{pmatrix} J_e(\theta) & 0 \\ L(\theta) & I \end{pmatrix}, a = \begin{pmatrix} \dot{r}_e \\ b \end{pmatrix}, \end{aligned}$$

and $\mu \in \Re^p$ is the newly introduced vector. Then the above problem can be solved by the neural network proposed in [28]:

$$\varepsilon \frac{d}{dt}\begin{pmatrix} x \\ u \end{pmatrix} = \begin{pmatrix} -x + P_{[\underline{\zeta},\overline{\zeta}]}(D^T u) \\ -Du + a \end{pmatrix}, \tag{8.15}$$

where $\varepsilon > 0$ is a scaling factor. For this neural network, the number of activation functions or amplifiers is $n+p$, the number of sates or integrators is $n+m+2p$, and the number of connections is of order $O[(m+p)\times(n+p)]$. However, the primal-dual neural network is restricted to solve strictly convex problems and thus cannot solve problem (8.1).

If we define

$$H(\theta)=\begin{pmatrix} J_e(\theta) \\ L(\theta) \\ I \end{pmatrix}, \quad \underline{\beta}=\begin{pmatrix} \dot{r}_e \\ -\infty \\ l \end{pmatrix}, \quad \overline{\beta}=\begin{pmatrix} \dot{r}_e \\ b \\ h \end{pmatrix},$$

the dual neural network (DNN) proposed in [33] can be utilized to solve problem (8.10) with dynamic equations:

- state equation

$$\varepsilon\frac{du}{dt}=P_{[\underline{\beta},\overline{\beta}]}(HH^Tu-u)-HH^Tu, \tag{8.16a}$$

- output equation

$$v=H^Tu, \tag{8.16b}$$

where $\varepsilon>0$. The output corresponds to the joint velocity $\dot{\theta}$ exactly. Apparently, for this neural network the number of amplifiers and the number of integrators are both $m+p+n$. A closer look at the definition of the box set $[\underline{\beta},\overline{\beta}]$ tells that actually only $p+n$ activation functions are needed, because a projection of any quantity to a point $\dot{r}_e$ is always equal to $\dot{r}_e$ and to ensure this happens there is no need to use an activation function. Therefore, realization of the neural network entails $p+n$ amplifiers only. The number of connections among the neural network is of order $O[(m+p+n)^2+n(m+p+n)]$ (the first term corresponds to the state equation and the second term corresponds to the output equation). Again, the dual neural network was designed for solving strictly convex QP problems and cannot solve problem (8.1).

Finally let's consider a simplified version of the above neural network, called a simplified dual neural network or SDNN, proposed in [17] for solving (8.10), which was even used in [16] for solving a similar problem to (8.10). By defining

$$F(\theta)=\begin{pmatrix} L(\theta) \\ I \end{pmatrix}, \underline{\xi}=\begin{pmatrix} -\infty \\ l \end{pmatrix}, \overline{\xi}=\begin{pmatrix} b \\ h \end{pmatrix}.$$

The SDNN for solving (8.10) can be written as

- state equation

$$\varepsilon\frac{du}{dt}=-FSF^Tu+P_{[\underline{\xi},\overline{\xi}]}(FSF^Tu-u+Fs)-Fs, \tag{8.17a}$$

- output equation

$$v=MF^Tu+s, \tag{8.17b}$$

where $\varepsilon>0, S=I-J_e^T(J_eJ_e^T)^{-1}J_e, s=J_e^T(J_eJ_e^T)^{-1}\dot{r}_e$. The output corresponds to the joint velocity $\dot{\theta}$ exactly. The number of amplifiers and the number of integrators of this neural network are both $p+n$, while the number of connections is in the order of $O[(p+n)^2+n(p+n)]$ (the first term corresponds to the state equation and the second term corresponds to the output equation). Again, the SDNN cannot solve problem (8.1).

Table 8.1 Comparisons of several salient neural networks in the literature for solving (8.10)

Model	Number of amplifiers	Number of integrators	Number of interconnections
PDNN (8.15)	$n+p$	$n+2p+m$	$O[(m+p)\times(n+p)]$
DNN (8.16)	$n+p$	$n+p+m$	$O[(m+p+n)^2+n(m+p+n)]$
SDNN (8.17)	$n+p$	$n+p$	$O[(n+p^2)+n(n+p)]$
IDNN (8.13)	$n+p$	$m+p$	$O[n(m+p)]$

The details of the above discussion about solving problem (8.10) are summarized in Table 8.1. As the Lagrange network (8.14) cannot solve (8.10), it is absent in the table. It is easy to conclude that overall the IDNN (8.13) is the simplest among these feasible neural networks by considering that m is always smaller than or equal to n. In addition, none of these neural networks solves problem (8.1). These facts strongly suggests the viability of the two neural networks, IDNN (8.13) and GPNN (8.12), in the motion planning and control of robot manipulators for obstacle avoidance.

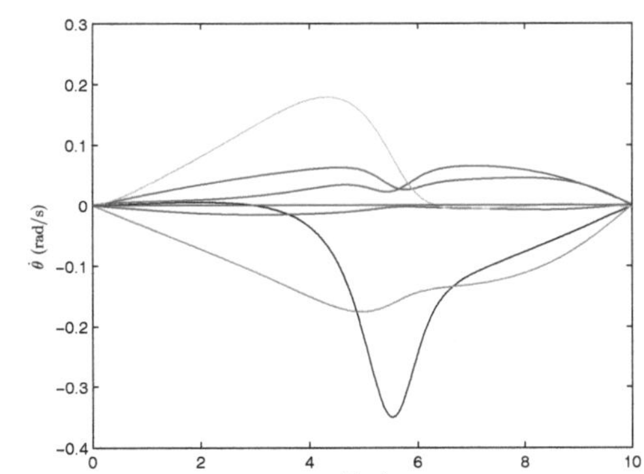

Fig. 8.4 Example 8.1: joint velocities

8.4 Simulation Results

In this section, the validity of the proposed obstacle avoidance scheme and the real-time solution capabilities of the proposed neural networks GPNN and IDNN are shown by using simulation results based on the Mitsubishi 7-DOF PA10-7C manipulator. The coordinate setup, structure parameters, joint limits, and joint velocity limits of this manipulator can be found in [27]. In this study, only the positioning of

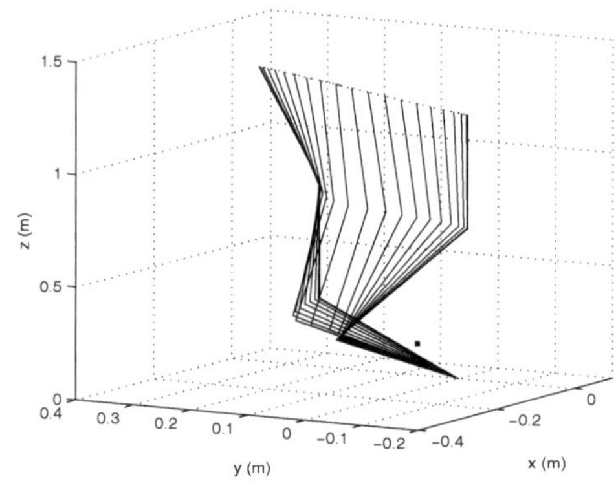

Fig. 8.5 Example 8.1: PA10 manipulator motion. The black dot denotes the obstacle point

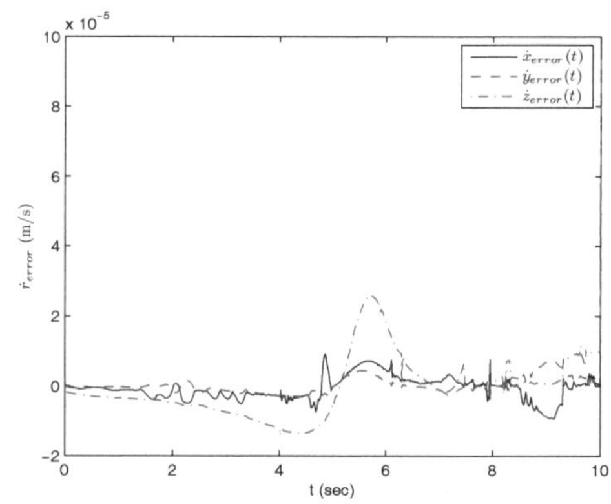

Fig. 8.6 Example 8.1: velocity error of the end-effector

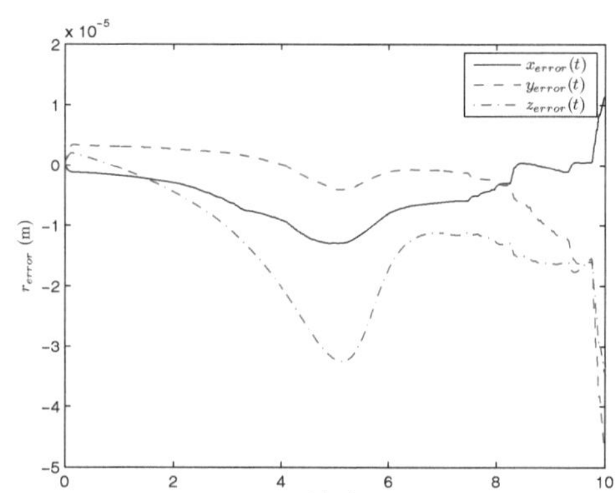

Fig. 8.7 Example 8.1: position error of the end-effector

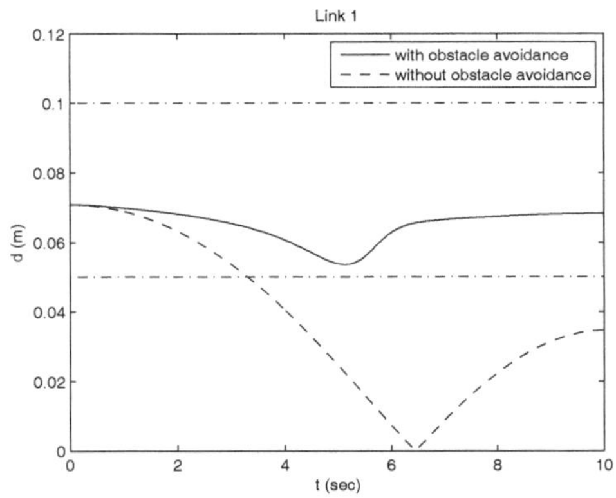

(a)

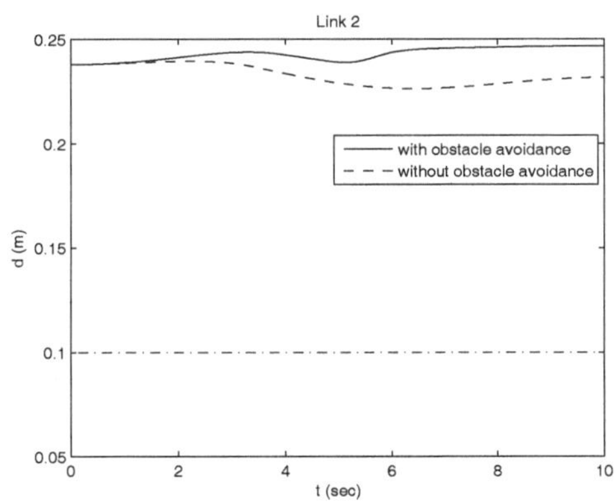

(b)

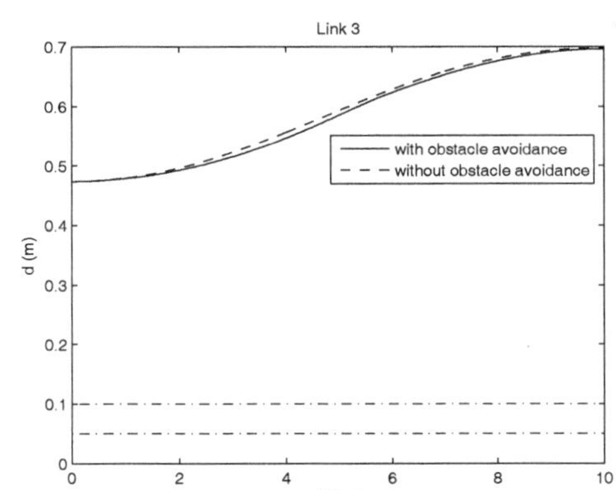

(c)

Fig. 8.8 Example 8.1: minimum distance between the obstacle point and (a) link 1; (b) link 2; and (c) link 3

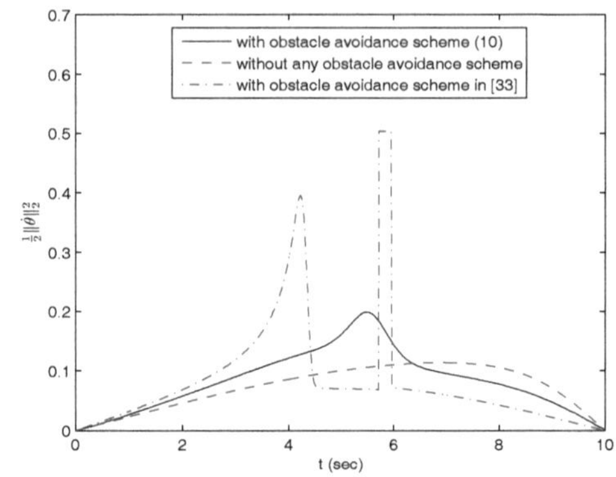

Fig. 8.9 Example 8.1: minimum of the objective function

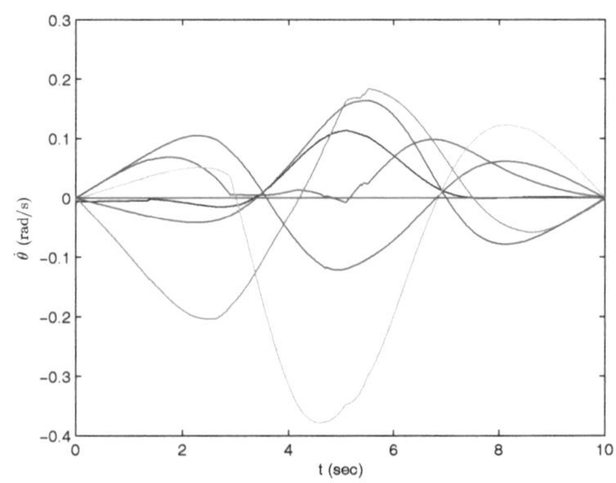

Fig. 8.10 Example 8.2: joint velocities

the end-point is concerned, then $m = 3$ and $n = 7$. The scaling factors ε that control the convergence rates of the neural networks (8.12) and (8.13) are both set to 10^{-4}. The inner and outer safety threshold d_1 and d_2 in (8.7) are set to 0.05m and 0.1m, respectively.

Example 8.1. (Linear Motion) In the first simulation example, the end-effector of the manipulator follows a straight line, which is determined by

$$x_e(t) = x_0 + \psi(t), y_e(t) = y_0 - 3\psi(t), z_e(t) = z_0 - 2\psi(t),$$

where $\psi(t) = -0.1 + 0.1\cos(\frac{\pi}{2}\sin\frac{\pi t}{20})$. The task time of the motion is 10s and the initial joint angles are set up as $\theta(0)$=$(0, -\pi/4, 0, \pi/2, 0, -\pi/4, 0)$. Then the initial position of the end-effector (x_0, y_0, z_0) can be calculated according to the configuration of the manipulator. The desired velocity $\dot{r}_e(t)$ is determined by taking the derivatives of $x_e(t), y_e(t), z_e(t)$ with respect to t. This example is used to illustrate

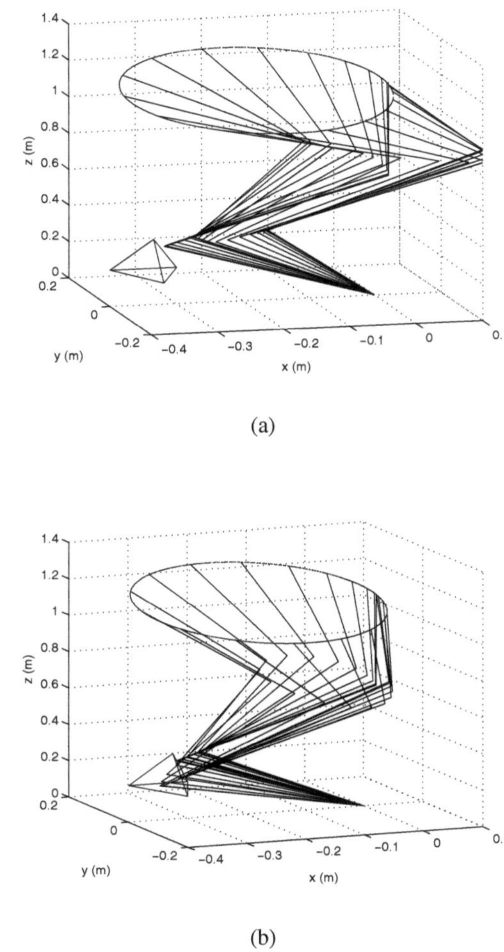

Fig. 8.11 Example 8.2: PA10 manipulator motion. The tetrahedron denotes the obstacle (a) with obstacle avoidance; and (b) without obstacle avoidance

the effectiveness of the QP formulation (8.10) and the suggested solution method IDNN (8.13), so the 3D obstacles are not considered. Instead, it is assumed that in the workspace there is one abstract obstacle point located at (-0.1m, 0.005, 0.2m). Then the calculation of the critical point C on each link can be done by either solving the QP problem (8.1) or adopting the approach described in [33]. We solve problems (8.1) and (8.10) by simulating the GPNN (8.12) and IDNN (8.13) concurrently. Figure 8.4 shows the continuous solution $\dot{\theta}(t)$, and Figure 8.5 illustrates the resulted motion of the manipulator in the 3D workspace. Figures 8.6 and 8.7 depict the tracking velocity error and position error, both of which are very small.

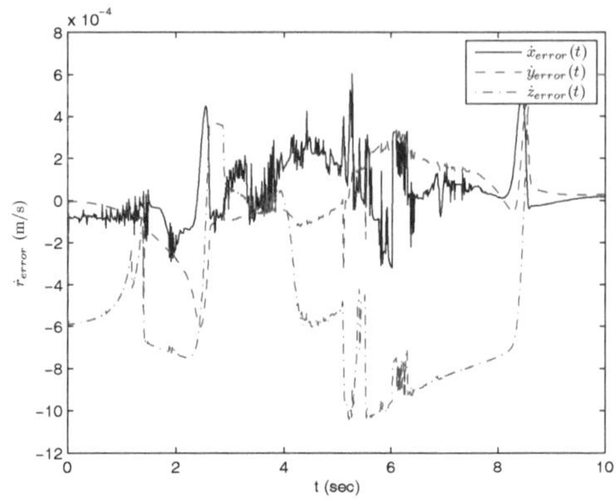

Fig. 8.12 Example 8.2: velocity error of the end-effector

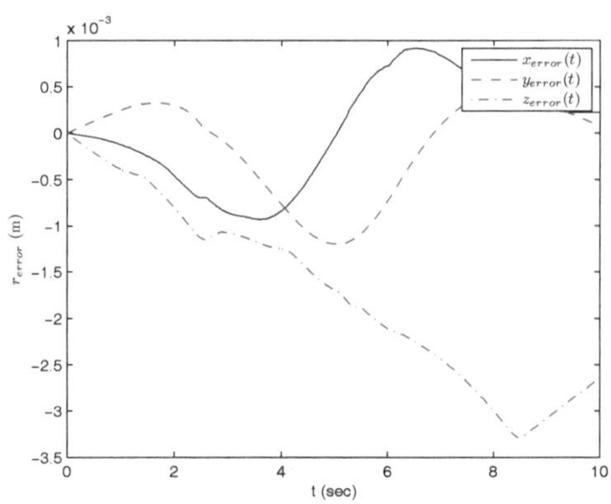

Fig. 8.13 Example 8.2: position error of the end-effector

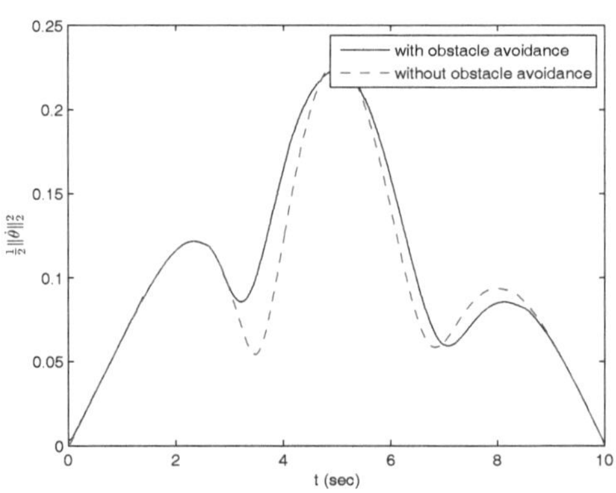

Fig. 8.14 Example 8.2: minimum of the objective function

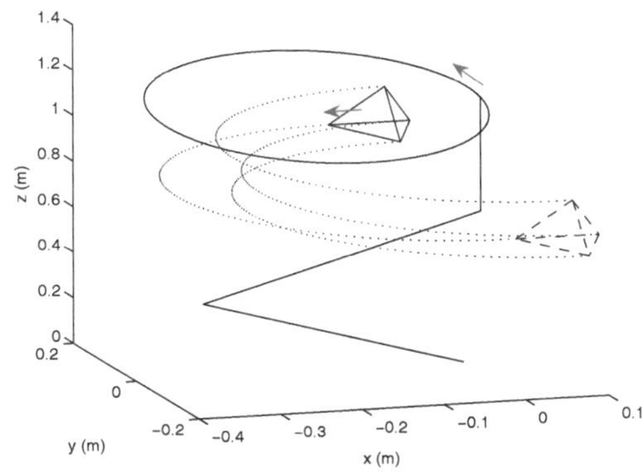

Fig. 8.15 Example 8.3: trajectories of the end-effector and the moving obstacle

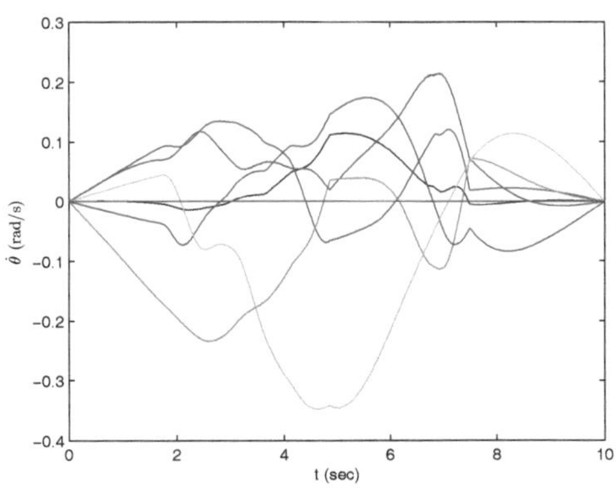

Fig. 8.16 Example 8.3: joint velocities

The minimum distances between the obstacle and links are shown in Figure 8.8 (see solid curves). It is seen from Figure 8.8(a) that Link 1 stays between the inner and outer safety thresholds from the beginning. The dashed curves shown in Figure 8.8 illustrate the results without the obstacle avoidance scheme, i.e., in (8.10) the inequality constraint $L(\theta(t))\dot{\theta}(t) \leq b(t)$ is absent. It is seen that in this case, Link 1 collides with the obstacle point. Figure 8.9 shows objective values of problem (8.10) with respect to time t with and without obstacle avoidance scheme. It is seen that the objective function value with the obstacle avoidance is greater than that without obstacle avoidance, which is reasonable.

Figure 8.9 also shows the time-varying objective function value of problem (8.10) with obstacle avoidance scheme proposed in [33]. The optimization performance of the present scheme (solid curve) is superior to that in the scheme in [33] (dash-dot curve) with better performance in the earlier phase of motion. In the later phase, an abrupt change for the scheme in [33] is observed though the smoothing technique is

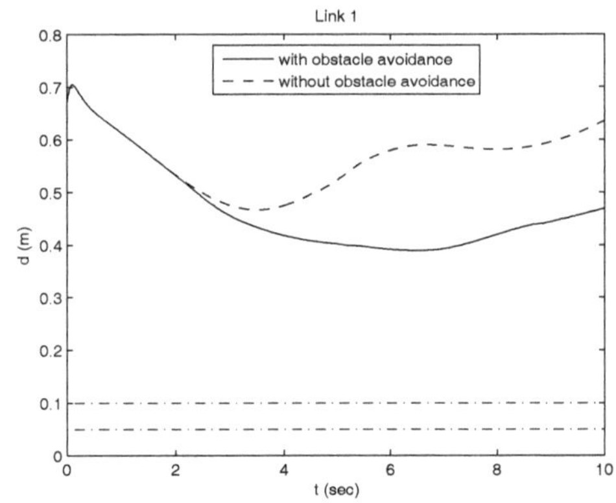

(a)

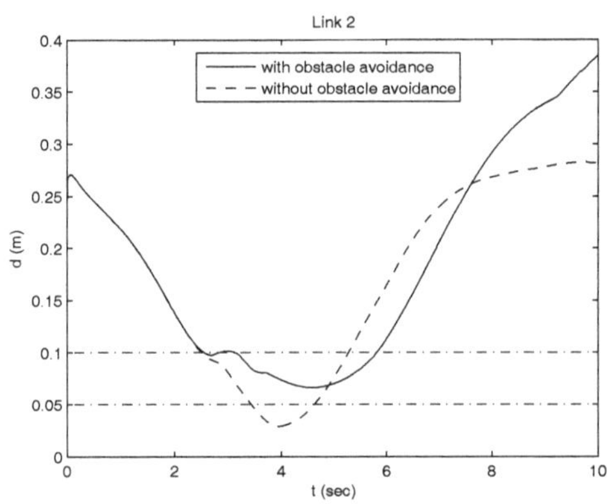

(b)

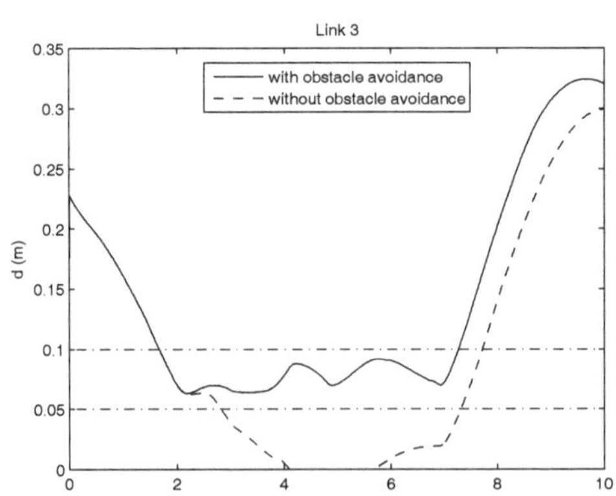

(c)

Fig. 8.17 Example 8.3: minimum distance between the obstacle and (a) link 1; (b) link 2; and (c) link 3

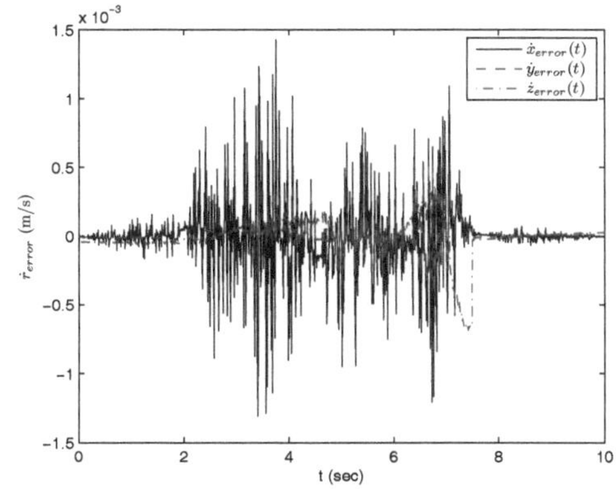

Fig. 8.18 Example 8.3: velocity error of the end-effector

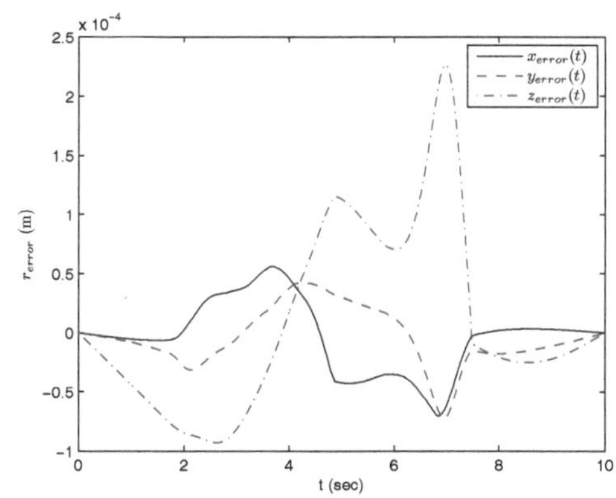

Fig. 8.19 Example 8.3: position error of the end-effector

adopted. This fact echoes the argument stated in Section 8.2.3 about the sizes of the feasible solution spaces of the two schemes.

Example 8.2. (Circular Motion) In this example, the desired motion of the end-effector is a circle of radius 20cm which is described by

$$\begin{cases} x_e(t) = x_0 - 0.2 + 0.2\cos(2\pi\sin^2\frac{\pi t}{20}) \\ y_e(t) = y_0 + 0.2\sin(2\pi\sin^2\frac{\pi t}{20})\cos\frac{\pi}{6} \\ z_e(t) = z_0 + 0.2\sin(2\pi\sin^2\frac{\pi t}{20})\sin\frac{\pi}{6}. \end{cases}$$

The task time of the motion is 10s and the joint variables are initialized as in Example 8.1. In the workspace, suppose that there is a tetrahedral obstacle whose vertices are respectively (-0.4m, 0, 0.2m), (-0.3m, 0, 0.2m), (-0.35m, -0.1m, 0.2m), (-0.35m, -0.05m, 0.4m). By some basic calculations, the parameters in (8.1) can be determined easily. We simultaneously solve problem (8.10) and problem (8.1) by simulating the IDNN (8.13) and GPNN (8.12). Figure 8.10 shows the continuous solution $\dot{\theta}(t)$ to problem (8.10), and Figure 11(a) illustrates the resulted motion of the manipulator in the 3D workspace. Figs. 8.12 and 8.13 depict the tracking velocity error and position error, which are very small.

In contrast with Figure 11(a), Figure 11(b) illustrates the simulated motion of the manipulator in the 3D workspace without the obstacle avoidance scheme. It is seen that in this case, the manipulator will definitely collide with the obstacle. Figure 8.14 shows the comparison of the minimized objective function with and without the obstacle avoidance scheme. It is reasonable to observe that the objective function value in the former case is greater than that in the latter case.

Example 8.3. (Moving Obstacle) Let the end-effector do the same circular motion as in Example 8.2 and suppose that there is a same shaped obstacle in the work space. The task time of the motion is also 10s and the joint variables are initialized as in Example 8.2 (as well as Example 8.1). But this time we suppose that the tetrahedral obstacle is moving while the manipulator is accomplishing the motion task. Let the obstacle move also along a circular path, which is described by

$$\begin{cases} x_o(t) = x_0 - 0.3\sin(\frac{\pi t}{10}) \\ y_o(t) = y_0 - 0.3 + 0.3\cos(\frac{\pi t}{10}) \\ z_o(t) = z_0, \end{cases}$$

where $(x_o, y_o, z_o)^T$ is any point on the obstacle and $(x_0, y_0, z_0)^T$ is the initial position of that point. Clearly, the path is a half circle with radius 30cm parallel to the x-y plane. The initial positions of the tetrahedron vertices are respectively (0.07m, 0.3m, 0.8m), (0.17m, 0.3m, 0.8m), (0.12m, 0.2m, 0.8m), (0.12m, 0.25m, 1m). Figure 8.15 illustrates the moving trajectory of the obstacle as well as the desired path of the end-effector.

We simultaneously solve problems (8.1) and (8.10) by coupling the GPNN (8.12) and IDNN (8.13). Figure 8.16 shows the continuous solution $\dot{\theta}(t)$ to problem (8.10). By virtue of the solution of problem (8.1) obtained by the GPNN (8.12) it is easy to calculate the minimum distances between the obstacle and manipulator arms, which are plotted in Figure 8.17. From the subfigure (c), when Link 3 enters the outer safety threshold, that is, the distance to the obstacle is smaller than 10cm, the obstacle avoidance scheme repels it away. Figure 8.17 also plots the distance between the obstacle and each link without the obstacle avoidance scheme. It is seen that in this case, Link 3 will penetrate into the obstacle (corresponding to zero distance). Figs. 8.18 and 8.19 depict the tracking velocity error and position error, both of which are very small.

8.5 Concluding Remarks

In the chapter, we present a novel scheme for real-time motion planning of kinematically redundant robot manipulators based on two recurrent neural networks.

Firstly, we propose a convex optimization problem formulation for determining the the critical points on the manipulator and the nearest points on the obstacles to the manipulator links by assuming convex-shape obstacles and a serial-link manipulator. In particular, if the obstacle is a polyhedron, the problem becomes a QP problem. For solving this problem in real time a GPNN is proposed. A new obstacle avoidance scheme for kinematically redundant manipulators control is proposed based on another QP formulation. Compared with existing QP formulations in the literature, the new scheme has larger feasible solution space and can thus yield better solutions in terms of objective functions. Two efficient neural networks, GPNN and IDNN, are adopted to solve the problems respectively in real time. Compared with existing counterparts for solving the two QP problems using neural networks, the selected GPNN and IDNN have low model complexity and good convergence property. The neural networks are simulated to plan and control the motion of the PA10-7C robot manipulator by solving the time-varying QP problems, which validates the effectiveness of the proposed approaches.

Acknowledgements The work described in the chapter was supported by the Research Grants Council of the Hong Kong Special Administrative Region, China, under Grants G_HK010/06 and CUHK417608E, by the National Natural Science Foundation of China under Grants 60805023, 60621062 and 60605003, by the National Key Foundation R&D Project under Grants 2003CB 317007, 2004CB318108 and 2007CB311003, by the China Postdoctoral Science Foundation, and by the Basic Research Foundation of Tsinghua National Laboratory for Information Science and Technology (TNList).

References

1. Al-Gallaf, E.A.: Multi-fingered robot hand optimal task force distribution - neural inverse kinematics approach. Robot. Auton. Syst. **54**(1), 34–51 (2006).
2. Allotta, B., Colla, V., Bioli, G.: Kinematic control of robots with joint constraints. J. Dyn. Syst. Meas. Control-Trans. ASME **121**(3), 433–442 (1999).
3. Cheng, F.T., Chen, T.H., Wang, Y.S., Sun, Y.Y.: Obstacle avoidance for redundant manipulators using the compact QP method. In: Proc. IEEE Int. Conf. Robotics and Automation (ICRA), vol. 3, pp. 262–269. Atlanta, Georgia, USA (1993).
4. Cheng, F.T., Lu, Y.T., Sun, Y.Y.: Window-shaped obstacle avoidance for a redundant manipulator. IEEE Trans. Syst., Man, Cybern. B **28**(6), 806–815 (1998).
5. Ding, H., Chan, S.P.: A real-time planning algorithm for obstacle avoidance of redundant robots. Journal of Intelligent and Robotic Systems **16**(3), 229–243 (1996).
6. Ding, H., Tso, S.K.: Redundancy resolution of robotic manipulators with neural computation. IEEE Trans. Industrial Electronics **46**(1), 230–233 (1999).
7. Ding, H., Wang, J.: Recurrent neural networks for minimum infinity-norm kinematic control of redundant manipulators. IEEE Trans. Syst., Man, Cybern. A **29**(3), 269–276 (1999).
8. Forti, M., Nistri, P., Quincampoix, M.: Generalized neural network for nonsmooth nonlinear programming problems. IEEE Trans. Circuits Syst. I **51**(9), 1741–1754 (2004).
9. Gao, X.B.: A neural network for a class of extended linear variational inequalities. Chinese Jounral of Electronics **10**(4), 471–475 (2001).
10. Glass, K., Colbaugh, R., Lim, D., Seraji, H.: Real-time collision avoidance for redundant manipulators. IEEE Trans. Robot. Autom. **11**(3), 448–457 (1995).

11. Guo, J., Hsia, T.C.: Joint trajectory generation for redundant robots in an environment with obstacles. Journal of Robotic Systems **10**(2), 199–215 (1993).
12. Hopfield, J.J., Tank, D.W.: Computing with neural circuits: a model. Scienc **233**(4764), 625–633 (1986).
13. Hu, X., Wang, J.: Design of general projection neural networks for solving monotone linear variational inequalities and linear and quadratic optimization problems. IEEE Trans. Syst., Man, Cybern. B **37**(5), 1414–1421 (2007).
14. Hu, X., Wang, J.: Solving generally constrained generalized linear variational inequalities using the general projection neural networks. IEEE Trans. Neural Netw. **18**(6), 1697–1708 (2007).
15. Hu, X., Wang, J.: An improved dual neural network for solving a class of quadratic programming problems and its k-winners-take-all application. IEEE Trans. Neural Netw. (2008). Accepted
16. Liu, S., Hu, X., Wang, J.: Obstacle avoidance for kinematically redundant manipulators based on an improved problem formulation and the simplified dual neural network. In: Proc. IEEE Three-Rivers Workshop on Soft Computing in Industrial Applications, pp. 67–72. Passau, Bavaria, Germany (2007).
17. Liu, S., Wang, J.: A simplified dual neural network for quadratic programming with its KWTA application. IEEE Trans. Neural Netw. **17**(6), 1500–1510 (2006).
18. Maciejewski, A.A., Klein, C.A.: Obstacle avoidance for kinematically redundant manipulators in dynamically varying environments. Intl. J. Robotics Research **4**(3), 109–117 (1985).
19. Mao, Z., Hsia, T.C.: Obstacle avoidance inverse kinematics solution of redundant robots by neural networks. Robotica **15**, 3–10 (1997).
20. Nakamura, Y., Hanafusa, H., Yoshikawa, T.: Task-priority based redundancy control of robot manipulators. Intl. J. Robotics Research **6**(2), 3–15 (1987).
21. Ohya, I., Kosaka, A., Ka, A.: Vision-based navigation by a mobile robot with obstacle avoidance using single-camera vision and ultrasonic sensing. IEEE Trans. Robot. Autom. **14**(6), 969–978 (1998).
22. Sciavicco, L., Siciliano, B.: Modeling and Control of Robot Manipulators. Springer-Verlag, London, U.K. (2000).
23. Shoval, S., Borenstein, J.: Using coded signals to benefit from ultrasonic sensor crosstalk in mobile robot obstacle avoidance. In: Proc. IEEE Int. Conf. Robotics and Automation (ICRA), vol. 3, pp. 2879–2884. Seoul, Korea (2001).
24. Tang, W.S., Lam, M., Wang, J.: Kinematic control and obstacle avoidance for redundant manipulators using a recurrent neural network. In: Proc. Intl. Conf. on Artificial Neural Networks, *Lecture Notes in Computer Science*, vol. 2130, pp. 922–929. Vienna, Austria (2001).
25. Tank, D.W., Hopfield, J.J.: Simple neural optimization networks: an A/D converter, signal decision circuit, and a linear programming circuit. IEEE Trans. Circuits Syst. **33**(5), 533–541 (1986).
26. Walker, I.D., Marcus, S.I.: Subtask performance by redundancy resolution for redundant robot manipulators. IEEE J. Robot. Autom. **4**(3), 350–354 (1988).
27. Wang, J., Hu, Q., Jiang, D.: A lagrangian network for kinematic control of redundant robot manipulators. IEEE Trans. Neural Netw. **10**(5), 1123–1132 (1999).
28. Xia, Y., Feng, G., Wang, J.: A primal-dual neural network for on-line resolving constrained kinematic redundancy in robot motion control. IEEE Trans. Syst., Man, Cybern. B **35**(1), 54–64 (2005).
29. Xia, Y., Wang, J.: A general methodology for designing globally convergent optimization neural networks. IEEE Trans. Neural Netw. **9**(6), 1331–1343 (1998).
30. Xia, Y., Wang, J.: A general projection neural network for solving monotone variational inequalities and related optimization problems. IEEE Trans. Neural Netw. **15**(2), 318–328 (2004).
31. Xia, Y., Wang, J., Fok, L.M.: Grasping force optimization of multi-fingered robotic hands using a recurrent neural network. IEEE Transactions on Robotics and Automation **20**(3), 549–554 (2004).

32. Zhang, Y., Ge, S.S., Lee, T.H.: A unified quadratic programming based dynamical system approach to joint torque optimization of physically constrained redundant manipulators. IEEE Trans. Syst., Man, Cybern. B **34**(5), 2126–2132 (2004).
33. Zhang, Y., Wang, J.: Obstacle avoidance for kinematically redundant manipulators using a dual neural network. IEEE Trans. Syst., Man, Cybern. B **34**(1), 752–759 (2004).

Chapter 9
Adaptive Neural Control of Uncertain Multi-variable Nonlinear Systems with Saturation and Dead-zone

Mou Chen, Shuzhi Sam Ge and Bernard Voon Ee How

Abstract In this chapter, adaptive neural control is developed for a class of uncertain MIMO nonlinear systems using neural networks. The MIMO system under study is a strict-feedback uncertain nonlinear system with non-symmetric input nonlinearities. Variable structure control (VSC) technique in combination with backstepping is proposed to tackle the input saturation and dead-zone. The spectral radius of the control coefficient matrix is introduced to design adaptive neural control in order to cancel the nonsingular assumption of the control coefficient matrix. Using the cascade property of system, the semi-global uniform ultimate boundedness of all signals in the closed-loop system is achieved. The tracking errors converge to small residual sets which are adjustable by updating design parameters. Finally, case study results are presented to illustrate the effectiveness of the proposed adaptive neural control.

9.1 Introduction

In the past decades, backstepping control has became one of the most popular robust and adaptive control system design techniques for some special classes of nonlinear systems. However, a number of works on backstepping design focus on single-input/single-output (SISO) nonlinear triangular systems [1–3], and the adaptive neu-

Mou Chen
Department of Electrical & Computer Engineering, National University of Singapore, Singapore 117576, e-mail: elechenm@nus.edu.sg

Shuzhi Sam Ge
Department of Electrical & Computer Engineering, National University of Singapore, Singapore 117576, e-mail: samge@nus.edu.sg

Bernard Voon Ee How
Department of Electrical & Computer Engineering, National University of Singapore, Singapore 117576, e-mail: bernardhow@nus.edu.sg

ral control of uncertain MIMO nonlinear systems with input nonlinearities need to be further investigated. In recent years, the universal approximation ability of neural networks makes it one of the effective tool in MIMO nonlinear system robust control design [4–8]. Direct adaptive control was investigated for a class of MIMO nonlinear systems using neural networks based on input-output discrete-time model with unknown interconnections between subsystems [4]. In [5], adaptive neural network (NN) control was proposed for a class of discrete-time MIMO nonlinear systems with triangular form inputs. Adaptive NN control was developed for a class of MIMO nonlinear systems with unknown bounded disturbances in discrete-time domain [6]. In [7], approximation-based control was presented for a class of MIMO nonlinear systems in block-triangular form with unknown state delays. By exploiting the special properties of the affine terms, adaptive neural control for two classes of uncertain MIMO nonlinear systems was proposed [8].

In practice, nonlinear systems usually possess time-varying disturbance, modeling error and other uncertainties [9–12]. It is important to develop effective control techniques for uncertain MIMO nonlinear systems which possess the above properties. One main challenge is the possible singularity of the control coefficient matrix which makes the control design complex. Existing research results mostly assume that the control coefficient matrix is known and nonsingular, or NN is used to approximate unknown control coefficient matrix and assume NN approximation nonsingular. In [13], robust backstepping control was developed for a class of nonlinear systems based on neural networks which requires the control coefficient matrix of each subsystem to be known and invertible. To relax the known restriction of the control coefficient matrix, robust tracking control for nonlinear MIMO systems with unknown control coefficient matrix via fuzzy approaches was investigated [14]. The proposed tracking control required the approximation of control coefficient matrix nonsingular, and the approximation error must satisfy a certain given condition. In [15], tracking control was proposed for affine MIMO nonlinear systems, and the nonsingularity requirement of control coefficient matrix was eliminated.

In this chapter, adaptive neural control is proposed for a class of uncertain MIMO nonlinear systems with saturation and dead-zone. Actuators cannot promptly respond to the command signal and saturation is unavoidable in most actuators as they can only provide a limited energy in a practical control system. There exist input dead-zone and saturation which are the most important non-smooth nonlinearities. Saturation and dead-zone can severely influence the control system performance. Thus, saturation and dead-zone as input nonlinearities have attracted a great amount of attention over the past years. The Takagi-Sugeno (T-S) fuzzy modeling approach was utilized to control the nonlinear systems with actuator saturation [16]. In [17], NN control was investigated for a class of nonlinear systems with actuator saturation. Globally stable adaptive control was proposed for minimum phase SISO plants with input saturation [18]. Dead-zone as a static nonsmooth input-output relationship gives zero output for a range of input values. To handle systems with unknown dead-zones, adaptive dead-zone inverse was proposed [19, 20]. Adaptive control was proposed for uncertain systems with unknown dead-zone nonlinearity by employing the dead-zone inverse [21]. In [22], after the introduction of practical dead-zone model, adaptive neural control was proposed for MIMO nonlinear state time-varying delay systems. Adaptive dynamic surface control was developed for a class of nonlinear systems with unknown dead-zone in pure feedback form [23]. In [24], nonlinear control was proposed to a three-axis stabilized flexible spacecraft with control input saturation/dead-zone. For uncertain nonlinear systems, integrating dead-zone with saturation needs to be further considered.

The motivation behind this work is to tackle the uncertain MIMO nonlinear systems with input saturation and dead-zone via adaptive neural control. The backstepping technique combining with neural networks is employed to design the robust control. The motivation of this chapter is three-fold. Firstly, NN-based methods can handle functional uncertainties and external disturbances which exist in the uncertain nonlinear system. Secondly, input nonlinearities such as saturation and dead-zone should be considered explicitly in the control design stage for stability and improved control performance. Thirdly, the nonsingular assumption of the control coefficient matrix of uncertain MIMO nonlinear system need to be relaxed. In this chapter, control coefficient matrix of each subsystem consists of both known and unknown matrices. The known matrix could be singular and the unknown matrix with bounded norm can be tackled via VSC. Thus, we propose the VSC technique in combination with backstepping for a class of cascade uncertain MIMO nonlinear systems with actuator dead-zone and saturation. The main contributions of the chapter are as follows:

1. Unlike existing works on uncertain MIMO nonlinear systems, the developed adaptive neural control takes into account non-symmetric input nonlinearities which are described using the practical model of saturation and dead-zone.
2. The matrix spectral radius is employed to design adaptive neural control for uncertain MIMO nonlinear systems, and the nonsingular assumption of control coefficient matrices have been eliminated.
3. Radial basis function neural networks (RBFNNs) are used to compensate for function uncertainties and external disturbances, which are commonly faced in uncertain MIMO nonlinear system control. Using the property of a cascaded system, the semi-global uniform ultimate boundedness of all signals in the closed-loop system is achieved.

The chapter is organized as follows: the problem formulation is presented for the control of uncertain MIMO nonlinear systems in Section 9.2. Section 9.3 investigate the adaptive neural control considering dead-zone and saturation for uncertain MIMO nonlinear systems. The simulation results are presented to demonstrate the effectiveness of proposed adaptive neural control in Section 9.4. Section 9.5 contains the conclusion.

We use the following notations in this chapter. $\lambda_{\max}(A)$ and $\lambda_{\min}(A)$ denote the largest and smallest eigenvalues of a square matrix A, respectively. $\|\cdot\|$ stands for Frobenius norm of matrices and Euclidean norm of vectors, i.e., given a matrix B and a vector ξ, the Frobenius norm and Euclidean norm are given by $\|B\|^2 = \mathrm{tr}(B^T B) = \sum_{i,j} b_{ij}^2$ and $\|\xi\|^2 = \sum_i \xi_i^2$. $\bar{x_i} = [x_1, x_2, \ldots, x_i]^T \in R^{i \times m}$ denotes the vector of partial state variables in nonlinear system.

9.2 Problem Formulation and Preliminaries

9.2.1 Problem Formulation

Consider the following uncertain MIMO nonlinear system:

$$\begin{aligned}
\dot{x}_1 &= F_1(x_1) + (G_1(x_1) + \Delta G_1(x_1))x_2 + D_1(x_1,t) \\
\dot{x}_2 &= F_2(\bar{x}_2) + (G_2(\bar{x}_2) + \Delta G_2(\bar{x}_2))x_3 + D_2(\bar{x}_2,t) \\
\dot{x}_3 &= F_3(\bar{x}_3) + (G_3(\bar{x}_3) + \Delta G_3(\bar{x}_3))x_4 + D_3(\bar{x}_3,t) \\
&\cdots \\
\dot{x}_n &= F_n(\bar{x}_n) + (G_n(\bar{x}_n) + \Delta G_n(\bar{x}_n))\Phi(u) + D_n(\bar{x}_n,t) \\
y &= x_1
\end{aligned} \tag{9.1}$$

where $x_i \in R^m, i = 1,2,\ldots,n$ are the state vectors; $y \in R^m$ is the system output vector; $u \in R^m$ is the control input vector; $\Phi(u) \in R^m$ denotes the input nonlinearity; $D_i \in R^m, i = 1,2,\ldots,n$ are unknown time-varying disturbances; $F_i \in R^{m\times 1}, i = 1,2,\ldots,n$ are unknown nonlinear functions which contain both parametric and nonparametric uncertainties; $G_i \in R^{m\times m}, i = 1,2,\ldots,n$ are known control coefficient matrices, and $\Delta G_i \in R^{m\times m}, i = 1,2,\ldots,n$ are unknown bounded perturbations of control coefficient matrices.

To facilitate control system design, the following assumptions and lemmas are presented and will be used in the subsequent developments.

Assumption 9.1 *Input nonlinearity $\Phi(u) = [\Phi(u_1),\Phi(u_2),\ldots,\Phi(u_m)]^T$ satisfies non-symmetric saturation and dead-zone as shown in Figure 9.1. That is, the control signal $u(t) = [u_1(t),u_2(t),\ldots,u_m(t)]^T$ is constrained by the saturation values $u_{l\max} = [u_{l1\max},u_{l2\max},\ldots,u_{lm\max}]^T$, $u_{r\max} = [u_{r1\max},u_{r2\max},\ldots,u_{rm\max}]^T$, dead-zone values $u_{l0} = [u_{l10},u_{l20},\ldots,u_{lm0}]^T$ and $u_{r0} = [u_{r10},u_{r20},\ldots,u_{rm0}]^T$. When $u_i > u_{ri1}$ or $u_i < u_{li1}$, the output of $\Phi(u)$ will be saturated, where $u_{li\max} < 0$, $u_{ri\max} > 0$, $u_{li1} < u_{li0} < 0$, $u_{ri1} > u_{ri0} > 0$, and the function $g_{ri}(u_i)$ and $g_{li}(u_i)$ are smooth nonlinear functions, $i = 1,2,\ldots,m$.*

Considering Assumption 9.1, the input nonlinearity $\Phi(u_i)$ can be expressed as

$$\Phi(u_i) = \begin{cases} \chi(u_i,u_{li\max},u_{ri\max},u_{li0},u_{ri0})(u_i - u_{ri0}), & \text{if } u_i \geq u_{ri0}(t) \\ \chi(u_i,u_{li\max},u_{ri\max},u_{li0},u_{ri0})(u_i - u_{li0}), & \text{if } u_i \leq u_{li0}(t) \\ 0, & \text{if } u_{li0}(t) \leq u_i \leq u_{ri0}(t). \end{cases} \tag{9.2}$$

According to the differential mean value theorem, we have that there exists $\psi_{ri}(u_i) \in [u_{ri0},u_{ri1}]$ and $\psi_{li}(u_i) \in [u_{li1},u_{li0}]$ such that

$$\chi(u_i,u_{li\max},u_{ri\max},u_{li0},u_{ri0}) = \begin{cases} \frac{u_{ri\max}}{u_i - u_{ri0}}, & \text{if } u_i > u_{ri1} \\ g'_{ri}(\psi_{ri}(u_i)), & \text{if } u_{ri0} \leq u_i \leq u_{ri1} \\ g'_{li}(\psi_{li}(u_i)), & \text{if } u_{li1} \leq u_i \leq u_{li0}, \\ \frac{u_{li\max}}{u_i - u_{li0}}, & \text{if } u_i < u_{li1}. \end{cases} \tag{9.3}$$

where $g'_{ri}(\psi_{ri}(u_i)) = \frac{dg_{ri}(z)}{dz}|_{z=\psi_{ri}(u_i)}$ and $g'_{li}(\psi_{li}(u_i)) = \frac{dg_{li}(z)}{dz}|_{z=\psi_{li}(u_i)}$.

Assumption 9.2 *[22, 23] The functions $g_{ri}(u_i)$ and $g_{li}(u_i)$ are smooth, and there exists unknown positive constants k_{li0}, k_{li1}, k_{ri0} and k_{ri1} such that*

$$0 \le k_{ri0} \le g'_{ri}(u_i) \le k_{ri1}, \quad u_i \in [u_{ri0}, u_{ri1}] \tag{9.4}$$

$$0 \le k_{li0} \le g'_{li}(u_i) \le k_{li1}, \quad u_i \in [u_{li1}, u_{li0}]. \tag{9.5}$$

According to (9.3), (9.4) and (9.5), we can know $\chi(u_i, u_{li\max}, u_{ri\max}, u_{li0}, u_{ri0}) > 0$.

Lemma 9.1. *[25,26] For bounded initial conditions, if there exists a C^1 continuous and positive definite Lyapunov function $V(x)$ satisfying $\pi_1(\|x\|) \le V(x) \le \pi_2(\|x\|)$, such that $\dot{V}(x) \le -\kappa V(x) + c$, where $\pi_1, \pi_2 : R^n \to R$ are class K functions and c is a positive constant, then the solution $x(t)$ is uniformly bounded.*

Lemma 9.2. *[25] For the continuous functions $D_{ij}(\bar{x_i}, t) : R^{i\times m} \times R \to R, i = 1,2,\ldots,n; j = 1,2,\ldots,m$, there exist positive, smooth, nondecreasing functions $p_{ij}(\bar{x_i}) : R^{i\times m} \to R^+$ and $q_{ij}(t) : R \to R^+$ such that*

$$|D_{ij}(\bar{x_i}, t)| \le p_{ij}(\bar{x_i}) + q_{ij}(t). \tag{9.6}$$

Assumption 9.3 *[25] For the time-dependent functions $q_{ij}(t), i = 1,2,\ldots,n; j = 1,2,\ldots,m$, there exist constants $\bar{q}_{ij} \in R^+, \forall t > t_0$, such that*

$$|q_{ij}(t)| \le \bar{q}_{ij}. \tag{9.7}$$

Assumption 9.4 *[27] For all known control coefficient matrices $G_i(\bar{x_i}), i = 1,2,\ldots,n$ of nonlinear system (9.1), there exist known positive constants $\zeta_i > 0$ such that $\|G_i(\bar{x_i})\| \le \zeta_i$, $\forall \bar{x_i} \in \Omega_i \subset R^{i\times m}$ with compact subset Ω_i containing the origin.*

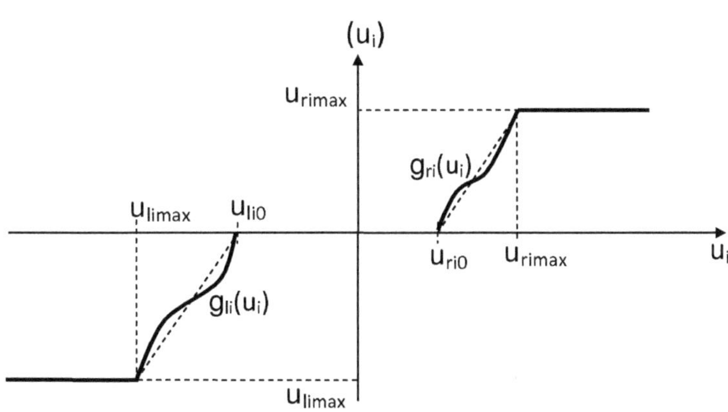

Fig. 9.1 Non-symmetric nonlinear saturation and dead-zone model

Assumption 9.5 *For $1 \leq i \leq n$, there exists known constants $\xi_{i1} \geq \xi_{i0} \geq 0, i = 1,2,\ldots,n$ such that $\xi_{i0} \leq \|\Delta G_i(\bar{x}_i)\| \leq \xi_{i1}$.*

Lemma 9.3. *[28] No eigenvalue of a matrix $A \in R^{m\times m}$ exceeds any of its norm in absolute value, that is*

$$|\lambda_i| \leq \|A\|, \quad i = 1,2,\ldots,m \tag{9.8}$$

where λ_i is a eigenvalue of matrix A.

Proof. See Appendix A1. □

Lemma 9.4. *Considering a matrix $B \in R^{m\times m}$ with spectral radius $\rho(B)$, there exists a positive constant $\tau > 0$ which makes matrix $B + (\rho(B) + \tau)I_{m\times m}$ nonsingular.*

Proof. See Appendix A2. □

Remark 9.1. Assumption 9.1 implies that the input nonlinearity $\Phi(u)$ satisfy the non-symmetric saturation and dead-zone. Namely, the control input $u(t)$ is constrained by saturation limits $u_{l\max}$, $u_{r\max}$, dead-zone values u_{l0} and u_{r0}. Lemma 9.1 allows one to separate the multivariable disturbance term $D_i(\bar{x}_i,t)$ into a bounded function in terms of $\bar{x}_i$, the internal states of nonlinear system, and a bounded function in terms of t, which generally includes exogenous effects and uncertainties. Assumption 9.3 is reasonable since the time-dependent component of the disturbance can be largely attributed to the exogenous effects, which have finite energy and, hence, are bounded [25]. Assumption 9.5 means that perturbations $\Delta G_i(\bar{x}_i)$ of control coefficient matrices $G_i(\bar{x}_i), i = 1,2,\ldots,n$ are bounded.

Remark 9.2. There are many practical systems that can be expressed as the nonlinear system form as shown in (9.1). For example, rigid robots and motors [29], ships [25], power converters [30], jet engines and aircrafts [31,32]. The nonlinear terms F_i and G_i depend on states $x_1, x_2, \ldots, x_i$. Thus, the system (9.1) is a strict-feedback uncertain MIMO nonlinear system. In this chapter, the matrix spectral radius is employed to design adaptive neural control for uncertain MIMO nonlinear systems (9.1). We need not assume that all control coefficient matrices $G_i(\bar{x}_i), i = 1,2,\ldots,n$ are invertible, and only require that the norm of control coefficient matrix is bounded. Considering Assumption 9.4 and Lemma 9.3, the spectral radius $\rho(G_i)$ of $G_i(\bar{x}_i)$ satisfies $\rho(G_i) \leq \zeta_i$. According to Lemma 9.4, we obtain $G_i(\bar{x}_i) + (\zeta_i + \tau_i)I_{m\times m}$ with $\tau_i > 0$, $i = 1,2,\ldots,n$ are nonsingular.

A adaptive neural control is developed to track the desired output of uncertain MIMO nonlinear system (9.1) in this chapter. RBFNNs are used to compensate for unknown nonlinear functions and unknown disturbances within the nonlinear system (9.1). The control objective is to make x_1 follow a certain desired trajectory x_{1d} in the presence of system uncertainties and disturbances. The proposed backstepping control will be rigorously shown to guarantee semiglobal uniform boundedness of the closed-loop system, and tracking errors converge to a very small neighborhood of the origin.

9.2.2 Neural Networks

In practical control, NNs have been successfully used as function approximators to solve different control problems [5]. The linear-in-parameters function approximator RBFNNs has been widely used to approximate the continuous function $f(\bar{Z}) : R^q \rightarrow R$ can be expressed as follows [33]:

$$f(\bar{Z}) = W^T S(\bar{Z}) + \varepsilon(\bar{Z}) \tag{9.9}$$

where $\bar{Z} = [\bar{z}_1, \bar{z}_2, \ldots, \bar{z}_q]^T \in R^q$ is the input vector of the approximator, $W \in R^p$ is a vector of updated weights, $S(\bar{Z}) = [s_1(\bar{Z}), s_2(\bar{Z}), \ldots, s_p(\bar{Z})]^T \in R^p$ is the vector of a known continuous (linear or nonlinear) basis function, $\varepsilon(\bar{Z})$ is the approximation error. According to the universal approximation property [33], the linearly parameterized NN can smoothly approximate any continuous function $f(\bar{Z})$ over the compact set $\Omega_{\bar{Z}} \in R^q$ to arbitrary any accuracy as

$$f(\bar{Z}) = W^{*T} S(\bar{Z}) + \varepsilon^*(\bar{Z}),\ \forall \bar{Z} \in \Omega_{\bar{Z}} \subset R^q \tag{9.10}$$

where W^* are the optimal weights, $\varepsilon^*(\bar{Z})$ is the smallest approximation error. The Gaussian RBFNNs is a particular network architecture which uses Gaussian functions of the form

$$s_i(\bar{Z}) = \exp[-(\bar{Z} - c_i)^T (\bar{Z} - c_i)/b_i^2),\ i = 1, 2, \ldots, l \tag{9.11}$$

where c_i and b_i are the center and width of the neural cell of the i-th hidden layer.

The optimal weight value of RBFNN is defined as

$$W^* = \arg \min_{\hat{W} \in \Omega_f} [\sup_{z \in S_{\bar{Z}}} |\hat{f}(\bar{Z}/\hat{W}) - f(\bar{Z})|] \tag{9.12}$$

where Ω_f is a valid field of the parameter and $S_{\bar{Z}} \subset R^n$ is an allowable set of the state vector.

Under the optimal weight value, there is

$$|f(\bar{Z}) - W^{*T} S(\bar{Z})| = |\varepsilon^*(\bar{Z})| \leq |\bar{\varepsilon}|. \tag{9.13}$$

9.3 Adaptive Neural Control and Stability Analysis

In this section, adaptive neural control is developed using RBFNNs in combination with the dynamics of MIMO nonlinear system (9.1).

Step 1: We define error variables $z_1 = x_1 - x_{1d}$, and $z_2 = x_2 - \alpha_1$, where $\alpha_1 \in R^m$ will be designed. Considering (9.1) and differentiating z_1 with respect to time, we obtain

$$\dot{z}_1 = F_1(x_1) + G_1(x_1)(z_2 + \alpha_1) + \Delta G_1(x_1) x_2 + D_1(x_1, t) - \dot{x}_{1d}. \tag{9.14}$$

Consider the Lyapunov function candidate

$$V_1^* = \frac{1}{2} z_1^T z_1. \tag{9.15}$$

The derivative of V_1^* along (9.14) is

$$\begin{aligned} \dot{V}_1^* &= z_1^T F_1(x_1) + z_1^T G_1(x_1)(z_2 + \alpha_1) + z_1^T \Delta G_1(x_1) x_2 + z_1^T D_1(x_1,t) - z_1^T \dot{x}_{1d} \\ &\le z_1^T F_1(x_1) + z_1^T G_1(x_1)(z_2 + \alpha_1) + \xi_{11} \|z_1\| \|x_2\| \\ &+ \sum_{j=1}^{m} |z_{1j}| (p_{1j}(x_1) + \bar{q}_{1j}) - z_1^T \dot{x}_{1d} \\ &= z_1^T F_1(x_1) + z_1^T G_1(x_1)(z_2 + \alpha_1) + \xi_{11} \|z_1\| \|x_2\| \\ &+ z_1^T \mathrm{Sgn}(z_1)(p_1(x_1) + \bar{q}_1) - z_1^T \dot{x}_{1d} \end{aligned} \tag{9.16}$$

where $\mathrm{Sgn}(z_1) := \mathrm{diag}\{\mathrm{sgn}(z_{1j})\}$, $p_1(x_1) = [p_{11}(x_1), p_{12}(x_1), \ldots, p_{1m}(x_1)]^T$, and $\bar{q}_1 = [\bar{q}_{11}, \bar{q}_{12}, \ldots, \bar{q}_{1m}]^T$.

We define

$$\rho_1(Z_1) = -F_1(x_1) - \mathrm{Sgn}(z_1)(p_1(x_1) + \bar{q}_1) + \dot{x}_{1d} \tag{9.17}$$

where $Z_1 = [x_1, z_1, \dot{x}_{1d}]^T$.

Considering (9.17), (9.16) can be rewritten as

$$\dot{V}_1^* \le z_1^T G_1(x_1)(z_2 + \alpha_1) + \xi_{11} \|z_1\| \|x_2\| - z_1^T \rho_1(Z_1). \tag{9.18}$$

Since $F_1(x_1)$, $p_1(x_1)$ and $\bar{q}_1$ are all unknown, then $\rho_1(Z_1)$ is also unknown which is approximated using the RBFNNs. Thus, we have

$$\hat{\rho}_1(Z_1) = \hat{W}_1^T S_1(Z_1) \tag{9.19}$$

where $\hat{W}_1 = \mathrm{diag}\{\hat{W}_{11}, \hat{W}_{12}, \ldots, \hat{W}_{1m}\}$ and $S_1(Z_1) = [S_{11}(Z_1), S_{12}(Z_1), \ldots, S_{1m}(Z_1)]^T$. $\hat{W}_1^T S_1(Z_1)$ approximates $W_1^{*T} S_1(Z_1)$ given by

$$\rho_1(Z_1) = W_1^{*T} S_1(Z_1) + \varepsilon_1 \tag{9.20}$$

where ε_1 is an approximation error.

Invoking Lemma 9.4, choose the following virtual control

$$\alpha_1 = (G_1(x_1) + \gamma_1 I_{m\times m})^{-1} \left(-K_1 z_1 + \hat{\rho}_1(Z_1) \right) \tag{9.21}$$

where $K_1 = K_1^T > 0$ and $\gamma_1 = \zeta_1 + \tau_1$.

Considering (9.20) and substituting (9.21) into (9.18), we obtain

$$
\begin{aligned}
\dot{V}_1^* \le & -z_1^T K_1 z_1 + z_1^T G_1(x_1) z_2 + z_1^T \hat{\rho}_1(Z_1) - z_1^T W_1^{*T} S_1(Z_1) - z_1^T \varepsilon_1 \\
& + \xi_{11}\|z_1\|\|x_2\| - \gamma_1 z_1^T \alpha_1 \\
= & -z_1^T K_1 z_1 + z_1^T G_1(x_1) z_2 + z_1^T \hat{W}_1^T S_1(Z_1) - z_1^T W_1^{*T} S_1(Z_1) - z_1^T \varepsilon_1 \\
& + \xi_{11}\|z_1\|\|x_2\| - \gamma_1 z_1^T \alpha_1 \\
= & -z_1^T K_1 z_1 + z_1^T G_1(x_1) z_2 + z_1^T \tilde{W}_1^T S_1(Z_1) - z_1^T \varepsilon_1 + \xi_{11}\|z_1\|\|x_2\| - \gamma_1 z_1^T \alpha_1
\end{aligned} \tag{9.22}
$$

where $\tilde{W}_1 := \hat{W}_1 - W_1^*$.

Considering the stability of error signal $\tilde{W}_1$, the augmented Lyapunov function candidate can be written as

$$
V_1 = \frac{1}{2} z_1^T z_1 + \frac{1}{2}\mathrm{tr}(\tilde{W}_1^T \Lambda_1^{-1} \tilde{W}_1) \tag{9.23}
$$

where $\Lambda_1^{-1} > 0$.

The time derivative of V_1 is

$$
\begin{aligned}
\dot{V}_1 \le & -z_1^T K_1 z_1 + z_1^T G_1(x_1) z_2 + z_1^T \tilde{W}_1^T S_1(Z_1) - z_1^T \varepsilon_1 \\
& + \xi_{11}\|z_1\|\|x_2\| - \gamma_1 z_1^T \alpha_1 + \mathrm{tr}(\tilde{W}_1^T \Lambda_1^{-1} \dot{\hat{W}}_1).
\end{aligned} \tag{9.24}
$$

Consider the adaptive law for $\hat{W}_1$ as

$$
\dot{\hat{W}}_1 = -\Lambda_1 (S_1(Z_1) z_1^T + \sigma_{m1} \hat{W}_1) \tag{9.25}
$$

where $\sigma_{m1} > 0$.

Noting that

$$
-\mathrm{tr}(\tilde{W}_1^T S_1(Z_1) z_1^T) = -\mathrm{tr}(z_1^T \tilde{W}_1^T S_1(Z_1)) = -z_1^T \tilde{W}_1^T S_1(Z_1) \tag{9.26}
$$

$$
2\mathrm{tr}(\tilde{W}_1^T \hat{W}_1) = \|\tilde{W}_1\|^2 + \|\hat{W}_1\|^2 - \|W_1^*\|^2 \ge \|\tilde{W}_1\|^2 - \|W_1^*\|^2, \tag{9.27}
$$

we have

$$
\begin{aligned}
\dot{V}_1 \le & -z_1^T (K_1 - \frac{1}{2} I_{m\times m}) z_1 + z_1^T G_1(x_1) z_2 + \xi_{11}\|z_1\|\|x_2\| - \gamma_1 z_1^T \alpha_1 \\
& + \frac{1}{2}\|\varepsilon_1\|^2 - \frac{\sigma_{m1}}{2}\|\tilde{W}_1\|^2 + \frac{\sigma_{m1}}{2}\|W_1^*\|^2.
\end{aligned} \tag{9.28}
$$

The first term on the right-hand side is stable if $K_1 - \frac{1}{2} I_{m\times m} > 0$, and the second term will be cancelled in the next step. The other terms will be considered in a stability analysis of the closed-loop system.

Step 2: Define the error variable $z_3 = x_3 - \alpha_2$. Considering (9.1) and differentiating z_2 with respect to time, we obtain

$$
\begin{aligned}
\dot{z}_2 &= F_2(\bar{x}_2) + G_2(\bar{x}_2) x_3 + \Delta G_2(\bar{x}_2) x_3 + D_2(\bar{x}_2, t) - \dot{\alpha}_1 \\
&= F_2(\bar{x}_2) + G_2(\bar{x}_2)(z_3 + \alpha_2) + \Delta G_2(\bar{x}_2) x_3 + D_2(\bar{x}_2, t) - \dot{\alpha}_1.
\end{aligned} \tag{9.29}
$$

Consider the Lyapunov function candidate

$$V_2^* = \frac{1}{2} z_2^T z_2. \tag{9.30}$$

The derivative of V_2^* along (9.29) is

$$\begin{aligned}
\dot{V}_2^* &= z_2^T F_2(\bar{x}_2) + z_2^T G_2(\bar{x}_2)(z_3 + \alpha_2) + z_2^T \Delta G_2(\bar{x}_2) x_3 + z_2^T D_2(\bar{x}_2, t) - z_2^T \dot{\alpha}_1 \\
&\le z_2^T F_2(\bar{x}_2) + z_2^T G_2(\bar{x}_2)(z_3 + \alpha_2) + \xi_{21} \|z_2\| \|x_3\| \\
&\quad + \sum_{j=1}^{m} |z_{2j}| (p_{2j}(\bar{x}_2) + \bar{q}_{2j}) - z_2^T \dot{\alpha}_1 \\
&= z_2^T F_2(\bar{x}_2) + z_2^T G_2(\bar{x}_2)(z_3 + \alpha_2) + \xi_{21} \|z_2\| \|x_3\| \\
&\quad + z_2^T \mathrm{Sgn}(z_2)(p_2(\bar{x}_2) + \bar{q}_2) - z_2^T \dot{\alpha}_1
\end{aligned} \tag{9.31}$$

where $\mathrm{Sgn}(z_2) := \mathrm{diag}\{\mathrm{sgn}(z_{2j})\}$, $p_2(\bar{x}_2) = [p_{21}(\bar{x}_2), p_{22}(\bar{x}_2), \ldots, p_{2m}(\bar{x}_2)]^T$, and $\bar{q}_2 = [\bar{q}_{21}, \bar{q}_{22}, \ldots, \bar{q}_{2m}]^T$.

We define

$$\rho_2(Z_2) = -F_2(\bar{x}_2) - \mathrm{Sgn}(z_2)(p_2(\bar{x}_2) + \bar{q}_2) + \dot{\alpha}_1 \tag{9.32}$$

where $Z_2 = [x_1, x_2, z_2, \dot{\alpha}_1]^T$.

Then, (9.31) can be expressed as

$$\dot{V}_2^* \le z_2^T G_2(\bar{x}_2)(z_3 + \alpha_2) - z_2^T \rho_2(Z_2) + \xi_{21} \|z_2\| \|x_3\|. \tag{9.33}$$

Since $F_2(x_1, x_2)$, $p_2(x_1, x_2)$ and $\bar{q}_2$ are all unknown, then $\rho_2(Z_2)$ is also unknown which is approximated using the RBFNNs. Thus, we obtain

$$\hat{\rho}_2(Z_2) = \hat{W}_2^T S_2(Z_2) \tag{9.34}$$

where $\hat{W}_2 = \mathrm{diag}\{\hat{W}_{21}, \hat{W}_{22}, \ldots, \hat{W}_{2m}\}$ and $S_2(Z_2) = [S_{21}(Z_2), S_{22}(Z_2), \ldots, S_{2m}(Z_2)]^T$. $\hat{W}_2^T S_2(Z_2)$ approximate $W_2^{*T} S_2(Z_2)$ given by

$$\rho_2(Z_2) = W_2^{*T} S_2(Z_2) + \varepsilon_2 \tag{9.35}$$

where ε_2 is an approximation error.

Evoking Lemma 9.4, design the following virtual control

$$\alpha_2 = (G_2(\bar{x}_2) + \gamma_2 I_{m\times m})^{-1} \left(-G_1^T(x_1) z_1 - K_2 z_2 + \hat{\rho}_2(Z_2) \right) \tag{9.36}$$

where $K_2 = K_2^T > 0$ and $\gamma_2 = \zeta_2 + \tau_2$.

Considering (9.35) and substituting (9.36) into (9.33), we have

$$\begin{aligned}\dot{V}_2^* &\leq -z_2^T K_2 z_2 + z_2^T G_2(\bar{x}_2) z_3 + z_2^T \hat{\rho}_2(Z_2) - z_2^T W_2^{*T} S_2(Z_2) \\ &\quad - z_2^T \varepsilon_2 - z_2^T G_1^T(x_1) z_1 - \gamma_2 z_2^T \alpha_2 + \xi_{21}\|z_2\|\|x_3\| \\ &= -z_2^T K_2 z_2 + z_2^T G_2(\bar{x}_2) z_3 + z_2^T \hat{W}_2^T S_2(Z_2) - z_2^T W_2^{*T} S_2(Z_2) \\ &\quad - z_2^T \varepsilon_2 - z_1^T G_1(x_1) z_2 - \gamma_2 z_2^T \alpha_2 + \xi_{21}\|z_2\|\|x_3\| \\ &= -z_2^T K_2 z_2 + z_2^T G_2(\bar{x}_2) z_3 + z_2^T \tilde{W}_2^T S_2(Z_2) - z_2^T \varepsilon_2 \\ &\quad - z_1^T G_1(x_1) z_2 - \gamma_2 z_2^T \alpha_2 + \xi_{21}\|z_2\|\|x_3\| \end{aligned} \tag{9.37}$$

where $\tilde{W}_2 := \hat{W}_2 - W_2^*$.

Considering the stability of error signal $\tilde{W}_2$, the augmented Lyapunov function can be expressed as

$$V_2 = V_1 + \frac{1}{2} z_2^T z_2 + \frac{1}{2}\mathrm{tr}(\tilde{W}_2^T \Lambda_2^{-1} \tilde{W}_2) \tag{9.38}$$

where $\Lambda_2^{-1} > 0$.

Invoking (9.28) and (9.37), the time derivative of V_2 is given by

$$\begin{aligned}\dot{V}_2 &\leq -z_1^T (K_1 - \frac{1}{2} I_{m\times m}) z_1 + z_1^T G_1(x) z_2 + \frac{1}{2}\|\varepsilon_1\|^2 - \frac{\sigma_{m1}}{2}\|\tilde{W}_1\|^2 + \frac{\sigma_{m1}}{2}\|W_1^*\|^2 \\ &\quad - \sum_{j=1}^{2} \gamma_j z_j^T \alpha_j + \sum_{j=1}^{2} \xi_{j1}\|z_j\|\|x_{j+1}\| - z_2^T K_2 z_2 + z_2^T G_2(\bar{x}_2) z_3 \\ &\quad + z_2^T \tilde{W}_2^T S_2(Z_2) - z_2^T \varepsilon_2 + \mathrm{tr}(\tilde{W}_2^T \Lambda_2^{-1} \dot{\hat{W}}_2) - z_1^T G_1(x_1) z_2 \\ &\leq -\sum_{j=1}^{2} z_j^T (K_j - \frac{1}{2} I_{m\times m}) z_j + \frac{1}{2}\sum_{j=1}^{2}\|\varepsilon_j\|^2 - \frac{\sigma_{m1}}{2}\|\tilde{W}_1\|^2 + \frac{\sigma_{m1}}{2}\|W_1^*\|^2 \\ &\quad - \sum_{j=1}^{2} \gamma_j z_j^T \alpha_j + \sum_{j=1}^{2} \xi_{j1}\|z_j\|\|x_{j+1}\| + z_2^T G_2(\bar{x}_2) z_3 \\ &\quad + z_2^T \tilde{W}_2^T S_2(Z_2) + \mathrm{tr}(\tilde{W}_2^T \Lambda_2^{-1} \dot{\hat{W}}_2). \end{aligned} \tag{9.39}$$

Consider the adaptive law for $\hat{W}_2$ as

$$\dot{\hat{W}}_2 = -\Lambda_2 (S_2(Z_2) z_2^T + \sigma_{m2} \hat{W}_2) \tag{9.40}$$

where $\sigma_{m2} > 0$.

Noting that

$$-\mathrm{tr}(\tilde{W}_2^T S_2(Z_2) z_2^T) = -\mathrm{tr}(z_2^T \tilde{W}_2^T S_2(Z_2)) = -z_2^T \tilde{W}_2^T S_2(Z_2) \tag{9.41}$$

$$2\mathrm{tr}(\tilde{W}_2^T \hat{W}_2) = \|\tilde{W}_2\|^2 + \|\hat{W}_2\|^2 - \|W_2^*\|^2 \geq \|\tilde{W}_2\|^2 - \|W_2^*\|^2, \tag{9.42}$$

we obtain

$$\begin{aligned}\dot{V}_2 \leq & -\sum_{j=1}^{2} z_j^T (K_j - \frac{1}{2} I_{m\times m}) z_j + z_2^T G_2(\bar{x}_2) z_3 - \sum_{j=1}^{2} \gamma_j z_j^T \alpha_j \\ & + \sum_{j=1}^{2} \xi_{j1} \|z_j\| \|x_{j+1}\| + \sum_{j=1}^{2} \frac{1}{2} \|\varepsilon_j\|^2 - \sum_{j=1}^{2} \frac{\sigma_{mj}}{2} \|\tilde{W}_j\|^2 + \sum_{j=1}^{2} \frac{\sigma_{mj}}{2} \|W_j^*\|^2 \end{aligned} \tag{9.43}$$

The first term on the right-hand side is stable if $\sum_{j=1}^{2}(K_j - \frac{1}{2} I_{m\times m}) > 0$, and the second term will be cancelled in the next step. The other terms will be considered in stability analysis of the closed-loop system.

Step i $(1 \leq i \leq n-1)$: Define the error variable $z_{i+1} = x_{i+1} - \alpha_i$. Considering (9.1) and differentiating z_i with respect to time, we have

$$\begin{aligned}\dot{z}_i &= F_i(\bar{x}_i) + G_2(\bar{x}_i) x_{i+1} + \Delta G_2(\bar{x}_i) x_{i+1} + D_i(\bar{x}_i, t) - \dot{\alpha}_{i-1} \\ &= F_i(\bar{x}_i) + G_2(\bar{x}_i)(z_{i+1} + \alpha_i) + \Delta G_2(\bar{x}_i) x_{i+1} + D_i(\bar{x}_i, t) - \dot{\alpha}_{i-1}. \end{aligned} \tag{9.44}$$

Consider the Lyapunov function candidate

$$V_i^* = \frac{1}{2} z_i^T z_i. \tag{9.45}$$

The derivative of V_i^* along (9.44) is

$$\begin{aligned}\dot{V}_i^* &= z_i^T F_i(\bar{x}_i) + z_i^T G_i(\bar{x}_i)(z_{i+1} + \alpha_i) + z_i^T \Delta G_2(\bar{x}_i) x_{i+1} + z_i^T D_i(\bar{x}_i, t) - z_i^T \dot{\alpha}_{i-1} \\ &\leq z_i^T F_i(\bar{x}_i) + z_i^T G_i(\bar{x}_i)(z_{i+1} + \alpha_i) + \xi_{i1} \|z_i\| \|x_{i+1}\| \\ &\quad + \sum_{j=1}^{m} |z_{ij}| (p_{ij}(\bar{x}_i) + \bar{q}_{ij}) - z_i^T \dot{\alpha}_{i-1} \\ &= z_i^T F_i(\bar{x}_i) + z_i^T G_i(\bar{x}_i)(z_{i+1} + \alpha_i) + \xi_{i1} \|z_i\| \|x_{i+1}\| \\ &\quad + z_i^T \mathrm{Sgn}(z_i)(p_i(\bar{x}_i) + \bar{q}_i) - z_i^T \dot{\alpha}_{i-1} \end{aligned} \tag{9.46}$$

where $\mathrm{Sgn}(z_i) := \mathrm{diag}\{\mathrm{sgn}(z_{ij})\}$, $p_i(\bar{x}_i) = [p_{i1}(\bar{x}_i), p_{i2}(\bar{x}_1), \ldots, p_{im}(\bar{x}_i)]^T$, and $\bar{q}_i = [\bar{q}_{i1}, \bar{q}_{i2}, \ldots, \bar{q}_{im}]^T$.

Define

$$\rho_i(Z_i) = -F_i(\bar{x}_i) - \mathrm{Sgn}(z_i)(p_i(\bar{x}_i) + \bar{q}_i) + \dot{\alpha}_{i-1} \tag{9.47}$$

where $Z_i = [\bar{x}_i, z_i, \dot{\alpha}_{i-1}]^T$.

Then, (9.46) can be expressed as

$$\dot{V}_i^* \leq z_i^T G_i(\bar{x}_i)(z_{i+1} + \alpha_i) + \xi_{i1} \|z_i\| \|x_{i+1}\| - z_i^T \rho_i(Z_i). \tag{9.48}$$

Since $F_i(\bar{x}_i)$, $p_i(\bar{x}_1)$ and $\bar{q}_i$ are all unknown, then $\rho_i(Z_i)$ is also unknown which is approximated using the RBFNNs. Thus, we have

$$\hat{\rho}_i(Z_i) = \hat{W}_i^T S_i(Z_i) \tag{9.49}$$

where $\hat{W}_i = \mathrm{diag}\{\hat{W}_{i1}, \hat{W}_{i2}, \dots, \hat{W}_{im}\}$ and $S_i(Z_i) = [S_{i1}(Z_i), S_{i2}(Z_i), \dots, S_{im}(Z_i)]^T$. $\hat{W}_i^T S_i(Z_i)$ approximate $W_i^{*T} S_i(Z_i)$ given by

$$\rho_i(Z_i) = W_i^{*T} S_i(Z_i) + \varepsilon_i \tag{9.50}$$

where ε_i is an approximation error.

Invoking Lemma 9.4, choose the following virtual control

$$\alpha_i = (G_i(\bar{x}_i) + \gamma_i I_{m\times m})^{-1} \left(-G_{i-1}^T(\bar{x}_{i-1}) z_{i-1} - K_i z_i + \hat{\rho}_i(Z_i) \right) \tag{9.51}$$

where $K_i = K_i^T > 0$ and $\gamma_i = \zeta_i + \tau_i$.

Considering (9.50) and substituting (9.51) into (9.48), we obtain

$$\begin{aligned}
\dot{V}_i^* \le & -z_i^T K_i z_i + z_i^T G_i(\bar{x}_i) z_{i+1} + z_i^T \hat{\rho}_i(Z_i) - z_i^T W_i^{*T} S_i(Z_i) \\
& - z_i^T \varepsilon_i - z_i^T G_{i-1}^T(\bar{x}_{i-1}) z_{i-1} - \gamma_i z_i^T \alpha_i + \xi_{i1} \|z_i\| \|x_{i+1}\| \\
= & -z_i^T K_i z_i + z_i^T G_i(\bar{x}_i) z_{i+1} + z_i^T \hat{W}_i^T S_i(Z_i) - z_i^T W_i^{*T} S_i(Z_i) \\
& - z_i^T \varepsilon_i - z_{i-1}^T G_{i-1}(\bar{x}_{i-1}) z_i - \gamma_i z_i^T \alpha_i + \xi_{i1} \|z_i\| \|x_{i+1}\| \\
= & -z_i^T K_i z_i + z_i^T G_i(\bar{x}_i) z_{i+1} + z_i^T \tilde{W}_i^T S_i(Z_i) - z_i^T \varepsilon_i \\
& - z_{i-1}^T G_{i-1}(\bar{x}_{i-1}) z_i - \gamma_i z_i^T \alpha_i + \xi_{i1} \|z_i\| \|x_{i+1}\|
\end{aligned} \tag{9.52}$$

where $\tilde{W}_i := \hat{W}_i - W_i^*$.

Considering the stability of error signal $\tilde{W}_i$, the augmented Lyapunov function candidate can be written as

$$V_i = V_{i-1} + \frac{1}{2} z_i^T z_i + \frac{1}{2} \mathrm{tr}(\tilde{W}_i^T \Lambda_i^{-1} \tilde{W}_i) \tag{9.53}$$

where $\Lambda_i^{-1} > 0$.

Invoking (9.43) and (9.52), the time derivative of V_i is given by

$$\begin{aligned}
\dot{V}_i \le & -\sum_{j=1}^{i-1} z_j^T (K_j - \frac{1}{2} I_{m\times m}) z_j + z_{i-1}^T G_{i-1}(\bar{x}_{i-1}) z_i + \frac{1}{2} \sum_{j=1}^{i-1} \|\varepsilon_j\|^2 - \sum_{j=1}^{i-1} \frac{\sigma_{mj}}{2} \|\tilde{W}_j\|^2 \\
& + \sum_{j=1}^{i-1} \frac{\sigma_{m1}}{2} \|W_j^*\|^2 - z_i^T K_i z_i + z_i^T G_i(\bar{x}_i) z_{i+1} + z_i^T \tilde{W}_i^T S_i(Z_i) - z_i^T \varepsilon_i \\
& + \mathrm{tr}(\tilde{W}_i^T \Lambda_i^{-1} \dot{\hat{W}}_i) - z_{i-1}^T G_{i-1}(\bar{x}_{i-1}) z_i - \sum_{j=1}^{i} \gamma_j z_j^T \alpha_j + \sum_{j=1}^{i} \xi_{j1} \|z_j\| \|x_{j+1}\| \\
\le & \sum_{j=1}^{i} -z_j^T (K_j - \frac{1}{2} I_{m\times m}) z_j + \sum_{j=1}^{i} \frac{1}{2} \|\varepsilon_j\|^2 - \sum_{j=1}^{i-1} \frac{\sigma_{mj}}{2} \|\tilde{W}_j\|^2 + \sum_{j=1}^{i-1} \frac{\sigma_{mj}}{2} \|W_j^*\|^2 \\
& - \sum_{j=1}^{i} \gamma_j z_j^T \alpha_j + \sum_{j=1}^{i} \xi_{j1} \|z_j\| \|x_{j+1}\| + z_i^T G_i(\bar{x}_i) z_{i+1} \\
& + z_i^T \tilde{W}_i^T S_i(Z_i) + \mathrm{tr}(\tilde{W}_i^T \Lambda_i^{-1} \dot{\hat{W}}_i).
\end{aligned} \tag{9.54}$$

Consider the adaptive law for $\hat{W}_i$ as

$$\dot{\hat{W}}_i = -\Lambda_i(S_i(Z_i)z_i^T + \sigma_{mi}\hat{W}_i) \tag{9.55}$$

where $\sigma_{mi} > 0$.

Noting that

$$-\mathrm{tr}(\tilde{W}_i^T S_i(Z_i)z_i^T) = -\mathrm{tr}(z_i^T \tilde{W}_i^T S_i(Z_i)) = -z_i^T \tilde{W}_i^T S_i(Z_i) \tag{9.56}$$

$$2\mathrm{tr}(\tilde{W}_i^T \hat{W}_i) = \|\tilde{W}_i\|^2 + \|\hat{W}_i\|^2 - \|W_i^*\|^2 \geq \|\tilde{W}_i\|^2 - \|W_i^*\|^2, \tag{9.57}$$

we obtain

$$\begin{aligned}\dot{V}_i \leq & -\sum_{j=1}^{i} z_j^T(K_j - \frac{1}{2}I_{m\times m})z_j + z_i^T G_i(\bar{x}_i)z_{i+1} - \sum_{j=1}^{i}\gamma_j z_j^T \alpha_j + \sum_{j=1}^{i}\xi_{j1}\|z_j\|\|x_{j+1}\| \\ & + \frac{1}{2}\sum_{j=1}^{i}\|\varepsilon_j\|^2 - \sum_{j=1}^{i}\frac{\sigma_{mj}}{2}\|\tilde{W}_j\|^2 + \sum_{j=1}^{i}\frac{\sigma_{mj}}{2}\|W_j^*\|^2.\end{aligned} \tag{9.58}$$

The first term on the right-hand side is stable if $\sum_{j=1}^{i}(K_j - \frac{1}{2}I_{m\times m}) > 0$, and the second term will be cancelled in the next step. The other terms will be considered in stability analysis of the closed-loop system.

Step n: By differentiating $z_n = x_n - \alpha_{n-1}$ with respect to time yields

$$\dot{z}_n = F_n(\bar{x}_n) + G_n(\bar{x}_n)\Phi(u) + \Delta G_n(\bar{x}_n)\Phi(u) + D_n(\bar{x}_n,t) - \dot{\alpha}_{n-1}. \tag{9.59}$$

Consider the Lyapunov function candidate

$$V_n^* = V_{n-1} + \frac{1}{2\delta}\sigma^T\sigma. \tag{9.60}$$

The sliding surface σ is defined as

$$\sigma = \vartheta_1 z_1 + \vartheta_2 z_2 + \ldots + \vartheta_{n-1}z_{n-1} + z_n \tag{9.61}$$

where $\vartheta_i > 0, i = 1,2,\ldots,n-1$ are the design parameters. The assistant design parameter δ is defined as follows

$$0 < \underline{\delta} \leq \delta \leq \bar{\delta} \leq \chi(u_i, u_{li\max}, u_{ri\max}, u_{li0}, u_{ri0}) \tag{9.62}$$

where $\bar{\delta} \geq \underline{\delta} > 0$. From the definition of δ, there exist constant $\delta_d > 0$ such that $|\dot{\delta}| \leq \delta_d$.

The derivative of V_n^* is given by

$$\dot{V}_n^* = \dot{V}_{n-1} - \frac{\sigma^T\sigma\dot{\delta}}{2\delta^2} + \frac{1}{\delta}\sigma^T\dot{\sigma}. \tag{9.63}$$

Considering (9.58), (9.61) and (9.62), we obtain

$$\begin{aligned}\dot{V}_n^* \leq & -\sum_{j=1}^{n-1} z_j^T(K_j - \frac{1}{2}I_{m\times m})z_j + \frac{1}{2}\sum_{j=1}^{n-1}\|\varepsilon_j\|^2 - \sum_{j=1}^{n-1}\frac{\sigma_{mj}}{2}\|\tilde{W}_j\|^2 + \sum_{j=1}^{n-1}\frac{\sigma_{mj}}{2}\|W_j^*\|^2 \\ & + \frac{\delta_d}{2\underline{\delta}^2}\sigma^T\sigma + z_{n-1}^T G_{n-1}(\bar{x}_{n-1})z_n + \frac{1}{\delta}\sigma^T(\sum_{j=1}^{n-1}\vartheta_j\dot{z}_j + \dot{z}_n) \\ & - \sum_{j=1}^{n-1}\gamma_j z_j^T\alpha_j + \sum_{j=1}^{n-1}\xi_{j1}\|z_j\|\|x_{j+1}\|. \end{aligned} \tag{9.64}$$

From (9.61), we have

$$z_n = \sigma - \vartheta_1 z_1 - \vartheta_2 z_2 - \ldots - \vartheta_{n-1}z_{n-1}. \tag{9.65}$$

Considering (9.59) and substituting (9.65) into (9.64), we obtain

$$\begin{aligned}\dot{V}_n^* \leq & -\sum_{j=1}^{n-1} z_j^T(K_j - \frac{1}{2}I_{m\times m})z_j + \frac{1}{2}\sum_{j=1}^{n-1}\|\varepsilon_j\|^2 - \sum_{j=1}^{n-1}\frac{\sigma_{mj}}{2}\|\tilde{W}_j\|^2 + \sum_{j=1}^{n-1}\frac{\sigma_{mj}}{2}\|W_j^*\|^2 \\ & + \frac{\delta_d}{2\underline{\delta}^2}\sigma^T\sigma + z_{n-1}^T G_{n-1}(\bar{x}_{n-1})(\sigma - \sum_{j=1}^{n-1}\vartheta_j z_j) + \frac{1}{\delta}\sigma^T(\sum_{j=1}^{n-1}\vartheta_j\dot{z}_j + \dot{z}_n) \\ & - \sum_{j=1}^{n-1}\gamma_j z_j^T\alpha_j + \sum_{j=1}^{n-1}\xi_{j1}\|z_j\|\|x_{j+1}\| \\ \leq & -\sum_{j=1}^{n-1} z_j^T(K_j - \frac{1}{2}I_{m\times m})z_j + \frac{1}{2}\sum_{j=1}^{n-1}\|\varepsilon_j\|^2 - \sum_{j=1}^{n-1}\frac{\sigma_{mj}}{2}\|\tilde{W}_j\|^2 + \sum_{j=1}^{n-1}\frac{\sigma_{mj}}{2}\|W_j^*\|^2 \\ & - \sum_{j=1}^{n-1}\gamma_j z_j^T\alpha_j + \sum_{j=1}^{n-1}\xi_{j1}\|z_j\|\|x_{j+1}\| - z_{n-1}^T G_{n-1}(\bar{x}_{n-1})\sum_{j=1}^{n-1}\vartheta_j z_j + \frac{\delta_d}{2\underline{\delta}^2}\sigma^T\sigma \\ & + \sigma^T G_{n-1}^T(\bar{x}_{n-1})z_{n-1} + \frac{1}{\delta}\sigma^T G_n(\bar{x}_n)\Phi(u) + \frac{1}{\delta}\sigma^T\Delta G_n(\bar{x}_n)\Phi(u) \\ & + \frac{1}{\delta}\sigma^T(\sum_{j=1}^{n-1}\vartheta_j\dot{z}_j + F_n(\bar{x}_n) + D_n(\bar{x}_n,t) - \dot{\alpha}_{n-1}). \end{aligned} \tag{9.66}$$

Define

$$h(Z) = \sum_{j=1}^{n-1}|\gamma_j||z_j^T\alpha_j| + \sum_{j=1}^{n-1}\xi_{j1}\|z_j\|\|x_{j+1}\| + \|\sigma\|\|G_{n-1}^T(\bar{x}_{n-1})z_{n-1}\| \tag{9.67}$$

$$\bar{\rho}_n(Z_n) = \frac{1}{\delta}\sigma^T(\sum_{j=1}^{n-1}\vartheta_j\dot{z}_j + F_n(\bar{x}_n) + D_n(\bar{x}_n,t) - \dot{\alpha}_{n-1}) \tag{9.68}$$

where $Z=[z_1,\ldots,z_{n-1},\alpha_1,\ldots,\alpha_{n-1}]$ and $Z_n=[\bar{x}_n,\alpha_1,\ldots,\alpha_{n-1},\dot{x}_{1d},\dot{\alpha}_1,\ldots,\dot{\alpha}_{n-1}]^T$.

According to (9.14), (9.29) and (9.44), we have

$$\begin{aligned}\bar{\rho}_n(Z_n) &= \frac{1}{\delta}\sigma^T(\vartheta_1(F_1(x_1)+G_1(x_1)(z_2+\alpha_1)-\dot{x}_{1d})\\ &\quad+\sum_{j=2}^{n-1}\vartheta_j(F_j(\bar{x}_j)+G_j(\bar{x}_j)(z_{j+1}+\alpha_j)-\dot{\alpha}_{j-1})\\ &\quad+F_n(\bar{x}_n)+D_n(\bar{x}_n,t)-\dot{\alpha}_{n-1}+\sum_{j=1}^{n-1}\vartheta_jD_j(\bar{x}_j,t))\\ &\le \frac{1}{\delta}\sigma^T(\vartheta_1(F_1(x_1)+G_1(x_1)(z_2+\alpha_1)-\dot{x}_{1d})\\ &\quad+\sum_{j=2}^{n-1}\vartheta_j(F_j(\bar{x}_j)+G_j(\bar{x}_j)(z_{j+1}+\alpha_j)-\dot{\alpha}_{j-1})+F_n(\bar{x}_n)\\ &\quad+\mathrm{Sgn}(\sigma)(P_n(\bar{x}_n)+\bar{q}_n)-\dot{\alpha}_{n-1}+\mathrm{Sgn}(\sigma)\sum_{j=1}^{n-1}\vartheta_j(P_j(\bar{x}_j)+\bar{q}_j))\end{aligned}\tag{9.69}$$

where $\mathrm{Sgn}(\sigma):=\mathrm{diag}\{\mathrm{sgn}(\sigma_j)\}$, $p_n(\bar{x}_n)=[p_{n1}(\bar{x}_n),p_{n2}(\bar{x}_n),\ldots,p_{nm}(\bar{x}_n)]^T$, $\bar{q}_n=[\bar{q}_{n1},\bar{q}_{n2},\ldots,\bar{q}_{nm}]^T$.

We define

$$\begin{aligned}\rho_n(Z_n) &= -\frac{1}{\delta}(\vartheta_1(F_1(x_1)+G_1(x_1)(z_2+\alpha_1)-\dot{x}_{1d})\\ &\quad+\sum_{j=2}^{n-1}\vartheta_j(F_j(\bar{x}_j)+G_j(\bar{x}_j)(z_{j+1}+\alpha_j)-\dot{\alpha}_{j-1})\\ &\quad+F_n(\bar{x}_n)+\mathrm{Sgn}(\sigma)(P_n(\bar{x}_n)+\bar{q}_n)-\dot{\alpha}_{n-1}+\mathrm{Sgn}(\sigma)\sum_{j=1}^{n-1}\vartheta_j(P_j(\bar{x}_j)+\bar{q}_j)).\end{aligned}$$

Since $\rho_n(Z_n)$ contains unknown terms, it is approximated using the RBFNNs. Thus, the above yields

$$\hat{\rho}_n(Z_n)=\hat{W}_n^TS_n(Z_n)\tag{9.70}$$

where $\hat{W}_n=\mathrm{diag}\{\hat{W}_{n1},\hat{W}_{n2},\ldots,\hat{W}_{nm}\}$ and $S_n(Z_n)=[S_{n1}(Z_n),S_{n2}(Z_n),\ldots,S_{nm}(Z_n)]^T$. $\hat{W}_n^TS_n(Z_n)$ approximate $W_n^{*T}S_n(Z_n)$ given by

$$\rho_n(Z_n)=W_n^{*T}S_n(Z_n)+\varepsilon_n\tag{9.71}$$

where ε_n is an approximation error.

Define

$$Q=\begin{bmatrix} K_1-\frac{1}{2}I_{m\times m} & 0 & \cdots & 0 \\ 0 & K_2-\frac{1}{2}I_{m\times m} & \cdots & 0 \\ \vdots & \vdots & \vdots & \vdots \\ \vartheta_1 G_{n-1} & \vartheta_2 G_{n-1} & \cdots & \vartheta_{n-1}G_{n-1}+K_{n-1}-\frac{1}{2}I_{m\times m} \end{bmatrix}. \tag{9.72}$$

We can easily choose appropriate design parameters $K_1, K_2, \ldots, K_{n-1}$ and ϑ_{n-1} which can make $Q > 0$.

Considering (9.67)-(9.72), (9.66) can be written as

$$\begin{aligned}\dot{V}_n^* \leq& -[z_1\ z_2\ \ldots,\ z_{n-1}]Q[z_1\ z_2\ \ldots,\ z_{n-1}]^T+\frac{1}{2}\sum_{j=1}^{n-1}\|\varepsilon_j\|^2-\sum_{j=1}^{n-1}\frac{\sigma_{mj}}{2}\|\tilde{W}_j\|^2\\ &+\sum_{j=1}^{n-1}\frac{\sigma_{mj}}{2}\|W_j^*\|^2+\frac{\delta_d}{2\underline{\delta}^2}\sigma^T\sigma+h(Z)+\frac{1}{\delta}\sigma^T G_n(\bar{x}_n)\Phi(u)\\ &+\frac{1}{\delta}\sigma^T\Delta G_n(\bar{x}_n)\Phi(u)-\sigma^T W_n^{*T}S_n(Z_n)-\sigma^T\varepsilon_n. \end{aligned}\tag{9.73}$$

Considering nonsingular matrix $G_n(\bar{x}_n)+\gamma_n I_{m\times m}$ with $\gamma_n=\zeta_n+\tau_n$, let $\theta = (G_n^T(\bar{x}_n)+\gamma_n I_{m\times m})\sigma=[\theta_1,\theta_2,\ldots,\theta_m]^T$. Then, the following VSC law with saturation /dead-zone is proposed:

$$u_i=\begin{cases} -(h(Z)+\sigma^T\Gamma\sigma+\|\sigma^T\|\|\hat{W}_n^T S(Z_n)\|)\frac{\theta_i}{\|\theta\|^2}+u_{ri0} & \text{if } \theta_i<0\\ 0 & \text{if } \theta_i=0\\ -(h(Z)+\sigma^T\Gamma\sigma+\|\sigma^T\|\|\hat{W}_n^T S(Z_n)\|)\frac{\theta_i}{\|\theta\|^2}+u_{li0} & \text{if } \theta_i>0 \end{cases}\tag{9.74}$$

where $\Gamma=\Gamma^T>0$ is an appropriate dimension positive matrix, u_{ri0} and u_{li0} are the dead-zone values.

To further design and analyze, the following lemma is required.

Lemma 9.5. *[24] For all the input nonlinearities $\Phi(u)$ satisfying Assumptions 9.1 and 9.2, the control law (9.74) satisfies the inequality*

$$\sigma^T(t)G_n(\bar{x})\Phi(u)\leq -\delta(h(Z)+\sigma^T\Gamma\sigma+\|\sigma^T\|\|\hat{W}_n^T S(Z_n)\|). \tag{9.75}$$

Proof. From (9.2) and (9.74), $u_i > u_{ri0}$ implies that $\theta_i(t)<0$ and thus

$$\begin{aligned}(u_i-u_{ri0})\Phi(u_i) &= -\frac{\theta_i}{\|\theta\|^2}(h(Z)+\sigma^T\Gamma\sigma+\|\sigma^T\|\|\hat{W}_n^T S(Z_n)\|)\Phi(u_i)\\ &\geq \delta(u_i-u_{ri0})^2\\ &= \delta(\frac{\theta_i}{\|\theta\|^2}(h(Z)+\sigma^T\Gamma\sigma+\|\sigma^T\|\|\hat{W}_n^T S(Z_n)\|))^2 \end{aligned}\tag{9.76}$$

whereas for $u_i < u_{li0}$, $\theta_i(t)>0$ and we have

$$
\begin{aligned}
(u_i - u_{li0})\Phi(u_i) &= -\frac{\theta_i}{\|\theta\|^2}(h(Z) + \sigma^T\Gamma\sigma + \|\sigma^T\|\|\hat{W}_n^T S(Z_n)\|)\Phi(u_i) \\
&\geq \delta(u_i - u_{li0})^2 \\
&= \delta(\frac{\theta_i}{\|\theta\|^2}(h(Z) + \sigma^T\Gamma\sigma + \|\sigma^T\|\|\hat{W}_n^T S(Z_n)\|))^2. \qquad (9.77)
\end{aligned}
$$

Considering (9.76) and (9.77) yields

$$
\theta_i\Phi(u_i) \leq -\delta\frac{\theta_i^2}{\|\theta\|^2}(h(Z) + \sigma^T\Gamma\sigma + \|\sigma^T\|\|\hat{W}_n^T S(Z_n)\|). \qquad (9.78)
$$

Therefore, the following inequality holds:

$$
\sigma^T G_n(\bar{x})\Phi(u) = \sum_{i=1}^{m}\theta_i\Phi(u_i) \leq -\delta(h(Z) + \sigma^T\Gamma\sigma + \|\sigma^T\|\|\hat{W}_n^T S(Z_n)\|). \, (9.79)
$$

This concludes the proof. This lemma shows that the input nonlinearity combining with the sliding surface satisfies (9.79). □

Considering the fact $\|\Phi(u)\| \leq \eta_n u_{\max}$ with $u_{\max} = \max\{u_{ri\max}, |u_{li\max}|\}, i = 1,2,3$ and $\eta_n > 0$ known constant, and substituting (9.79) into (9.73) yields

$$
\begin{aligned}
\dot{V}_n^* \leq & -[z_1\ z_2\ \ldots,\ z_{n-1}]Q[z_1\ z_2\ \ldots,\ z_{n-1}]^T + \frac{1}{2}\sum_{j=1}^{n-1}\|\varepsilon_j\|^2 - \sum_{j=1}^{n-1}\frac{\sigma_{mj}}{2}\|\tilde{W}_j\|^2 \\
& + \sum_{j=1}^{n-1}\frac{\sigma_{mj}}{2}\|W_j^*\|^2 + \frac{\delta_d}{2\underline{\delta}^2}\sigma^T\sigma + h(Z) - \frac{\gamma_n}{\delta}\sigma^T\Phi(u) + \frac{1}{\delta}\sigma^T\Delta G_n(\bar{x}_n)\Phi(u) \\
& - h(Z) - \sigma^T\Gamma\sigma - \|\sigma^T\|\|\hat{W}_n^T S(Z_n)\| - \sigma^T\hat{W}_n^T S(Z_n) \\
& + \sigma^T\hat{W}_n S(Z_n) - \sigma^T W_n^{*T} S_n(Z_n) - \sigma^T\varepsilon_n \\
\leq & -[z_1\ z_2\ \ldots,\ z_{n-1}]Q[z_1\ z_2\ \ldots,\ z_{n-1}]^T + \frac{1}{2}\sum_{j=1}^{n-1}\|\varepsilon_j\|^2 - \sum_{j=1}^{n-1}\frac{\sigma_{mj}}{2}\|\tilde{W}_j\|^2 \\
& + \sum_{j=1}^{n-1}\frac{\sigma_{mj}}{2}\|W_j^*\|^2\frac{\delta_d}{2\underline{\delta}^2}\sigma^T\sigma + \frac{\gamma_n + \xi_{n1}}{2\underline{\delta}}\sigma^T\sigma + \frac{\gamma_n + \xi_{n1}}{2\underline{\delta}}\eta_n^2 u_{\max}^2 \\
& - \sigma^T\Gamma\sigma + \sigma^T\tilde{W}_n^T S(Z_n) - \sigma^T\varepsilon_n. \qquad (9.80)
\end{aligned}
$$

where $\tilde{W}_n := \hat{W}_n - W_n^*$.

Considering the stability of error signal $\tilde{W}_n$, the augmented Lyapunov function candidate can be expressed as

$$
V_n = V_n^* + \frac{1}{2}\text{tr}(\tilde{W}_n^T\Lambda_n^{-1}\tilde{W}_n) \qquad (9.81)
$$

where $\Lambda_n^{-1} > 0$.

Invoking (9.80) and $-\sigma^T\varepsilon_n \leq \frac{1}{2}(\sigma^T\sigma + \varepsilon_n^T\varepsilon_n)$, the derivative of V_n is given by

$$\dot{V}_n \le -[z_1\ z_2\ \ldots,\ z_{n-1}]Q[z_1\ z_2\ \ldots,\ z_{n-1}]^T + \frac{1}{2}\sum_{j=1}^{n}\|\varepsilon_j\|^2 - \sum_{j=1}^{n-1}\frac{\sigma_{mj}}{2}\|\tilde{W}_j\|^2$$
$$+ \sum_{j=1}^{n-1}\frac{\sigma_{mj}}{2}\|W_j^*\|^2 - \sigma^T(\Gamma - (\frac{\delta_d}{2\underline{\delta}^2} + \frac{\gamma_n + \xi_{n1}}{2\underline{\delta}} + \frac{1}{2})I_{m\times m})\sigma$$
$$+ \frac{\gamma_n + \xi_{n1}}{2\underline{\delta}}\eta_n^2 u_{\max}^2 + \sigma^T \tilde{W}_n^T S(Z_n) + \mathrm{tr}(\tilde{W}_n^T \Lambda_n^{-1}\dot{\hat{W}}_n). \tag{9.82}$$

Consider the adaptive law for $\hat{W}_n$ as

$$\dot{\hat{W}}_n = -\Lambda_n(S_n(Z_n)\sigma^T + \sigma_{mn}\hat{W}_n) \tag{9.83}$$

where $\sigma_{mn} > 0$.

Noting that

$$-\mathrm{tr}(\tilde{W}_n^T S_n(Z_n)\sigma^T) = -\mathrm{tr}(\sigma^T \tilde{W}_n^T S_n(Z_n)) = -\sigma^T \tilde{W}_n^T S_n(Z_n) \tag{9.84}$$

$$2\mathrm{tr}(\tilde{W}_n^T \hat{W}_n) = \|\tilde{W}_n\|^2 + \|\hat{W}_n\|^2 - \|W_n^*\|^2 \ge \|\tilde{W}_n\|^2 - \|W_n^*\|^2, \tag{9.85}$$

we have

$$\dot{V}_n \le -[z_1\ z_2\ \ldots,\ z_{n-1}]Q[z_1\ z_2\ \ldots,\ z_{n-1}]^T + \frac{1}{2}\sum_{j=1}^{n}\|\varepsilon_j\|^2 - \sum_{j=1}^{n}\frac{\sigma_{mj}}{2}\|\tilde{W}_j\|^2$$
$$+ \sum_{j=1}^{n}\frac{\sigma_{mj}}{2}\|W_j^*\|^2 - \sigma^T(\Gamma - (\frac{\delta_d}{2\underline{\delta}^2} + \frac{\gamma_n + \xi_{n1}}{2\underline{\delta}} + \frac{1}{2})I_{m\times m})\sigma + \frac{\gamma_n + \xi_{n1}}{2\underline{\delta}}\eta_n^2 u_{\max}^2$$
$$\le -\kappa V_n + C \tag{9.86}$$

where

$$\kappa := \min\left(\lambda_{\min}(Q), 2\delta\lambda_{\min}(Q_1), \min_{j=1,2,\ldots,n}\left(\frac{\sigma_{mj}}{\lambda_{\max}(\Lambda_j^{-1})}\right)\right)$$
$$C := \sum_{j=1}^{m}\frac{\sigma_{mj}}{2}\|W_j^*\|^2 + \frac{1}{2}\sum_{j=1}^{m}\|\bar{\varepsilon}_j\|^2 + \frac{\gamma_n + \xi_{n1}}{2\underline{\delta}}\eta_n^2 u_{\max}^2$$
$$Q_1 := \Gamma - (\frac{\delta_d}{2\underline{\delta}^2} + \frac{\gamma_n + \xi_{n1}}{2\underline{\delta}} + \frac{1}{2})I_{m\times m}. \tag{9.87}$$

The above design procedure can be summarized in the following theorem, which contains the results of adaptive neural control for uncertain MIMO nonlinear system (9.1).

Theorem 9.1. *Considering the strict-feedback nonlinear system (9.1) with known coefficient matrices satisfies Assumptions 9.1-9.5. Given that the system initial conditions are bounded, and that full state information are available. Under the control laws (9.74), and parameter updated laws (9.25), (9.40), (9.55), and (9.83), the trajectories of the closed-loop system are semiglobally uniformly bounded. The track-*

ing error z_1 converges to a compact set $\Omega_{zs} := \{z_1 \in R^m \mid \|z_1\| \le \sqrt{(2C/\kappa)}\}$ asymptotically, where C and κ are defined in (9.87).

Proof. According to (9.86) and Lemma 9.1, it can directly show that the signals z_1, z_2, ..., z_n , σ and $\tilde{W}_j, j = 1,2,\ldots,n$ are semiglobally uniformly bounded. For completeness, the details of the proof are provided here. Multiplying (9.86) by $e^{\kappa t}$, there can yield $\frac{d}{dt}(V_n)e^{\kappa t}$. Integration of this equation yields

$$V_n \le \left(V_n(0) - \tfrac{C}{\kappa}\right)e^{-\kappa t} + \frac{C}{\kappa} \le V_n(0) + \frac{C}{\kappa}. \tag{9.88}$$

Considering (9.60) and (9.81), we obtain

$$\frac{1}{2}\|z_1\|^2 \le V_n(0) + \frac{C}{\kappa} \;\Rightarrow\; \|z_1\| \le \sqrt{2\left(V_n(0) + \tfrac{C}{\kappa}\right)}. \tag{9.89}$$

The bounds of z_i and $\tilde{W}_i$, $i = 1,2,\ldots,n$ can be similarly shown. This concludes the proof. □

Remark 9.3. In this chapter, VSC in combination with backstepping and Lyapunov synthesis is proposed for a class of uncertain MIMO nonlinear system with non-symmetric input nonlinearities of saturation and dead-zone. To the best of our knowledge, the input nonlinearity expressed as (9.2) is to capture the most realistic situation, and as such it is different from the existing description [24]. The parameters u_{l0} and u_{r0} of dead-zones need to be known in the design of VSC which is the same as the assumption in [24, 34, 35].

Remark 9.4. In the proposed adaptive neural control, the principle of designing the neural network $\hat{W}_i^T S(Z_i)$, $i = 1,2,\ldots,n$ as such is to use as few neurons as possible to approximate the unknown function $\rho_i(Z_i)$. As shown in [8], the weight estimates $\hat{W}_1,\ldots,\hat{W}_{i-1}$ are not recommend to be taken as inputs to NN because of the curse of dimensionality of RBFNN. The minimal number of RBFNN input Z_i can be decided following the method in [8].

Remark 9.5. For uncertain nonlinear system (9.1), the uniform boundedness of the closed-loop system established in Theorem 9.1 is semiglobal due to the use of approximation-based control, which are only valid within a compact set. In the design process of adaptive neural control, the assistant design parameter δ is used which satisfies $0 < \delta \le \chi(u_i, u_{li\max}, u_{ri\max}, u_{li0}, u_{ri0})$. We treat it as an unknown part of system compound uncertainty $\rho_n(Z_n)$, and is approximated by neural networks. It does not appear in the design of parameter adaptive law and control law. Hence, similar to [26], it is not necessary to choose an exact value for δ.

9.4 Simulation Results

Consider the following an uncertain MIMO nonlinear system with saturation and dead-zone:

$$\begin{aligned}\dot{x}_1 &= F_1(x_1)+(G_1(x_1)+\Delta G_1(x_1))x_2+D_1(x_1,t)\\ \dot{x}_2 &= F_2(\bar{x}_2)+(G_2(\bar{x}_2)+\Delta G_2(\bar{x}_2))\Phi(u)+D_2(\bar{x}_2,t)\\ y &= x_1\end{aligned} \tag{9.90}$$

where

$$x_1=[x_{11},x_{12}]^T,\ x_2=[x_{21},x_{22}]^T,\ F_1(x_1)=\begin{bmatrix}-0.2\sin(x_{11})\cos(x_{11}x_{12})\\ 0.1x_{11}x_{12}\end{bmatrix},$$

$$G_1(x_1)=\begin{bmatrix}1.2+\cos(x_{11})\sin(x_{12}) & 0\\ 0 & 1.5-\cos(x_{12})\sin(x_{11})\end{bmatrix},$$

$$F_2(\bar{x}_2)=\begin{bmatrix}-0.5x_{12}x_{21}\\ 0.8x_{11}x_{22}\end{bmatrix},\ \Delta G_1(x_1)=\begin{bmatrix}0\\0\end{bmatrix},\ G_2(\bar{x}_2)=\begin{bmatrix}\cos(x_{21}) & -\sin(x_{22})\\ \sin(x_{22}) & \cos(x_{21})\end{bmatrix},$$

$$\Delta G_2(\bar{x}_2)=\begin{bmatrix}0.1\sin(x_{11}x_{21}) & 0.1\cos(x_{11}x_{21})\\ 0.1\cos(x_{11}x_{21}) & 0.1\sin(x_{21}x_{22})\end{bmatrix},$$

$$D_1(x_1,t)=\begin{bmatrix}0.2\sin(x_{11})\cos(x_{12})+0.04\cos(0.3t)\\ 0.1\cos(x_{11})\sin(x_{12})+0.03\cos(0.2t)\end{bmatrix},$$

$$D_2(\bar{x}_2,t)=\begin{bmatrix}0.15\sin(x_{21})\cos(x_{22})+0.05\cos(0.2t)\\ 0.12\sin(x_{22})\cos(x_{21})+0.21\cos(0.3t)\end{bmatrix}.$$

Now, we design adaptive neural control for an uncertain MIMO nonlinear system (9.90). The control objective is to design robust control for system (9.90) such that the system output $y=x_1$ follows the desired trajectory x_{1d}, where the desired trajectories are taken as $x_{11d}=0.5[\sin(t)+0.5\sin(0.5t)]$ and $x_{12d}=0.5\sin(1.5t)+\sin(0.5t)$.

The robust control is designed as follows:

$$\alpha_1=(G_1(x_1)+\gamma_1 I_{m\times m})^{-1}\left(-K_1z_1+\hat{\rho}_1(Z_1)\right) \tag{9.91}$$

$$u_i=\begin{cases}-(h(Z)+\sigma^T\Gamma\sigma+\|\sigma^T\|\|\hat{W}_n^TS(Z_n)\|)\frac{\theta_i}{\|\theta\|^2}+u_{ri0} & \text{if } \theta_i<0\\ 0 & \text{if } \theta_i=0\\ -(h(Z)+\sigma^T\Gamma\sigma+\|\sigma^T\|\|\hat{W}_n^TS(Z_n)\|)\frac{\theta_i}{\|\theta\|^2}+u_{li0} & \text{if } \theta_i>0\end{cases} \tag{9.92}$$

where $K_1=K_1^T>0$, $\sigma=\vartheta_1z_1+z_2$, $\theta=G_n^T(\bar{x}_n)\sigma$, $h(Z)=|\gamma_1||z_1^T\alpha_1|+\xi_{11}\|z_1\|\|x_2\|+\|\sigma\|\|G_1^T(\bar{x}_1)z_1\|$ and $i=1,2$. $\hat{\rho}_1(Z_1)$ is the approximation of F_1, and $\hat{W}_2^TS(Z_2)$ is the approximation of F_2.

The design parameters of the control are chosen as $K_1 = \mathrm{diag}\{2.5, 1.0\}$, $\vartheta_1 = 15$ and $\Gamma = \mathrm{diag}\{250, 250\}$. The dead-zone values are taken as $u_{r10} = 0.2$, $u_{l10} = -0.3$, $u_{r20} = 0.25$, $= u_{l20} = -0.4$, the dead-zone values are chosen as $u_{r1\max} = 7.0$, $u_{r2\max} = 7.5$, $u_{l1\max} = -6.0$, $u_{l2\max} = -8.5$, and $k_{r11} = k_{l11} = k_{r21} = k_{l21} = 1$.

The simulation results for the tracking output are shown in Figures 9.2 and 9.3 under initial states $x_{11} = 0.01$ and $x_{12} = -0.01$. It can be observed that the output x_{11} and x_{12} follows the desired trajectory x_{11d} and x_{12d} despite the unknown system dynamics, perturbation to the control coefficient matrices and input nonlinearities. The tracking errors for e_{11} and e_{12} are shown in Figures 9.4 and 9.5 respectively. From Figures 9.6 and 9.7, the control inputs are saturated in the transient phase and chattering is observed as a result of the variable structure control. These simulation results show that good tracking performance can be obtained under the adaptive neural control.

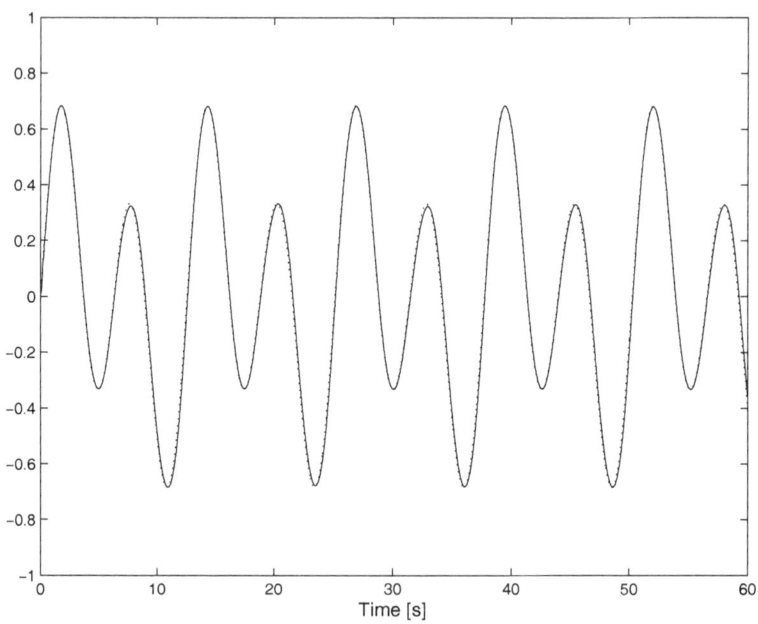

Fig. 9.2 Output x_{11}(solid line) follows desired trajectory x_{11d} (dashed line)

9.5 Conclusions

Adaptive neural control has been proposed for uncertain MIMO nonlinear systems with saturation and dead-zone in this chapter. Considering non-symmetric saturation and dead-zone nonlinearities of actuator, the VSC control in combination with

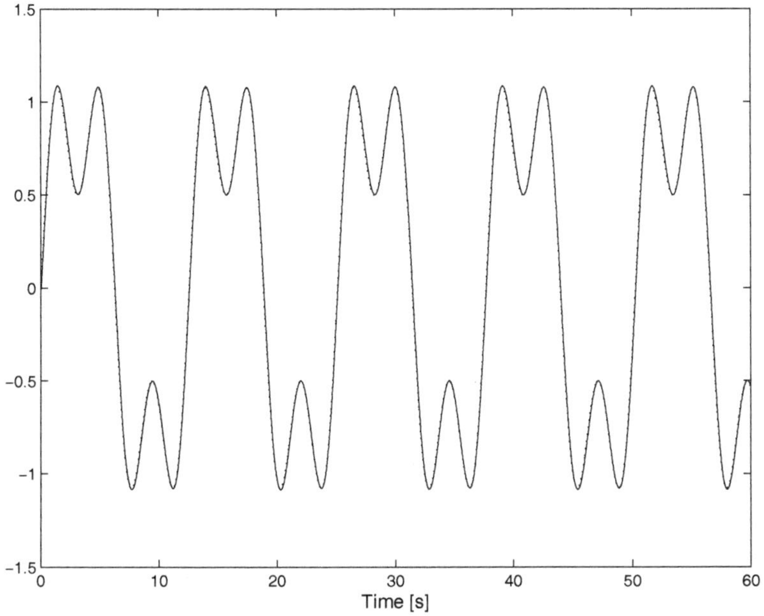

Fig. 9.3 Output x_{12}(solid line) follows desired trajectory x_{12d} (dashed line)

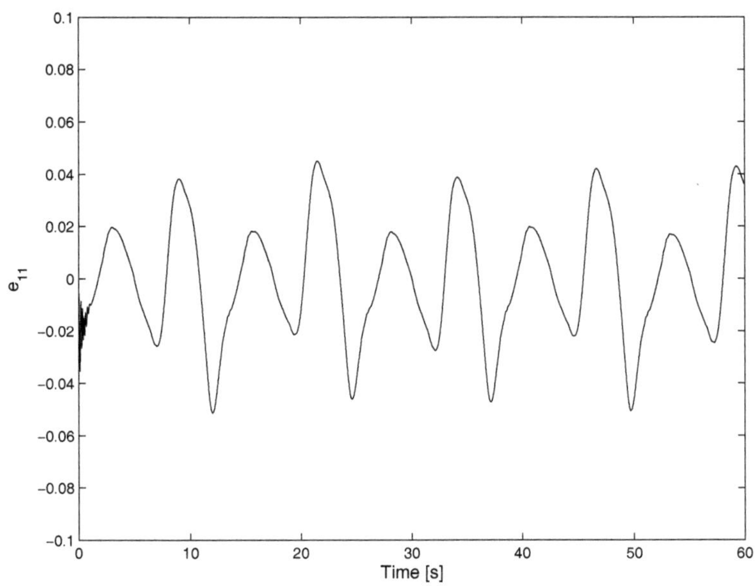

Fig. 9.4 Tracking error of e_{11}

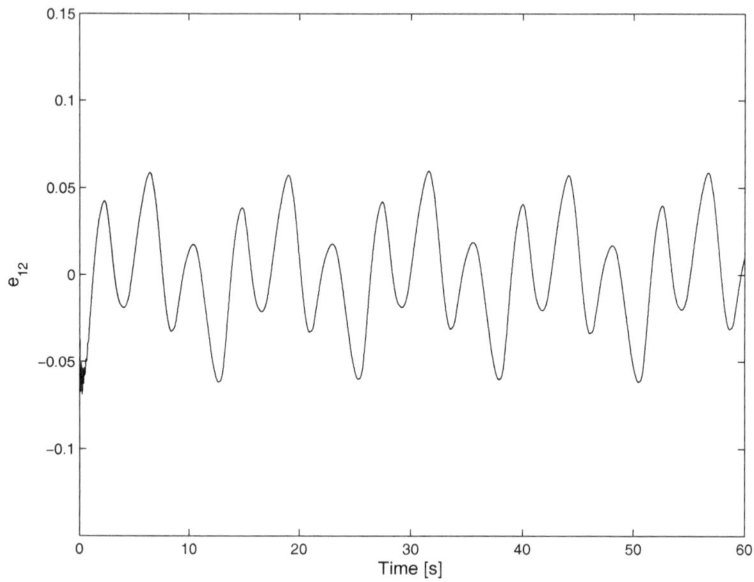

Fig. 9.5 Tracking error of e_{12}

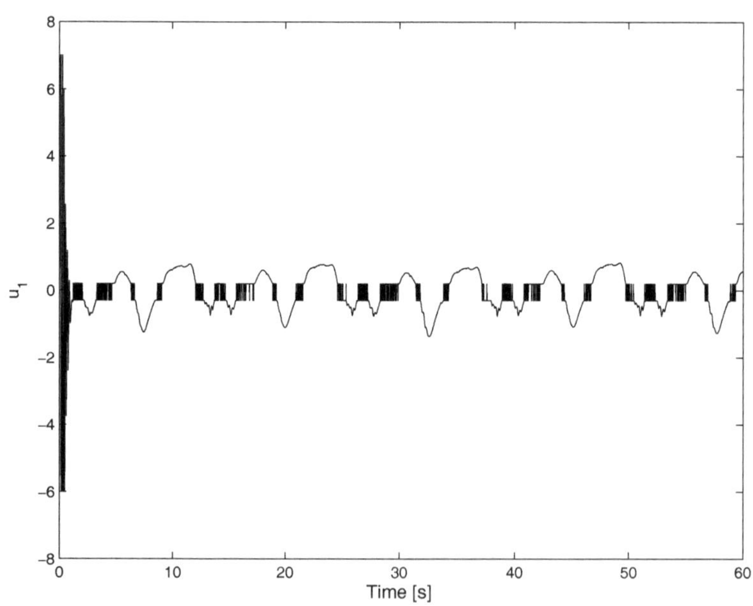

Fig. 9.6 Control signal u_1

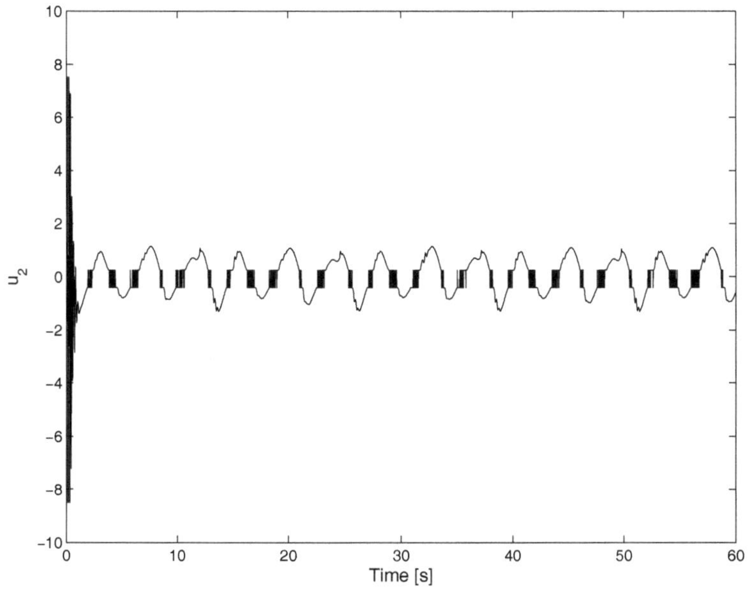

Fig. 9.7 Control signal u_2

backstepping and Lyapunov synthesis has been investigated. The cascade property of the studied systems has been fully utilized in developing the control structure and neural weight learning laws. It is proved that the proposed adaptive neural control is able to guarantee semi-global uniform ultimate boundedness of all signals in the closed-loop system. Finally, simulation studies are presented to illustrate the effectiveness of the proposed control.

Appendix 1

Proof of Lemma 9.3:

Proof. Let $\|A\| = \bar{\nu}$, and define $B = [1/(\bar{\nu}+\bar{\varepsilon})]A$ with $\bar{\nu} > 0$ and $\bar{\varepsilon} > 0$. We obtain

$$\|B\| \leq [1/(\bar{\nu}+\bar{\varepsilon})]\|A\| = \bar{\nu}/(\bar{\nu}+\bar{\varepsilon}) < 1.$$

Thus, $B^n \to 0$ as $n \to \infty$ for $\|B\| < 1$. We can also know that the eigenvalues μ_i of the matrix B satisfy $|\mu_i| < 1,\ i = 1,2,\ldots,m$. Moreover, there is the following fact

$$\mu_i = [1/(\bar{\nu}+\bar{\varepsilon})]\lambda_i.$$

Therefore, we have

$$|\lambda_i| < \bar{v} + \bar{\varepsilon}.$$

Since $\bar{\varepsilon}$ can be chosen arbitrarily small, it follows that $|\lambda_i| \leq \bar{v}$. This concludes the proof. □

Appendix 2

Proof of Lemma 9.4:

Proof. For any matrix B, there exists a nonsingular matrix $T \in R^{m\times m}$ such that

$$T^{-1}BT = J = \text{diag}\{J_1, J_2, \ldots, J_t\} \tag{9.93}$$

where $J_i = \begin{bmatrix} \lambda_i & 1 & 0 & \ldots & 0 \\ 0 & \lambda_i & 1 & \ddots & \vdots \\ \vdots & \ddots & \ddots & \ddots & \vdots \\ \vdots & \ddots & \ddots & \ddots & 1 \\ 0 & \ldots & \ldots & 0 & \lambda_i \end{bmatrix} \in R^{m_i\times m_i}$ is a Jordan block, λ_i is a eigenvalue of matrix B, $i = 1, 2, \ldots, t$, $m_1 + m_2 + \ldots + m_t = n$ and the matrix J is called a Jordan canonical form.

Considering (9.93), we obtain

$$B = T\text{diag}\{J_1, J_2, \ldots, J_t\}T^{-1}. \tag{9.94}$$

Hence, we have

$$B + (\rho(B) + \Delta)I = T(\text{diag}\{J_1, J_2, \ldots, J_t\} + (\rho(B) + \Delta)I)T^{-1}. \tag{9.95}$$

Invoking the definition of matrix spectral radius and $\Delta > 0$, we can know that each diagonally element of matrix $B + (\rho(B) + \Delta)I$ is larger than zero. Thus, this concludes the proof. □

References

1. Zhang T, Ge SS, Hang CC (2000) Adaptive neural network control for strict-feedback nonlinear systems using backstepping design. Automatica 36:1835-1846.
2. Gong JQ, Yao B (2001) Neural network adaptive robust control of nonlinear systems in semi-strict feedback form. Automatica 37:1149-1160.
3. Wang D, Huang J (2005) Neural network-based adaptive dynamic surface control for a class of uncertain nonlinear systems in strict-feedback form. IEEE Trans Neural Netw 16:195-202.
4. Ge SS, Li GY, Zhang J *et al.* (2004) Direct adaptive control for a class of MIMO nonlinear systems using neural networks. IEEE Trans Automat Control 49:2001-2006.

5. Zhang J, Ge SS, Lee TH (2005) Output feedback control of a class of discrete MIMO nonlinear systems with triangular form inputs. IEEE Trans Neural Netw 16:1491-1503.
6. Ge SS, Zhang J, Lee TH (2004) Adaptive neural network control for a class of MIMO nonlinear systems with disturbances in discrete-time. IEEE Trans Sys Man Cybern B 34:1630-1644.
7. Ge SS, Tee KP (2007) Approximation-based control of nonlinear MIMO time-delay systems. Automatica 43:31-43.
8. Ge SS, Wang C (2004) Adaptive neural control of uncertain MIMO nonlinear systems. IEEE Trans Neural Netw 15:674-692.
9. Ge SS (1998) Advanced control techniques of robotic manipulators. Proc of the American Control Conf, June 1998:2185-2199.
10. Chang YC (2001) An adaptive H_∞ tracking control for a class of nonlinear multiple-input-multiple-output (MIMO) systems. IEEE Trans Automat Contr 46:1432-1437.
11. Lee CY, Lee JJ (2004) Adaptive control for uncertain nonlinear systems based on multiple neural networks. IEEE Trans Sys Man Cybern B 34:325-333.
12. Chang YC, Yen HM (2005) Adaptive output feedback tracking control for a class of uncertain nonlinear systems using neural networks. IEEE Trans Sys Man Cybern B 35:1311-1316.
13. Kwan C, Lewis FL (2000) Robust backstepping control of nonlinear systems using neural networks. IEEE Trans Sys Man Cybern A 30:753-766.
14. Chang YC (2000) Robust tracking control for nonlinear MIMO systems via fuzzy approaches. Automatica 36:1535-1545.
15. Rovithakis GA (1999) Tracking control of multi-input affine nonlinear dynamical systems with unknown nonlinearities using dynamical neural networks. IEEE Trans Sys Man Cybern B 29:179-189.
16. Cao YY, Lin ZL (2003) Robust stability analysis and fuzzy-scheduling control for nonlinear systems subject to actuator saturation. IEEE Trans Fuzzy Syst 11:57-67.
17. Gao WZ, Selmic RR (2006) Neural network control of a class of nonlinear systems with actuator saturation. IEEE Trans Neural Netw 17:1539-1547.
18. Zhong YS (2005) Globally stable adaptive system design for minimum phase SISO plants with input saturation. Automatica 41:674-692.
19. Tao G, Kokotovic PV (1994) Adaptive sliding control of plants with unknown dead-zone. IEEE Trans Automat Contr 39:59-68.
20. Tao G, Kokotovic PV (1995) Discrete-time adaptive control of plants with unknown output dead-zones. Automatica 31:287-291.
21. Zhou J, Wen C, Zhang Y (2006) Adaptive output control of nonlinear systems with uncertain dead-zone nonlinearity. IEEE Trans Automat Contr 51:504-510.
22. Zhang TP, Ge SS (2007) Adaptive neural control of MIMO nonlinear state time-varying delay systems with unknown dead-zones and gain signs. Automatica. 43:1021-1033
23. Zhang TP, Ge SS (2008) Adaptive dynamic surface control of nonlinear systems with unknown dead zone in pure feedback form. Automatica 44:1895-1903.
24. Hu QL, Ma GF, Xie LH (2008) Robust and adaptive variable structure output feedback control of uncertain systems with input nonlinearity. Automatica. 44:552-559
25. Tee KP, Ge SS (2006) Control of fully actuated ocean surface vessels using a class of feedforward approximators. IEEE Trans Contr Syst Tech 14:750-7560.
26. Ge SS, Wang C (2002) Direct adaptive NN control of a class of nonlinear systems. IEEE Trans Neural Netw 13:214-221.
27. Kim YH, Ha IJ (2000) Asymptotic state tracking in a class of nonlinear systems via learning-based inversion. IEEE Trans Automat Contr 45:2011-2017.
28. Usmani RA (1987) Applied linear algebra. Marcel Dekker, New York.
29. Dawson D M, Carroll JJ, Schneider M (1994) Integrator backstepping control of a brush DC motor turning a robotic load. IEEE Trans Contr Syst Tech 2:233-244.
30. Sabanovic A, Sabanovic N, Ohnishi K (1993) Sliding modes in power converters and motion control systems. Int J Contr 57:1237-1259.
31. Diao YX, Passino KM (2001) Stable fault-tolerant adaptive fuzzy/neural control for a turbine engine. IEEE Trans Contr Syst Tech 9:494-509.

32. Tee KP, Ge SS, Tay FEH (2008) Adaptive neural network control for helicopters in vertical flight. IEEE Trans Contr Syst Tech 16:753-762.
33. Ge SS, Hang CC, Lee TH *et al.* (2001) Stable adaptive neural network control. Kluwer Academic, Norwell, USA.
34. Wang XS, Hong H, Su CY (2003) Model reference adaptive control of continous time systems with an unknown dead-zone. IEE proc Contr Theo & Appl 150:261-266.
35. Shyu KK, Liu WJ, Hsu KC (2005) Design of large-scale time-delayed systems with dead-zone input via variable structure control. Automatica 41:1239-1246.

Part III
Fuzzy Neural Control

Chapter 10
An Online Self-constructing Fuzzy Neural Network with Restrictive Growth

Ning Wang, Meng Joo Er and Xianyao Meng

Abstract In this chapter, a novel paradigm, termed online self-constructing fuzzy neural network with restrictive growth (OSFNNRG), which incorporates a pruning strategy into some restrictive growing criteria, is proposed. The proposed approach not only speeds up the online learning process but also results in a more parsimonious fuzzy neural network while maintaining comparable performance and accuracy by virtue of the growing and pruning mechanism. The OSFNNRG starts with no hidden neurons and parsimoniously generates new hidden units according to the new growing criteria as learning proceeds. In the second learning phase, all the free parameters of hidden units, regardless of whether they are new or originally existing, are updated by the extended Kalman filter (EKF) method. The performance of the OSFNNRG algorithm is compared with other popular algorithms in the areas of function approximation, nonlinear dynamic system identification and chaotic time series prediction, etc. Simulation results demonstrate that the proposed OSFNNRG algorithm is faster in learning speed and the network structure is more compact while maintaining comparable generalization performance and accuracy.

10.1 Introduction

Fuzzy systems and neural networks play an important role in the fields of artificial intelligence and machine learning. As an effective technique to express imprecision, uncertainty, and knowledge of domain experts, fuzzy logic is used to deal with problems that classical approaches find difficult or impossible to tackle since

Ning Wang
Dalian Maritime University, Dalian 116026, P.R. China, e-mail: n.wang.dmu.cn@gmail.com

Meng Joo Er
Nanyang Technological University, 639798, Singapore, e-mail: emjer@ntu.edu.sg

Xianyao Meng
Dalian Maritime University, Dalian 116026, P.R. China, e-mail: n.wang.dmu.cn@gmail.com

a fuzzy system can approximate any continuous function on a compact set to any accuracy [20]. However, the main challenge on how to extract a suitable collection of fuzzy rules from the available data set and expert knowledge is still an open problem. To cope with this difficulty, many researchers have been working tirelessly to find various automatic methods of designing a fuzzy system by combining with neural networks [10], since the learning algorithms of neural networks can be utilized to identify the structure and parameters of a fuzzy system which in turn enhance the interpretation of the resulting fuzzy neural network. Innovative merger of a fuzzy system and neural networks has resulted in a fruitful area where there are many effective and typical approaches [7, 8, 12, 15–17, 21, 22] besides the well known ANFIS [9]. A significant contribution was made by Platt [17] through the development of an algorithm named resource-allocating network (RAN) that adds hidden units to the network based on the novelty of new data in the sequential learning process. An improved approach called RAN via the extended Kalman filter (RANEKF) [12] was provided to enhance the performance of the RAN by adopting an EKF instead of the least-mean-square (LMS) method for adjusting network parameters. They all start with no hidden neurons and allocate the new hidden units when some criteria are satisfied. However, once the hidden neuron is generated, it will never be removed anymore. To circumvent this drawback, the minimal resource allocation network (MRAN) [15, 16] was developed by using the pruning method in which inactive hidden neurons can be detected and removed during the learning process. Hence, a compact network can be implemented. Other improvements of the RAN developed in [18] and [19] take into consideration the pseudo-Gaussian (PG) function and orthogonal techniques including QR factorization (also called QR decomposition) and singular value decomposition (SVD), and have been applied to the problem of time series analysis. Recently, a growing and pruning RBF (GAP-RBF) approach [7] has been proposed by simplifying the Gaussian function for the significance of each hidden neuron and directly linking the required learning accuracy to the significance in the sequential learning process. The generalized GAP-RBF (GGAP-RBF) algorithm of [8] can be used for any arbitrary sampling density for training samples. It is well known that we should obtain a good range and/or sampling distribution of the input space in advance. An online sequential extreme learning machine (OS-ELM) [14] has been proposed for learning data one-by-one and/or chunk-by-chunk with fixed or varying chunk size. However, the parameters of hidden nodes are randomly selected and only the output weights are analytically updated based on the sequentially arriving data. It should be noted that the number of hidden neurons in the OS-ELM is fixed in advance by users. On the contrary, a major development was made by Chen *et al.* [2] who proposed an orthogonal least square (OLS) learning algorithm in which both structure and parameter identification are made. Similar to [17], the OLS approach also produces a network smaller than the randomly selected one. In [3], a hierarchically self-organizing approach, whereby the structure was identified by input-output pairs, was developed. An online self-constructing paradigm was proposed in [11], which is inherently a modified Takagi-Sugeno (T-S) fuzzy rule-based model possessing the learning ability of neural networks. The structure of the premise as well as the consequent part are determined by clustering the input via an online self-organizing approach and by assigning a singleton to each rule initially and adding some additional significant terms (forming a linear equation of input variables) incrementally in the learning process, respectively. For parameter identification, the consequent parameters were tuned by the LMS while the precondition parameters were updated by the backpropagation (BP) algorithm which is known to be slow and easy to be entrapped into local minima. Lately, a hierarchical online self-organizing learning algorithm for the dynamic fuzzy neural network (DFNN) based on RBF neural network has been developed in [21], in which not only the parameters can be adjusted by the linear least square (LLS) but also the structure can be self-adaptive via growing and pruning criteria. To be precise, a generalized DFNN (GDFNN) based on the ellipsoidal basis function (EBF) has been developed in [22], in which a novel online parameter allocation mechanism based on ε-completeness was developed to allocate the initial widths for each dimension of the input variables. It should be noted that the improved performance of the GDFNN is at the cost of lowering learning speed. The resulting approaches termed DFNN and GDFNN have been applied to a large area of engineering problems [4]- [6]. Similar to the GDFNN, a self-organizing fuzzy neural network (SOFNN) [13] with a pruning strategy using the optimal brain surgeon (OBS) approach has been proposed to extract fuzzy rules online. Unfortunately, like other online learning algorithms, it also encounters the problem that the learning speed is slower down due to complicated growing and pruning criteria and it makes the network topology difficult to understand although the required learning accuracy can be obtained. In this chapter, we propose an OSFNNRG that attempts to speed up the sequential learning process by incorporating the pruning strategy into the growing criteria leading to a novel growth mechanism. The OSFNNRG starts with no hidden neurons and restrictively generates neurons based on the proposed growing criteria so that a more parsimonious structure of the fuzzy neural network can be obtained. The EKF is adapted to tune all the free parameters in the parameter identification phase for sequentially arriving training data pairs. The high performance of the OSFNNRG learning algorithm is demonstrated on some benchmark problems in the areas of function approximation, nonlinear dynamic system identification and chaotic time series prediction. Comprehensive comparisons with other well-known learning algorithms have been conducted. Simulation results show that the proposed OSFNNRG algorithm can provide faster learning speed and more compact network structure with comparable generalization performance and accuracy.

The chapter is organized as follows. Section 10.2 briefly describes the architecture of the OSFNNRG, and the learning algorithm is presented in Section 10.3 in detail. Section 10.4 shows simulation studies of the OSFNNRG algorithm, including quantitative and qualitative performance comparisons with other popular algorithms. Section 10.5 concludes the chapter.

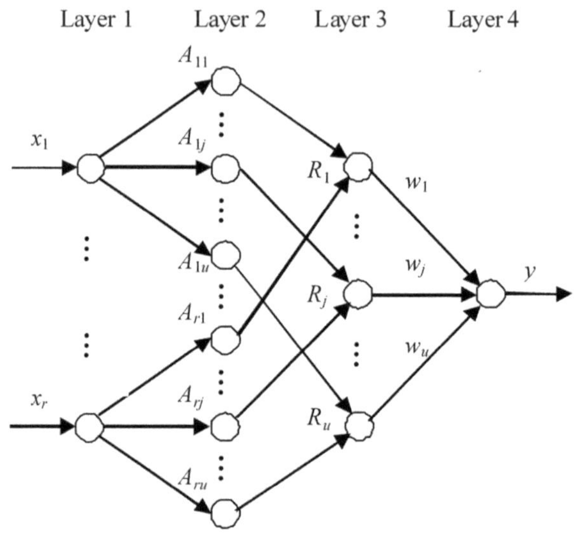

Fig. 10.1 Architecture of the OSFNNRG

10.2 Architecture of the OSFNNRG

In this section, the structure of the OSFNNRG, shown in Figure 10.1, will be introduced. The four-layered network similar to [21, 22] realizes a Sugeno fuzzy inference system, which could be described as follows:

$$\text{Rule } j: \text{ IF } x_1 \text{ is } A_{1j} \text{ and ... and } x_r \text{ is } A_{rj} \text{ THEN } y \text{ is } w_j, \; j = 1, 2, ..., u\,. \tag{10.1}$$

Let r be the number of input variables, and each variable x_i, $i = 1, 2, ..., r$ in layer 1 has u fuzzy subsets A_{ij}, $j = 1, 2, ..., u$ shown in layer 2 of which the corresponding membership functions are defined in the form of Gaussian functions given by

$$\mu_{ij}(x_i) = \exp\left(-\frac{(x_i - c_{ij})^2}{\sigma_{ij}^2}\right), \; i = 1, 2, ..., r, \; j = 1, 2, ..., u\,, \tag{10.2}$$

where μ_{ij} is the jth membership function of x_i, c_{ij} and σ_{ij} are the center and width of the jth Gaussian membership function of x_i, respectively. Each node in layer 3 represents a possible IF-part of fuzzy rules. If multiplication is selected as T-norm to calculate each rule's firing strength, the output of the jth rule R_j $(j = 1, 2, ..., u)$ could be computed by

$$\varphi_j(x_1, x_2, ..., x_r) = \exp\left(-\sum_{i=1}^{r} \frac{(x_i - c_{ij})^2}{\sigma_{ij}^2}\right), \; j = 1, 2, ..., u\,. \tag{10.3}$$

As the output layer, layer 4 has single output node for multi-input and single-output (MISO) systems. However, the results could be readily applied to multi-input and multi-output (MIMO) systems since a special MIMO system could be decomposed into several MISO systems. The output is the weighted summation of incoming signals given by

$$y(x_1, x_2, ..., x_r) = \sum_{j=1}^{u} w_j \varphi_j \,, \tag{10.4}$$

where w_j is the consequent parameter in the THEN-part of the jth rule.

10.3 Learning Algorithm of the OSFNNRG

In this section, the main ideas behind the OSFNNRG will be presented. For each observation (X^k, t^k), $k = 1, 2, ..., n$, where n is the number of total training data pairs, $X^k \in R^r$ and $t^k \in R$ are the kth input vector and desired output, respectively. The overall OSFNNRG output y^k of the existing structure could be obtained by (10.1)-(10.4). In the learning process, suppose that the fuzzy neural network has generated u hidden neurons in layer 3 for the kth observation. The proposed online learning algorithm in this chapter also includes the structure learning and parameter learning. However, this novel algorithm leads to a more compact and faster fuzzy neural network. For structure identification, only the growing criteria is considered in that it will be effective and fast if the pruning criteria could be incorporated into the process of rule generation. Hence, a more parsimonious fuzzy neural network can be implemented by the resulting growing criteria. In the parameter learning, the popular EKF is used to adjust the parameters of the premise and the consequent parts of fuzzy rules simultaneously.

10.3.1 Criteria of Rule Generation

10.3.1.1 System Errors

Usually, the rule generation critically depends on the output error of a system with regard to the reference signal [17]. When the kth observation coming, calculate the system error as follows:

$$\|e^k\| = \|t^k - y^k\|, \; k = 1, 2, ..., n \,. \tag{10.5}$$

If

$$\|e^k\| > k_e, \; k_e = \max\{e_{\max}\beta^{k-1}, \; e_{\min}\} \,, \tag{10.6}$$

the performance of the fuzzy neural network is poor and a new rule should be recruited if other criteria have been satisfied. Otherwise, no new rule will be generated. Here, k_e is a predefined threshold that decays during the learning process, where $e_{\max}$ is the maximum error chosen, $e_{\min}$ is the desired accuracy and $\beta \in (0,1)$ is the convergence constant.

10.3.1.2 Input Partitioning

To some extent, the structure learning of fuzzy neural networks means the effective and economical partitioning of the input space. The performance and structure of the resulting fuzzy neural network is therefore strongly correlative to the location and the size of the input membership function. Many methods have been presented in the literatures [1, 2, 7, 8, 12, 13, 15–17, 21, 22], of which there is a minimum distance [12, 15–17] between the new observation and the existing membership functions, ε-completeness [13, 21, 22] and various concepts of significance [1, 2, 7, 8], etc. For the sake of computational simplicity the distance criterion is considered in this chapter. For the kth observation $X^k = [x_{1k}, x_{2k}, ..., x_{rk}]^{\mathrm{T}}$, the distances from the existing cluster centers could be described as follows:

$$d_{kj} = \|X^k - C_j\|, \ k = 1, 2, ..., n, \ j = 1, 2, ..., u \,, \tag{10.7}$$

where $C_j = [c_{1j}, c_{2j}, ..., c_{rj}]^{\mathrm{T}}$ is the center of the jth cluster. The minimum distance between the kth observation and the nearest center could also be obtained by

$$d_{k,\min} = \min_j d_{kj}, \ j = 1, 2, ..., u \,. \tag{10.8}$$

If

$$d_{k,\min} > k_d, \ k_d = \max\{d_{\max}\gamma^{k-1}, \ d_{\min}\} \,, \tag{10.9}$$

the existing input membership functions cannot partition the input space well. Hence, it needs a new cluster or the centers and widths of the existing membership functions need adjustments. Where $d_{\max}$ and $d_{\min}$ are the maximum and minimum distances chosen, respectively, and $\gamma \in (0,1)$ is the decay constant.

10.3.1.3 Generalization Capability

Essentially, the error reduction ratio (ERR) method [2] is a special kind of the OLS. Therefore, the ERR method is usually used to calculate the sensitivity and the significance of input variables and fuzzy rules in order to check which variable or rule would be deleted. In this chapter, the ERR method is introduced to be a growing criterion rather than pruning since pruning after growing will consequentially increase the computational burden while a synthetic growth without pruning would results in a faster and more compact training process which is necessary for an online self-constructing fuzzy neural network.

Given n input-output data pairs, (X^k, t^k), $k = 1,2,...,n$, consider (10.7) as a special case of the linear regression model,

$$t^k = \sum_{j=1}^{u} w_j \varphi_{jk}(X^k) + e^k . \tag{10.10}$$

The above described model could be rewritten in the following compact form:

$$T = \Phi W + E , \tag{10.11}$$

where $T = [t^1, t^2, ..., t^n]^{\mathrm{T}} \in R^n$ is the desired output vector, $W = [w_1, w_2, ..., w_u]^{\mathrm{T}} \in R^u$ is the weight vector, $E = [e^1, e^2, ..., e^n]^{\mathrm{T}} \in R^n$ is the error vector which is assumed to be uncorrelated with the regressors, and $\Phi = [\psi_1, \psi_2, ..., \psi_u] \in R^{n\times u}$ is the output matrix of layer 3 given by

$$\Phi = \begin{bmatrix} \varphi_{11} & \cdots & \varphi_{u1} \\ \vdots & \ddots & \vdots \\ \varphi_{1n} & \cdots & \varphi_{un} \end{bmatrix} . \tag{10.12}$$

For matrix Φ, if its row number is larger than the column number, we can transform it into a set of orthogonal basis vectors by QR decomposition,

$$\Phi = PQ , \tag{10.13}$$

where $P = [p_1, p_2, ..., p_u] \in R^{n\times u}$ has the same dimension as Φ with orthogonal columns and $Q \in R^{u\times u}$ is an upper triangular matrix. The orthogonality makes it feasible to compute the individual contribution of each rule to the desired output energy from each vector. Substituting (10.13) into (10.11) yields

$$T = PQW + E = PG + E , \tag{10.14}$$

where $G = [g_1, g_2, ..., g_u]^{\mathrm{T}} = (P^{\mathrm{T}}P)^{-1}P^{\mathrm{T}}T \in R^u$ could be obtained by the LLS method, and equivalently,

$$g_i = \frac{p_i^{\mathrm{T}} T}{p_i^{\mathrm{T}} p_i}, \; i = 1,2,...,u . \tag{10.15}$$

An ERR due to p_i as defined in [2], is given by

$$err_i = \frac{g_i^2 p_i^{\mathrm{T}} p_i}{T^{\mathrm{T}} T}, \; i = 1,2,...,u . \tag{10.16}$$

Substituting (10.15) into (10.16) yields

$$err_i = \frac{(p_i^{\mathrm{T}} T)^2}{p_i^{\mathrm{T}} p_i T^{\mathrm{T}} T}, \; i = 1,2,...,u . \tag{10.17}$$

Originally, the ERR offers a simple and effective manner to seek a subset of significant regressors. Alternatively, the ERR is used to define a restrained growing criterion named generalization factor (GF) for checking the generalization capability of the OSFNNRG, furthermore simplifying and accelerating the learning process. We define

$$GF = \sum_{i=1}^{u} err_i . \tag{10.18}$$

If $GF < k_{GF}$, the generalization capability is poor and therefore, the fuzzy neural network either needs more hidden neurons or adjusts the free parameters to achieve high generalization performance. Otherwise, no hidden nodes will be created, where k_{GF} is the threshold chosen.

10.3.2 Parameter Adjustment

Note that the following parameter learning phase is performed on the entire network after structure learning, regardless of whether the nodes are newly generated or are existent originally. In addition, for the sequential learning, the methods of LLS [2, 11, 21, 22], the Kalman filter (KF) [13] and the EKF algorithm [12, 14–16] have been popularly applied to the parameter identification in the field of self-constructing fuzzy neural networks. Among them, the LLS method could achieve fast computation and obtain the optimal resolution in one computing step. However, this method is sensitive to signal to noise ratio and it is difficult to identify parameters when the resolution matrix is ill-conditioned. The KF and EKF method would not be so fast as the LLS method since the adaptation is much more complex. Nevertheless, they could implement a robust online learning algorithm and not be sensitive to noise. When no neuron is added, the network parameter vector $W_{EKF} = [w_1, C_1^{\mathrm{T}}, \sigma_1, \ldots, w_u, C_u^{\mathrm{T}}, \sigma_u]$ is adapted using the following EKF algorithm [12],

$$W_{EKF}(k) = W_{EKF}(k-1) + e^k \kappa_k , \tag{10.19}$$

where κ_k is the Kalman gain vector given by

$$\kappa_k = [R_k + a_k^{\mathrm{T}} P_{k-1} a_k]^{-1} P_{k-1} a_k , \tag{10.20}$$

where a_k is the gradient vector and has the following form:

$$\begin{aligned} a_k = \Big[& \varphi_1(X^k), \varphi_1(X^k)\frac{2w_1}{\sigma_1^2}(X^k - C_1)^{\mathrm{T}}, \varphi_1(X^k)\frac{2w_1}{\sigma_1^3}\|X^k - C_1\|^2, \cdots, \\ & \varphi_u(X^k), \varphi_u(X^k)\frac{2w_u}{\sigma_u^2}(X^k - C_u)^{\mathrm{T}}, \varphi_u(X^k)\frac{2w_u}{\sigma_u^3}\|X^k - C_u\|^2\Big]^{\mathrm{T}} , \end{aligned} \tag{10.21}$$

where R_k is the variance of the measurement noise, and P_k is the error covariance matrix which is updated by

$$P_k = [I - \kappa_k a_k^{\mathrm{T}}]P_{k-1} + Q_0 I \,, \tag{10.22}$$

where Q_0 is a scalar which determines the allowed random step in the direction of the gradient vector and I is the identity matrix. When a new hidden neuron is allocated, the dimensionality of the P_k increases to

$$P_k = \begin{bmatrix} P_{k-1} & 0 \\ 0 & P_0 I \end{bmatrix} \,, \tag{10.23}$$

where P_0 is an estimate of the uncertainty in the initial values assigned to the parameters. The dimension of the identify matrix I is equal to the number of new parameters introduced by the new hidden unit.

10.3.3 Complete OSFNNRG Algorithm

In this subsection, the proposed OSFNNRG learning algorithm is summarized:

10.3.3.1 Initialization

When the first observation (X^1, t^1) comes, some initializations should be conducted since there is no hidden neuron at the beginning of the learning process.

$$\begin{aligned} C_0 &= C_1 = X^1 \,, \\ \sigma_0 &= \sigma_1 = d_{\max} \,, \\ w_0 &= w_1 = t^1 \,. \end{aligned} \tag{10.24}$$

10.3.3.2 Growth Procedure

For the kth observation (X^k, t^k), suppose that there exist u hidden neurons in layer 3, calculate e^k, $d_{k,\min}$ and GF according to (10.6), (10.8) and (10.18), respectively.

If

$$\|e^k\| > k_e, \; d_{k,\min} > k_d \text{ and } GF < k_{GF} \,, \tag{10.25}$$

then a new hidden neuron should be generated.

Here the decaying parameters k_e and k_d have been given in (10.6) and (10.9), respectively. The center, width and weight for the newly created unit are set as follows:

$$C_{u+1} = X^k \,,$$
$$\sigma(u+1) = k_0 d_{k,\min} \,,$$
$$w_{u+1} = e^k \,. \tag{10.26}$$

where k_0 is a predefined parameter which determines the overlapping degree. Then, adjust the dimension of $W_{EKF}(k)$ and P_{k-1}, respectively.

10.3.3.3 Parameter Adjustment

When the above process of generation is finished, no matter whether there are newly generated units or not, the parameters of all the existing neurons will be updated by the EKF algorithm which is described by (10.19)-(10.23). All the involving parameter vectors are updated, including κ_k, a_k, $W_{EKF}(k)$ and P_k according to (10.19)-(10.22), respectively and sequentially.

10.4 Simulation Studies

In this section, the effectiveness and superiority of the proposed algorithm is demonstrated in function approximation, nonlinear dynamic system identification and Mackey-Glass time series prediction. Some comparisons are made with other significant works such as RAN [17], RANEKF [12], MRAN [15, 16], ANFIS [9], OLS [2], RBF-AFS [3], DFNN [21], GDFNN [22], GGAP-RBF [8], OS-ELM [14] and SOFNN [3], and so on. All the simulations are carried out in the unified environment running in Matlab 7.1, with a Pentium 4, 3.0 GHz CPU.

Example 10.1. (Hermite Function Approximation) The underlying function which needs to be approximated in the first experiment is the Hermite polynomial given by

$$f(x) \;=\; 1.1\,(1-x+2x^2)\exp(-0.5x^2)\,. \tag{10.27}$$

A random sampling of the interval $[-4,4]$ is used in obtaining 200 input-output data pairs for the training set. And the parameters used in this experimental example

Table 10.1 Comparisons of the OSFNNRG with other algorithms (Example 10.1)

Algorithms	Rule number	RMSE	Training time (sec.)
RANEKF	13	0.0262	0.48
MRAN	7	0.0438	0.39
DFNN	6	0.0054	0.87
GDFNN	6	0.0097	0.93
OSFNNRG	5	0.0089	0.26

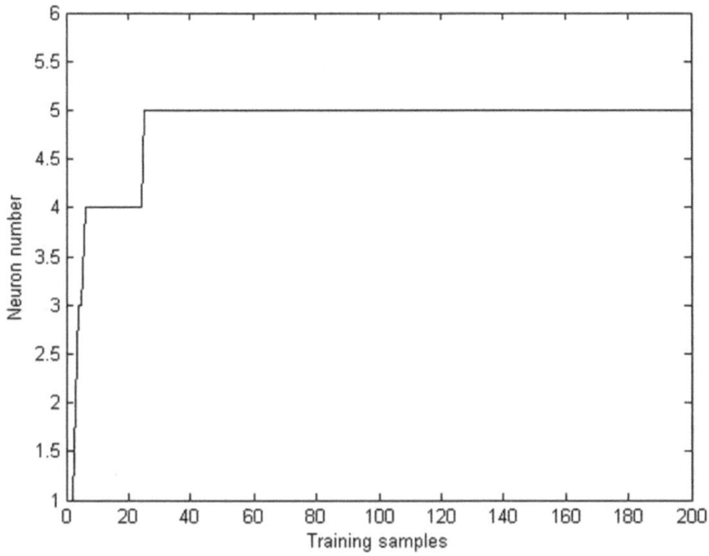

Fig. 10.2 Growth of neurons

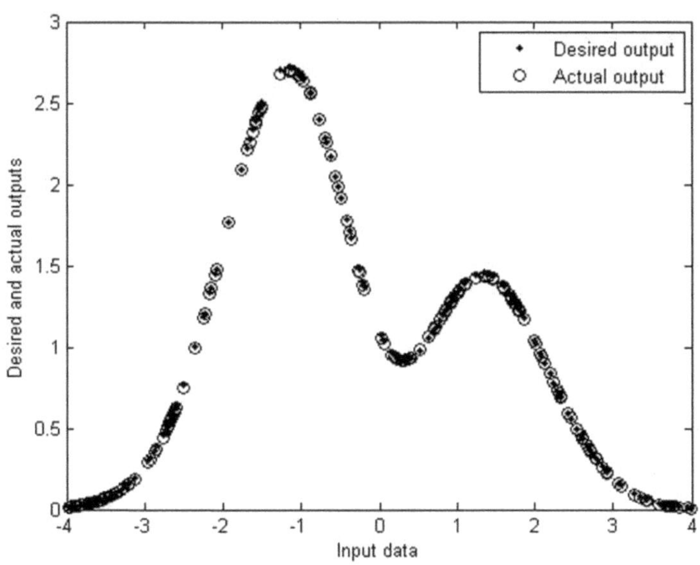

Fig. 10.3 The Hermite function and the SOFNNRG approximation

are as follows: $d_{max} = 1.0$, $d_{min} = 0.1$, $e_{max} = 0.5$, $e_{min} = 0.02$, $k_0 = 0.5$, $\beta = 0.97$, $\gamma = 0.97$, $GF = 0.99$, $P_0 = 1.0$, $Q_0 = 0.5$ and $R_k = 1.0$. The growth of rule is shown

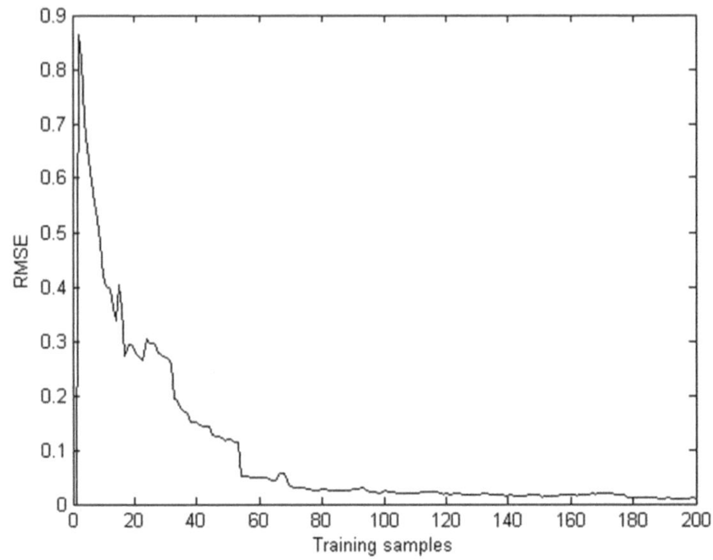

Fig. 10.4 Root mean squared error (RMSE) during training

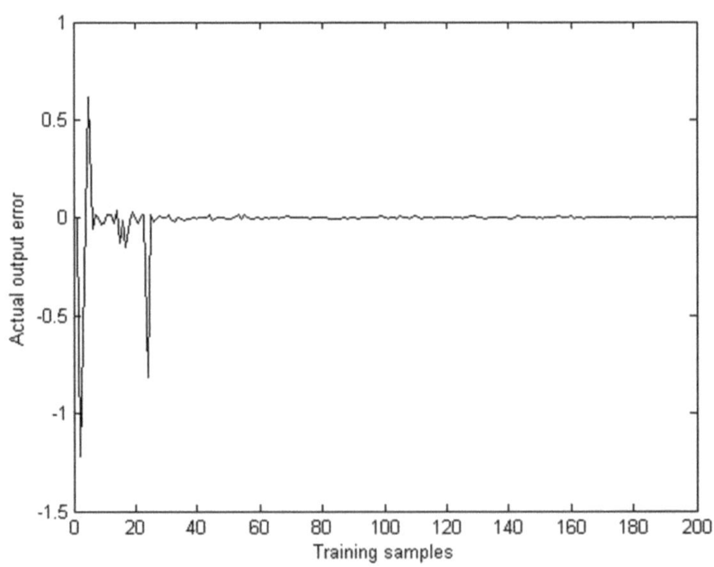

Fig. 10.5 Actual output error during training

in Figure 10.2, in which we see that there are only 5 fuzzy rules finally. Figure 10.3 shows the underlying function curve and the OSFNNRG approximation curve, from which we can see that the resulting fuzzy neural network can approximate well to the original function. Figures 10.4 and Figure 10.5 plot the root mean squared error (RMSE) and actual output error during the training process, respectively. The results illustrate the good approximation performance. Comparisons of the OSFNNRG with other algorithms [12, 15, 16, 21, 22] are listed in Table 10.1. It can be seen that the OSFNNRG has a slightly larger training RMSE than that of the DFNN and the GDFNN, in which the LLS method has been used to obtain the optimal weights and the consequent of each fuzzy rule adopts the polynomial function of input variables. However, the OSFNNRG obtains the most compact structure and the fastest learning process while the approximation error is nearly as good as that of the DFNN and the GDFNN. It should be noted that the superiority to the early approaches of RANEKF and MRAN is remarkable.

Example 10.2. (Modeling a Three-dimensional Nonlinear Function) In this example, a multi-dimensional nonlinear function, to be approximated, which is widely used to verify the algorithms adopted in [9, 13, 22] is given by

$$f(x_1,x_2,x_3) = (1+x_1^{0.5}+x_2^{-1}+x_3^{-1.5})^2 . \tag{10.28}$$

A total of 216 training data are randomly sampled from the input space $[1,6]^3$. The parameters are chosen as follows: $d_{\max}=1.0$, $d_{\min}=0.1$, $e_{\max}=1.1$, $e_{\min}=0.02$, $k_0=1.15$, $\beta=0.97$, $\gamma=0.97$, $GF=0.99$, $P_0=1.0$, $Q_0=0.1$ and $R_k=1.0$. To compare the performance with other approaches, the performance index is chosen to be the same as that in [9], which is given by

$$APE = \frac{1}{n}\sum_{k=1}^{n}\frac{|t^k-y^k|}{|t^k|}\times 100\% , \tag{10.29}$$

where n denotes the number of training data pairs, t^k and y^k are the kth desired and actual output of the fuzzy neural network, respectively. Another 125 data pairs are randomly selected from the same discourse of universe to check the generalization of the resulting fuzzy neural network. The results are shown in Figures 10.6-10.8, in which we see that there are only 7 fuzzy rules in the OSFNNRG system which could model satisfactorily the underlying multi-dimensional function. The RMSE approaches to 0.28 and the actual output error during the learning process is also contained in a very small range. Comparisons of the proposed algorithm with ANFIS, GDFNN and SOFNN are listed in Table 10.2, which shows that the OSFNNRG obtains the most parsimonious structure while the generalization performance is worse than ANFIS and SOFNN, but nearly same as the GDFNN. On the other hand, all of the other three approaches adopt polynomial functions in their consequents of fuzzy rules, which would result in a high computational complexity in the case that the input dimensions are high although the approximation surface is smoother. In addition, it is well known that the ANFIS is a batch learning approach using the BP method rather than the sequential learning. Therefore, the OSFNNRG

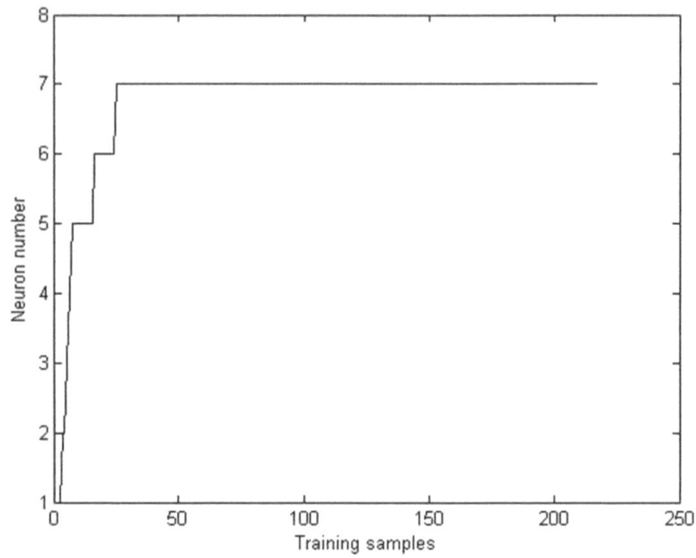

Fig. 10.6 Growth of neurons

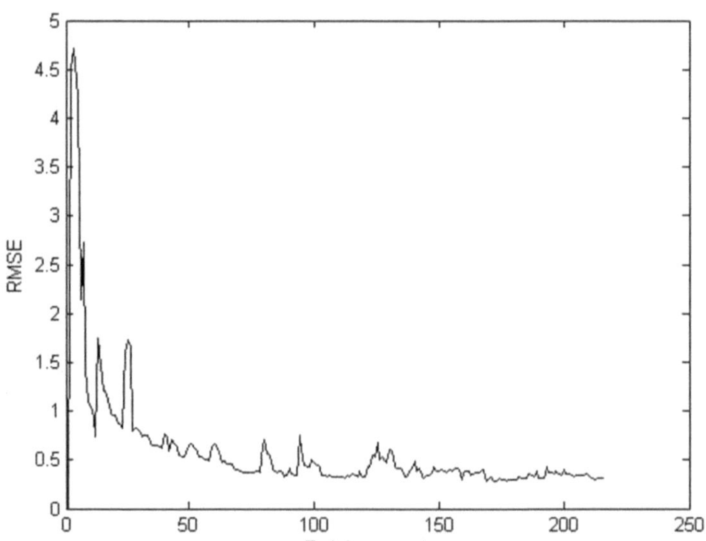

Fig. 10.7 RMSE during training

provides the best performance in the sense of compact structure and fast computational speed, while the comparable approximation performance is obtained.

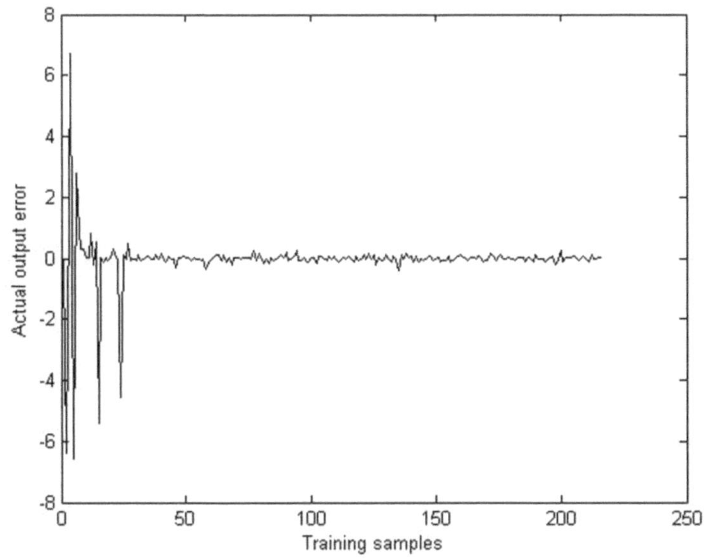

Fig. 10.8 Actual output error during training

Table 10.2 Comparisons of the OSFNNRG with other algorithms (Example 10.2)

Algorithms	Rule number	$APE_{trn}(\%)$	$APE_{chk}(\%)$	Training time (sec.)
ANFIS	8	0.043	1.066	-*
GDFNN	10	2.11	1.54(e_m=0.8781)*	1.86
SOFNN	9	1.1380	1.1244	-
OSFNNRG	7	1.89	2.95(e_m=1.0515)	0.53

* The results are not listed in the original papers.

Example 10.3. (Nonlinear Dynamic System Identification) In the third experiment, the plant to be identified is described as follows:

$$y(k+1) = \frac{y(k)y(k-1)[y(k)+2.5]}{1+y^2(k)+y^2(k-1)} + u(k) . \quad (10.30)$$

To identify the plant, which is also used in [2,3,21,22], a series-parallel identification model is governed by

$$y(k+1) = f(y(k), y(k-1)) + u(k) , \quad (10.31)$$

where f is the function implemented by the OSFNNRG with three inputs and one output, and the input is given by

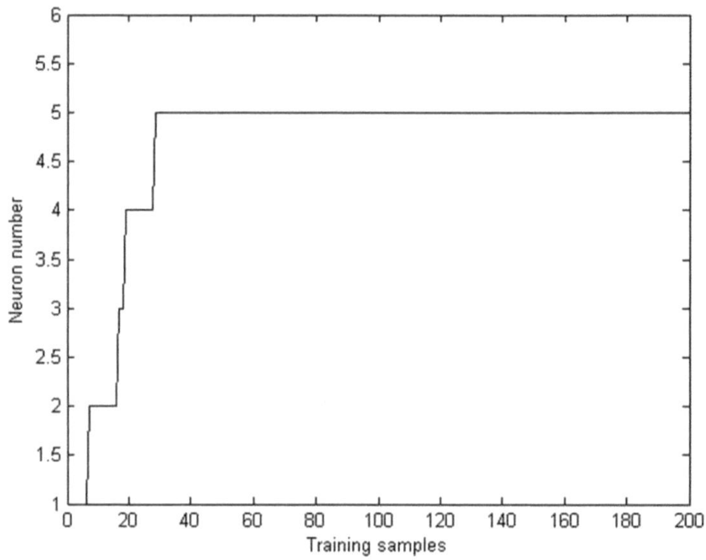

Fig. 10.9 Growth of neurons

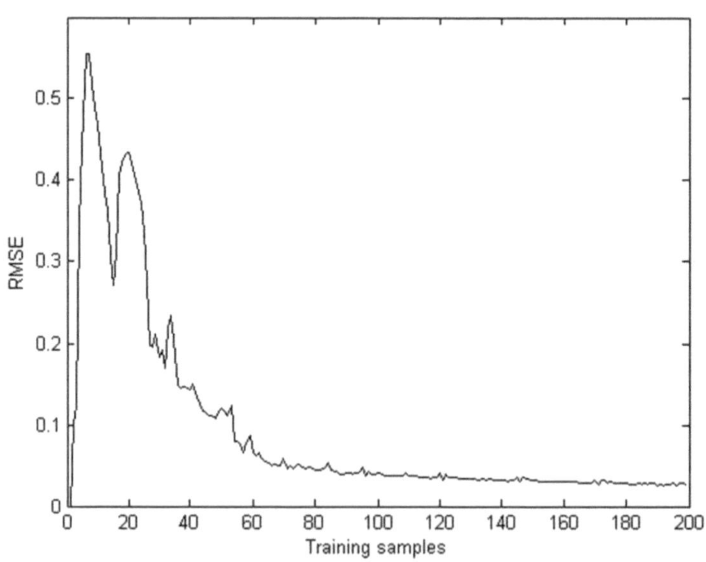

Fig. 10.10 RMSE during training

$$u(k) = \sin(2\pi k/25)\,. \tag{10.32}$$

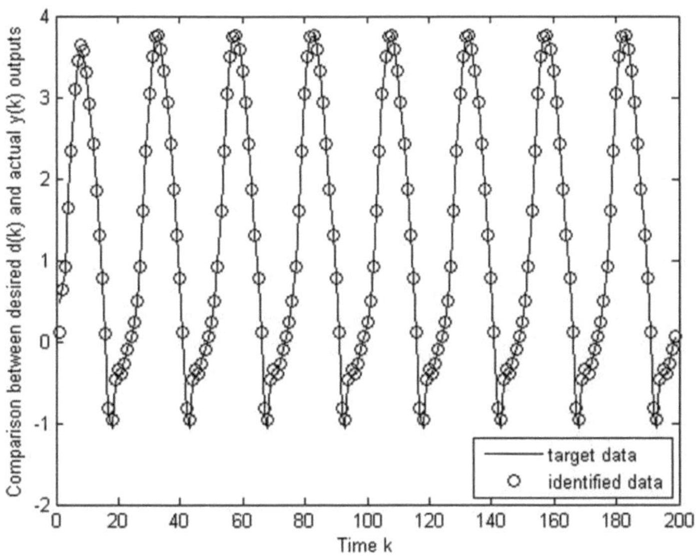

Fig. 10.11 Identification result

Table 10.3 Comparisons of the OSFNNRG with other algorithms (Example 10.3)

Algorithms	Rule number	RMSE	Parameter number	Training time (sec.)
RBF-AFS	35	0.1384	280	-*
OLS	65	0.0288	326	-
DFNN	6	0.0283	48	0.99
GDFNN	8	0.0108	56	1.14
OSFNNRG	5	0.0252	25	0.31

* The results are not listed in the original papers.

The parameters are set as follows: $d_{\max} = 2.2$, $d_{\min} = 0.2$, $e_{\max} = 1.15$, $e_{\min} = 0.02$, $k_0 = 0.25$, $\beta = 0.97$, $\gamma = 0.97$, $GF = 0.98$, $P_0 = 1.0$, $Q_0 = 0.4$ and $R_k = 1.0$. Simulation results are shown in Figures 10.9-10.11, from which it can be seen that the identification performance of the OSFNNRG is perfect and the number of involving fuzzy rules is comparative small. The performance comparisons of different methods are listed in Table 10.3, from which it can be concluded that the OSFNNRG provides the best performance on a more compact fuzzy neural network and a fast and effective online learning algorithm can be achieved.

Example 10.4. (Mackey-Glass Time Series Prediction) One of the classical benchmark problems in literatures [2,3,8,12,14–17,21] is the chaotic Mackey-Glass time series prediction. Suppose that the time series is generated by the discrete version given by

Table 10.4 Comparisons of the OSFNNRG with other algorithms (Example 10.4)

Algorithms	Rule number	Training RMSE	Testing RMSE	Training time (sec.)
RAN	39	0.1006	0.0466	58.127*
RANEKF	23	0.0726	0.0240	62.674*
MRAN	16	0.1101	0.0337	57.205*
GGAP-RBF	13	0.0700	0.0368	24.326*
OS-ELM	120	0.0184	0.0186	10.0603*
OLS	13	0.0158	0.0162	-**
RBF-AFS	21	0.0107	0.0128	-**
DFNN	5	0.0132	0.0131	93.2169
OSFNNRG	11	0.0073	0.0147	7.6195

* The number of training samples is selected as 4000.
** The results are not listed in the original papers.

$$x(t+1) = (1-a)x(t) + \frac{bx(t-\tau)}{1+x^{10}(t-\tau)}, \qquad (10.33)$$

where the parameters $a = 0.1$, $b = 0.2$, $\tau = 17$ and the initial condition $x(0) = 1.2$ are chosen to be the same as those in [21]. The problem is to predict the value $x(t+P)$ in the future from the past values, $\{x(t), x(t-\Delta t), ..., x(t-(n-1)\Delta t)\}$. As given by [3], setting $P = \Delta t = 6$ and $n = 4$, we can express the prediction model as follows:

$$x(t+6) = f[x(t), x(t-6), x(t-12), x(t-18)]. \qquad (10.34)$$

For the training, we have extracted 1000 data points between $t = 124$ and $t = 1123$ and have used them for preparing the input and output sample data in the structure of (34). The parameters are set as follows: $d_{\max} = 2$, $d_{\min} = 0.2$, $e_{\max} = 0.9$, $e_{\min} = 0.02$, $k_0 = 1.2$, $\beta = 0.97$, $\gamma = 0.97$, $GF = 0.9978$, $P_0 = 1.1$, $Q_0 = 0.003$ and $R_k = 1.1$. The training results plotted in Figures 10.12-10.15 show that the proposed approach can approximate the underlying model. In order to demonstrate the prediction capability of the learned OSFNNRG, another 1000 data points between $t = 1124$ and $t = 2123$ are tested. The testing results shown in Figures 10.16-10.17 indicate that the prediction performance of the resulting OSFNNRG is wonderful. Comprehensive comparisons of the OSFNNRG with other popular algorithms are listed in Table 10.4, from which the DFNN obtains the most compact structure while the learning time is considerably long. Comparing with the others, the OSFNNRG has more compact structure and shorter training time. Furthermore, the training and testing RMSE of the proposed approach can approach to 0.0073 and 0.0127, respectively. It is considerably remarkable compared with the others. It should be noted that the methods like RAN, RANEKF, MRAN, GGAP-RBF and OS-ELM need 4000 training samples rather than 1000 for learning, although the performances of the resulting systems are worse than that of the OSFNNRG. In terms of the training time, the ones marked by † are conducted in an ordinary PC with the same CPU

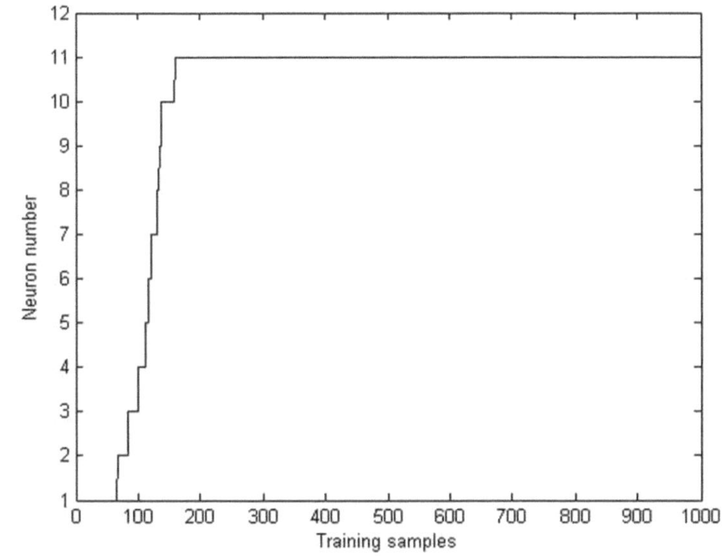

Fig. 10.12 Identification result

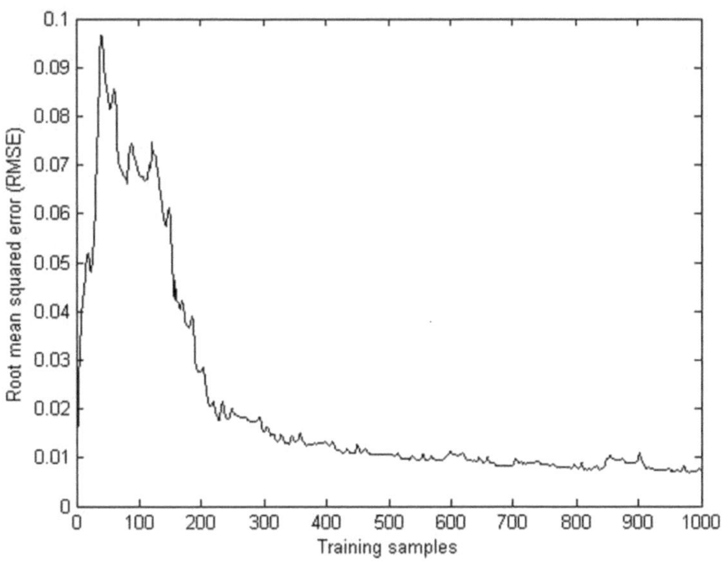

Fig. 10.13 Identification result

(3.0 GHz). Note that the traditional BP method is used in the OLS and RBF-AFS.

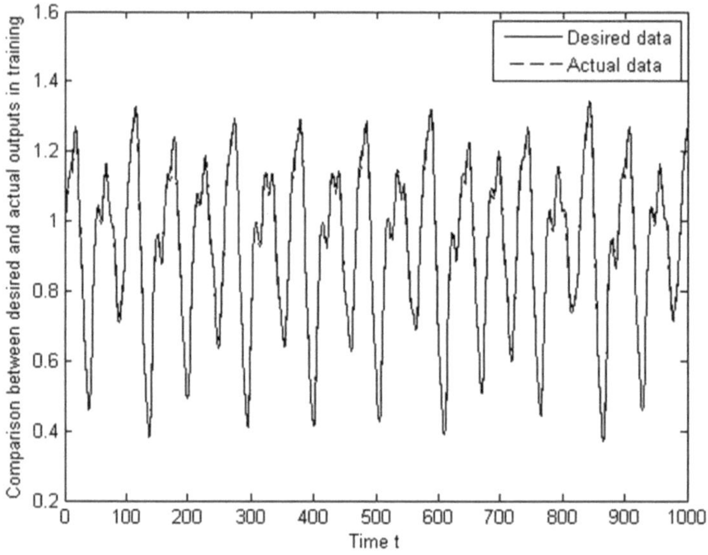

Fig. 10.14 Identification result

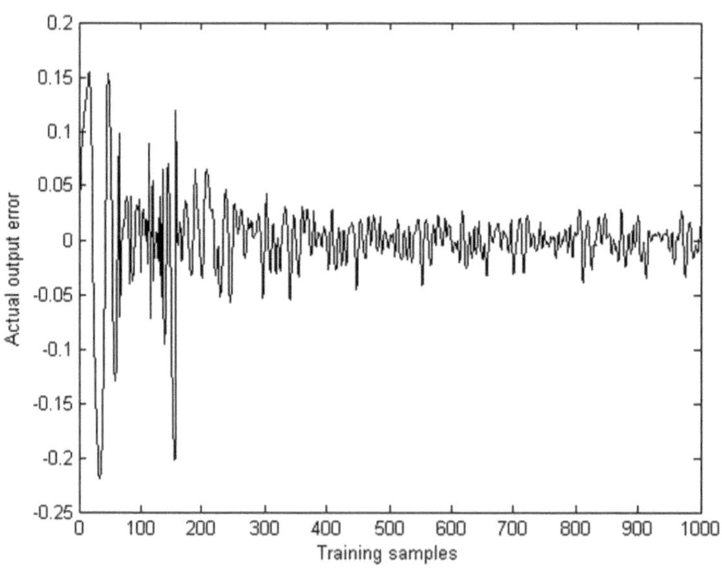

Fig. 10.15 Identification result

Hence, the results are comparable and the OSFNNRG can be considered as the fastest approach.

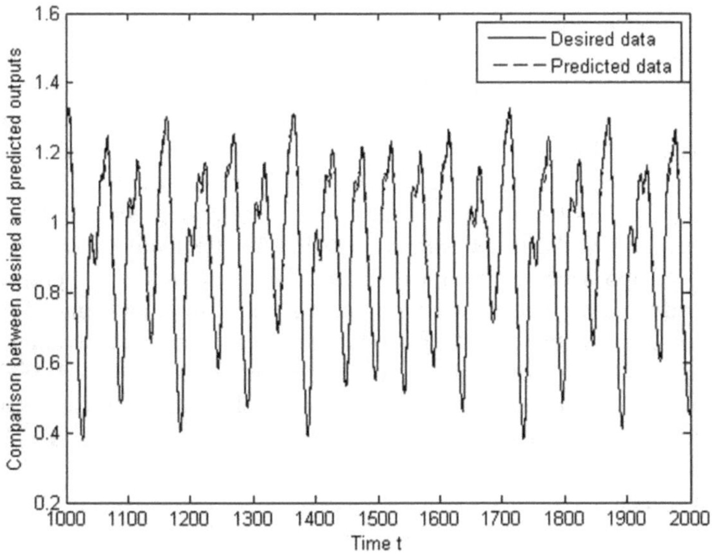

Fig. 10.16 Desired and predicted outputs during testing

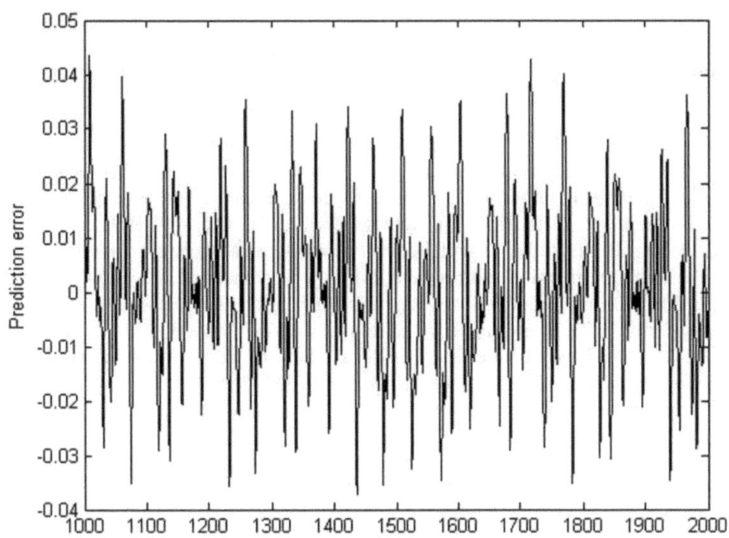

Fig. 10.17 Prediction error during testing

10.5 Conclusions

In this chapter, an online self-constructing algorithm for realizing a fast and parsimonious fuzzy neural network has been developed. Unlike some of the existing sequential learning approaches, the proposed algorithm does not adopt the pruning technology which would increase the computational burden and consequently slow down the learning process. A novel growing criterion has been proposed by incorporating a simple growth criterion and the pruning process which results in a restricted growth strategy. Besides the system error and the input partitioning, the error reduction ratio (ERR) has been used to define a generalization factor which is a new growing criterion that facilitates smooth and parsimonious structure growth. After structure learning, the EKF has been applied to parameter identification of the resulting fuzzy neural network in each learning epoch. The effectiveness and superiority of the proposed approach has been demonstrated in static function approximation, nonlinear modeling, nonlinear dynamic system identification and chaotic time series prediction. Simulation results demonstrate that a faster and more compact fuzzy neural network with high performance can be achieved by the proposed OSFNNRG algorithm online. Comprehensive comparisons with other latest approaches indicate that the overall performance of the proposed approach is superior to the others in terms of learning speed and resulting system structure, while comparable performance is obtained.

Acknowledgements The authors would like to thank the China Scholarship Council (CSC) for partially supporting the joint-training research in Singapore.

References

1. Chao, C.T., Chen, Y.J., Teng, C.C.: Simplification of fuzzy-neural systems using similarity analysis. IEEE Trans. on Syst. Man and Cybern. Part B Cybern. 26, (2), 344–354 (1996).
2. Chen, S., Cowan, C.F.N., Grant, P.M.: Orthogonal least squares learning algorithm for radial basis function network. IEEE Trans. on Neural Netw. 2, (2), 302–309 (1991).
3. Cho, K.B., Wang, B.H.: Radial basis function based adaptive fuzzy systems and their applications to system identification and prediction. Fuzzy Sets and Syst. 83, 325–339 (1996).
4. Gao, Y., Er, M.J.: Online adaptive fuzzy neural identification and control of a class of MIMO nonlinear systems. IEEE Trans. on Fuzzy Syst. 11, (4), 12–32 (2003).
5. Gao, Y., Er, M.J.: NARMAX time series model prediction: feedforward and recurrent fuzzy neural network approaches. Fuzzy Sets and Syst. 150, 331–350 (2005).
6. Gao, Y., Er, M.J.: An intelligent adaptive control scheme for postsurgical blood pressure regulation. IEEE Trans. on Neural Netw. 16, (2), 475–483 (2005).
7. Huang, G.B., Saratchandran, P., Sundararajan, N.: An efficient sequential learning algorithm for growing and pruning RBF (GAP-RBF) networks. IEEE Trans. on Syst. Man and Cybern. Part B Cybern. 34, (6), 2284–2292 (2004).
8. Huang, G.B., Saratchandran, P., Sundararajan, N.: A generalized growing and pruning RBF (GGAP-RBF) neural network for function approximation. IEEE Trans. on Neural Netw. 16, (1), 57–67 (2005).
9. Jang, J.-S.R.: ANFIS: Adaptive-network-based fuzzy inference system. IEEE Trans. on Syst. Man and Cybern. 23, 665–684 (1993).
10. Jang, J.-S.R., Sun, C.T., Mizutani, E.: Neuro-Fuzzy and Soft Computing. Prentice-Hall, Englewood Cliffs (1997)
11. Juang, C.-F., Lin, C.-T.: An on-line self-constructing neural fuzzy inference network and its applications. IEEE Trans. on Fuzzy Syst. 6, (1), 12–32 (1998).

12. Kadirkamanathan, V., Niranjan, M.: A function estimation approach to sequential learning with neural networks. Neural Comput. 5, 954–975 (1993).
13. Leng, G., McGinnity, T.M., Prasad, G.: An approach for on-line extraction of fuzzy rules using a self-organising fuzzy neural network. Fuzzy Sets and Syst. 150, 211–243 (2005)
14. Liang, N.Y., Huang, G.B., Saratchandran, P., Sundararajan, N.: A fast and accurate online sequential learning algorithm for feedforward networks. IEEE Trans. on Neural Netw. 17, (6), 1411–1423 (2006).
15. Lu, Y.W., Sundararajan, N., Saratchandran, P.: A sequential learning scheme for function approximation using minimal radial basis function (RBF) neural networks. Neural Comput. 9, 461–478 (1997).
16. Lu, Y.W., Sundararajan, N., Saratchandran, P.: Performance evaluation of a sequential minimal radial basis function (RBF) neural network learning algorithm. IEEE Trans. on Neural Netw. 9, (2), 308–318 (1998).
17. Platt, J.: A resource-allocating network for function interpolation. Neural Comput. 3, 213–225 (1991).
18. Rojas, I., Pomares, H., Bernier, J.L., Ortega, J., Pino, B., Pelayo, F.J., Prieto, A.: Time series analysis using normalized PG-RBF network with regression weights. Neurocomput. 42, 267–285 (2002).
19. Salmeron, M., Ortega, J., Puntonet, C.G., Prieto, A.: Improved RAN sequential prediction using orthogonal techniques. Neurocomput. 41, 153–172 (2001).
20. Wang L.X.: Adaptive Fuzzy Systems and Control: Design and Stability Analysis. Prentice-Hall, Englewood Cliffs (1994).
21. Wu, S.Q., Er, M.J.: Dynamic fuzzy neural networks - a novel approach to function approximation. IEEE Trans. on Syst. Man and Cybern. Part B Cybern. 30, (2), 358–364 (2000).
22. Wu, S.Q., Er, M.J., Gao, Y.: A fast approach for automatic generation of fuzzy rules by generalized dynamic fuzzy neural networks. IEEE Trans. on Fuzzy Syst. 9, (4), 578–594 (2001).

Chapter 11
Nonlinear System Control Using Functional-link-based Neuro-fuzzy Networks

Chin-Teng Lin, Cheng-Hung Chen and Cheng-Jian Lin

Abstract This study presents a functional-link-based neuro-fuzzy network (FLNFN) structure for nonlinear system control. The proposed FLNFN model uses a functional link neural network (FLNN) to the consequent part of the fuzzy rules. This study uses orthogonal polynomials and linearly independent functions in a functional expansion of the FLNN. Thus, the consequent part of the proposed FLNFN model is a nonlinear combination of input variables. An online learning algorithm, which consists of structure learning and parameter learning, is also presented. The structure learning depends on the entropy measure to determine the number of fuzzy rules. The parameter learning, based on the gradient descent method, can adjust the shape of the membership function and the corresponding weights of the FLNN. Finally, the FLNFN model is applied in various simulations. Results of this study demonstrate the effectiveness of the proposed FLNFN model.

11.1 Introduction

Nonlinear system control is an important tool, which is adopted to improve control performance and achieve robust fault-tolerant behavior. Among nonlinear control techniques, those based on artificial neural networks and fuzzy systems have become popular topics of research in recent years [1,2] because classical control theory

Chin-Teng Lin
Department of Electrical and Control Engineering, National Chiao-Tung University, Hsinchu 300, Taiwan, R.O.C., e-mail: ctlin@nctu.edu.tw

Cheng-Hung Chen
Department of Electrical and Control Engineering, National Chiao-Tung University, Hsinchu 300, Taiwan, R.O.C., e-mail: chchen.ece93g@nctu.edu.tw

Cheng-Jian Lin
Department of Computer Science and Information Engineering, National Chin-Yi University of Technology, Taiping City, Taichung County 411, Taiwan, R.O.C., e-mail: cjlin@ncut.edu.tw

usually requires that a mathematical model be used in designing a controller. However, the inaccuracy of the mathematical modeling of plants usually degrades the performance of the controller, especially for nonlinear and complex control problems [3]. On the contrary, both the fuzzy system controller and the artificial neural network controller provide key advantages over traditional adaptive control systems. Although traditional neural networks can learn from data and feedback, the meaning associated with each neuron and each weight in the network is not easily interpreted. Alternatively, the fuzzy systems are easily appreciated, because they use linguistic terms and the structure of if-then rules. However, the learning capacity of fuzzy systems is less than that of neural networks. According to the literature review mentioned above, neuro-fuzzy networks (NFN) [4–13] provide the advantages of both neural networks and fuzzy systems, unlike pure neural networks or fuzzy systems alone. A NFN bring the low-level learning and computational power of neural networks into fuzzy systems and give the high-level human-like thinking and reasoning of fuzzy systems to neural networks.

Two typical types of NFN are the Mamdani-type and the TSK-type. For Mamdani-type NFN [7–9], the minimum fuzzy implication is adopted in fuzzy reasoning. For TSK-type NFN [10–13], the consequence part of each rule is a linear combination of input variables. Many researchers [12, 13] have shown that TSK-type NFN offer better network size and learning accuracy than Mamdani-type NFN. In the typical TSK-type neural fuzzy network (TSK-type NFN), which is a linear polynomial of input variables, the model output is approximated locally by the rule hyper-planes. Nevertheless, the traditional TSK-type neural fuzzy network does not take full advantage of the mapping capabilities that may be offered by the consequent part. Introducing a nonlinear function, especially a neural structure, to the consequent part of the fuzzy rules has yielded the NARA [14] and the CANFIS [15] models. These models [14, 15] apply multilayer neural networks to the consequent part of the fuzzy rules. Although the interpretability of the model is reduced, the representational capability of the model is markedly improved. However, the multilayer neural network has such disadvantages as slower convergence and greater computational complexity. Therefore, this study uses the FLNN [16, 17] to the consequent part of the fuzzy rules, called FLNFN. The consequent part of the proposed FLNFN model is a nonlinear combination of input variables, which differs from the other existing models [8, 12, 13]. The FLNN is a single layer neural structure capable of forming arbitrarily complex decision regions by generating nonlinear decision boundaries with nonlinear functional expansion. The FLNN [18] was conveniently used for function approximation and pattern classification with faster convergence rate and less computational loading than a multilayer neural network. Moreover, using the functional expansion can effectively increase the dimensionality of the input vector, so the hyper-planes generated by the FLNN will provide a good discrimination capability in input data space.

This study presents a FLNFN structure for nonlinear system control. The FLNFN model, which combines a NFN with a FLNN, is designed to improve the accuracy of functional approximation. Each fuzzy rule that corresponds to a FLNN consists of a functional expansion of input variables. The orthogonal polynomials and linearly independent functions are adopted as FLNN bases. An online learning algorithm, consisting of structure learning and parameter learning, is proposed to construct the FLNFN model automatically. The structure learning algorithm determines whether or not to add a new node which satisfies the fuzzy partition of input variables. Initially, the FLNFN model has no rules. The rules are automatically generated from training data by entropy measure. The parameter learning algorithm is based on back-propagation to tune the free parameters in the FLNFN model simultaneously to minimize an output error function. The advantages of the proposed FLNFN model are summarized below.

1. The consequent of the fuzzy rules of the proposed model is a nonlinear combination of input variables. This study uses the FLNN to the consequent part of the fuzzy rules. The local properties of the consequent part in the FLNFN model enable a nonlinear combination of input variables to be approximated more effectively.
2. The online learning algorithm can automatically construct the FLNFN model. No rules or memberships exist initially. They are created automatically as learning proceeds, as online incoming training data are received and as structure and parameter learning are performed.
3. As demonstrated in Section 11.4, the proposed FLNFN model is a more adaptive and effective controller than the other models.

This study is organized as follows. Section 11.2 describes the structure of the suggested model. Section 11.3 presents the online structure and parameter learning algorithms. Next, Section 11.4 presents the results of simulations of various problems. Finally, the last section ion draws conclusions and future works.

11.2 Structure of Functional-link-based Neuro-fuzzy Network

This section describes the structure of FLNN and the structure of the FLNFN model. In FLNN, the input data usually incorporate high order effects and thus artificially increases the dimensions of the input space using a functional expansion. Accordingly, the input representation is enhanced and linear separability is achieved in the extended space. The FLNFN model adopted the FLNN generating complex nonlinear combination of input variables to the consequent part of the fuzzy rules. The rest of this section details these structures.

11.2.1 Functional Link Neural Networks

FLNN is a single layer network in which the need for hidden layers is removed. While the input variables generated by the linear links of neural networks are linearly weighted, the functional link acts on an element of input variables by generating a set of linearly independent functions (i.e., the use of suitable orthogonal polynomials for a functional expansion), and then evaluating these functions with the variables as the arguments. Therefore, the FLNN structure considers trigonometric functions. For example, for a two-dimensional input $\mathbf{X}=[\mathrm{x}_1, \mathrm{x}_2]^T$, the enhanced input is obtained using trigonometric functions in $\Phi=[1, \mathrm{x}_1, \sin(\pi\ \mathrm{x}_1), \cos(\pi\ \mathrm{x}_1), \ldots, \mathrm{x}_2, \sin(\pi\ \mathrm{x}_2), \cos(\pi\ \mathrm{x}_2), \ldots]^T$. Thus, the input variables can be separated in the enhanced space [16]. In the FLNN structure with reference to Figure 11.1, a set of basis functions Φ and a fixed number of weight parameters $\mathbf{W}$ represent $f_{\mathbf{W}}(x)$ The theory behind the FLNN for multidimensional function approximation has been discussed elsewhere [19] and is analyzed below.

Consider a set of basis functions $\mathbf{B}=\{\phi_k \in \Phi(\mathrm{A})\}_{k\in\mathbf{K}}$, $\mathbf{K}=1, 2,\ldots$, with the following properties; 1) $\phi_1=1$, 2) the subset $B_j=\{\phi_k \in \mathbf{B}\}_{k=1}^{M}$ is a linearly independent set, meaning that if $\sum_{k=1}^{M} w_k\phi_k=0$, then $w_k=0$ for all $k=1, 2,\ldots,M$, and 3) $\sup_j[\sum_{k=1}^{j}\|\phi_k\|_A^2]^{1/2}<\infty$.

Let $\mathbf{B}=\{\phi_k\}_{k=1}^{M}$ be a set of basis functions to be considered, as shown in Figure 11.1. The FLNN comprises M basis functions $\{\phi_1, \phi_2,\ldots,\phi_M\}\in\mathbf{B}_M$. The linear sum of the jth node is given by

$$\hat{y}_j = \sum_{k=1}^{M} w_{kj}\phi_k(\mathbf{X}) \tag{11.1}$$

where $\mathbf{X}\in A\subset\Re^N$, $\mathbf{X}=[x_1, x_2,\ldots, x_N]^T$ is the input vector and $\mathbf{W}_j=[w_{j1}, w_{j2},\ldots, w_{jM}]^T$ is the weight vector associated with the jth output of the FLNN. $\hat{\mathbf{y}}_j$ denotes the local output of the FLNN structure and the consequent part of the jth fuzzy rule in the FLNFN model. Thus, (11.1) can be expressed in matrix form as $\hat{\mathbf{y}}_j=\mathbf{W}_j\Phi$, where $\Phi=[\phi_1(x), \phi_2(x),\ldots,\phi_M(x)]^T$ is the basis function vector, which is the output of the functional expansion block. The m-dimensional linear output may be given by $\hat{\mathbf{y}}=\mathbf{W}\Phi$, where $\hat{\mathbf{y}}=[\hat{y}_1, \hat{y}_2,\ldots, \hat{y}_m]^T$, m denotes the number of functional link bases, which equals the number of fuzzy rules in the FLNFN model, and $\mathbf{W}$ is a $(m\times M)$-dimensional weight matrix of the FLNN given by $\mathbf{W}=[\mathbf{w}_1, \mathbf{w}_2,\ldots, \mathbf{w}_M]^T$. The jth output of the FLNN is given by $\hat{y}_j'=\rho(\hat{y}_j)$, where the nonlinear function $\rho(\cdot)=\tanh(\cdot)$. Thus, the m-dimensional output vector is given by

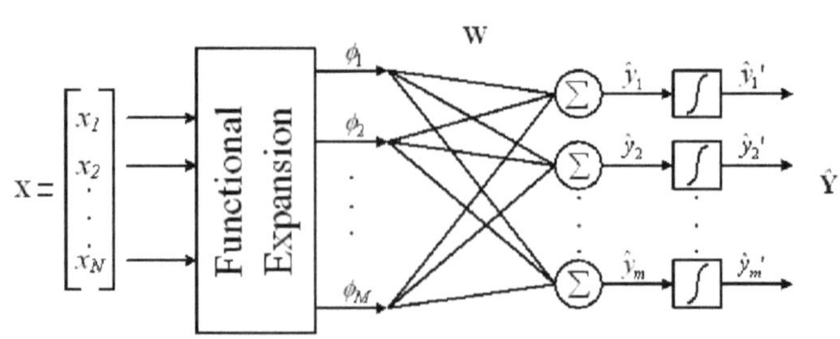

Fig. 11.1 Structure of FLNN

$$\hat{\mathbf{Y}} = \rho(\hat{y}) = f_{\mathbf{W}}(x) \tag{11.2}$$

where $\hat{\mathbf{Y}}$ denotes the output of the FLNN. In the FLNFN model, the corresponding weights of functional link bases do not exist in the initial state, and the amount of the corresponding weights of functional link bases generated by the online learning algorithm is consistent with the number of fuzzy rules. Section 11.3 details the online learning algorithm.

11.2.2 Structure of the FLNFN Model

This subsection describes the FLNFN model, which uses a nonlinear combination of input variables. Each fuzzy rule corresponds to a sub-FLNN, comprising a functional link. Figure 11.2 presents the structure of the proposed FLNFN model. Nodes in layer 1 are input nodes, which represent input variables. Nodes in layer 2 are called membership function nodes and act as membership functions, which express the input fuzzy linguistic variables. Nodes in this layer are adopted to determine Gaussian membership values. Each node in layer 3 is called a rule node. Nodes in layer 3 are equal to the number of fuzzy sets that correspond to each external linguistic input variable. Links before layer 3 represent the preconditions of the rules, and links after layer 3 represent the consequences of the rule nodes. Nodes in layer 4 are called consequent nodes, each of which is a nonlinear combination of the input variables. The node in layer 5 is called the output node; it is recommended by layers 3 and 4, and acts as a defuzzifier.

The FLNFN model realizes a fuzzy if-then rule $Rule-j$ in the following form.

$$\begin{aligned}&\text{IF } x_1 \text{ is } A_{1j} \text{ and } \cdots x_i \text{ is } A_{ij} \text{ and } \cdots x_N \text{ is } A_{Nj}\\ &\text{THEN } \hat{y}_j = \textstyle\sum_{k=1}^{M} w_{kj}\phi_k \;= w_{1j}\phi_1 + w_{2j}\phi_2 + \ldots + w_{Mj}\phi_M\end{aligned} \tag{11.3}$$

where x_i and $\hat{y}_j$ are the input and local output variables, respectively; A_{ij} is the linguistic term of the precondition part with Gaussian membership function; N is the number of input variables; w_{kj} is the link weight of the local output; ϕ_k is the basis trigonometric function of the input variables; M is the number of basis function, and $Rule-j$ is the jth fuzzy rule.

The operation functions of the nodes in each layer of the FLNFN model are now described. In the following description, $u^{(l)}$ denotes the output of a node in the lth layer.

Layer 1 (Input Node): No computation is performed in this layer. Each node in this layer is an input node, which corresponds to one input variable, and only transmits input values to the next layer directly:

$$u_i^{(1)} = x_i. \tag{11.4}$$

Layer 2 (Membership Function Node): Nodes in this layer correspond to a single linguistic label of the input variables in Layer 1. Therefore, the calculated membership value specifies the degree to which an input value belongs to a fuzzy set in Layer 2. The implemented Gaussian membership function in Layer 2 is

$$u_{ij}^{(2)} = exp(-\frac{[u_i^{(1)} - m_{ij}]^2}{\sigma_{ij}^2}) \tag{11.5}$$

where m_{ij} and σ_{ij} are the mean and variance of the Gaussian membership function, respectively, of the jth term of the ith input variable x_i.

Layer 3 (Rule Node): Nodes in this layer represent the preconditioned part of a fuzzy logic rule. They receive one-dimensional membership degrees of the associated rule from the nodes of a set in Layer 2. Here, the product operator described above is adopted to perform the IF-condition matching of the fuzzy rules. As a result, the output function of each inference node is

$$u_j^{(3)} = (\prod_i u_{ij}^{(2)}) \tag{11.6}$$

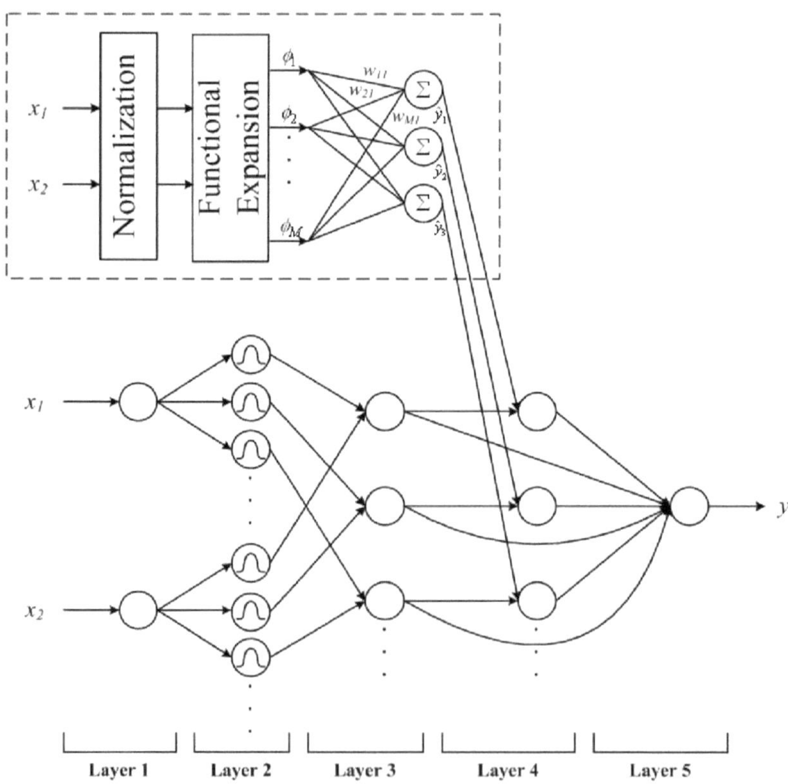

Fig. 11.2 Structure of the proposed FLNFN model

where the $\prod_i u_{ij}^{(2)}$ of a rule node represents the firing strength of its corresponding rule.

Layer 4 (Consequent Node): Nodes in this layer are called consequent nodes. The input to a node in Layer 4 is the output from Layer 3, and the other inputs are nonlinear combinations of input variables from a FLNN, where the nonlinear combination function has not used the function tanh(·), as shown in Figure 11.2. For such a node,

$$u^{(4)} = u_j^{(3)} \cdot \sum_{k=1}^{M} w_{kj}\phi_k \tag{11.7}$$

where w_{kj} is the corresponding link weight of FLNN and ϕ_k is the functional expansion of input variables. The functional expansion uses a trigonometric polynomial basis function, given by $[x_1, \sin(\pi x_1), \cos(\pi x_1), x_2, \sin(\pi x_2), \cos(\pi x_2)]$ for two-dimensional input variables. Therefore, M is the number of basis functions, M=3×N, where N is the number of input variables.

Layer 5 (Output Node): Each node in this layer corresponds to a single output variable. The node integrates all of the actions recommended by Layers 3 and 4 and acts as a defuzzifier with,

$$y = u^{(5)} = \frac{\sum_{j=1}^{R} u_j^{(4)}}{\sum_{j=1}^{R} u_j^{(3)}} \tag{11.8}$$

where R is the number of fuzzy rules, and y is the output of the FLNFN model. As described above, the number of tuning parameters for the FLNFN model is known to be 5×N×R, where N and R denote the number of inputs and existing rules, respectively.

11.3 Learning Algorithms of the FLNFN Model

This section presents an online learning algorithm for constructing the FLNFN model. The proposed learning algorithm comprises a structure learning phase and a parameter learning phase. Figure 11.3 presents a flow diagram of the learning scheme for the FLNFN model. Structure learning is based on the entropy measure used to determine whether a new rule should be added to satisfy the fuzzy partitioning of input variables. Parameter learning is based on supervised learning algorithms. The back-propagation algorithm minimizes a given cost function by adjusting the link weights in the consequent part and the parameters of the membership functions. Initially, there are no nodes in the network except the input-output nodes, i.e., there are no any nodes in the FLNFN model. The nodes are created automatically as learning proceeds, upon the reception of online incoming training data in the structure and parameter learning processes. The rest of this section details the structure learning phase and the parameter learning phase.

11.3.1 Structure Learning Phase

The first step in structure learning is to determine whether a new rule from should be extracted from the training data and to determine the number of fuzzy sets in the universal of discourse of each input variable, since one cluster in the input space corresponds to one potential fuzzy logic rule, which represent the mean and variance of that cluster, respectively. For each incoming pattern x_i, the rule firing strength can be regarded as the degree to which the incoming pattern belongs to the corresponding cluster. Entropy measure between each data point and each membership function is calculated based on a similarity measure. A data point of closed mean will have a lower entropy. Therefore, the entropy values between data points and current membership functions are calculated to determine whether or not to add a new rule. For computational efficiency, the entropy measure can be calculated using the firing strength from $u_{ij}^{(2)}$ as follows,

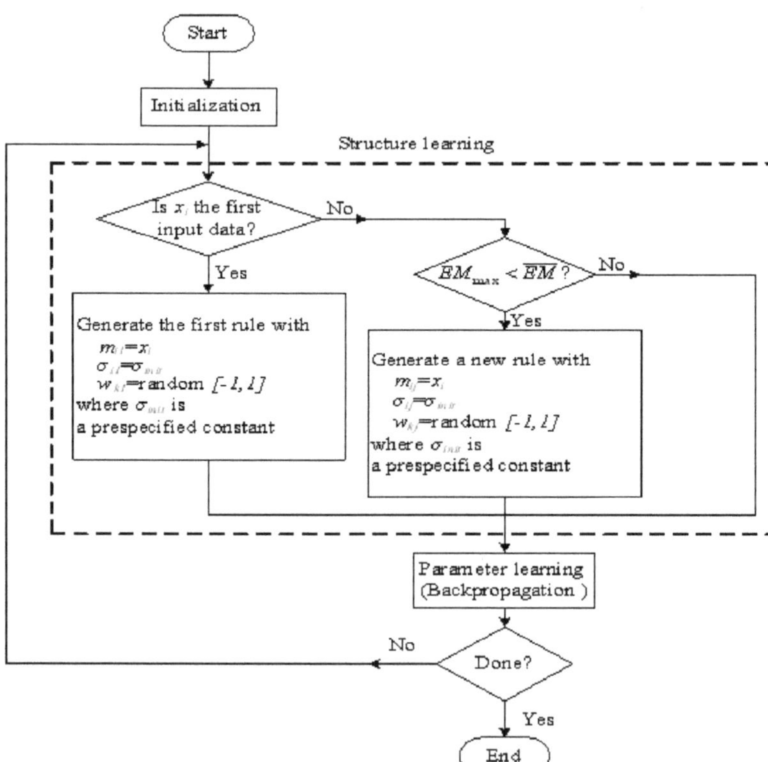

Fig. 11.3 Flow diagram of the structure/parameter learning for the FLNFN model

$$EM_j = -\sum_{i=1}^{N} D_{ij} \log_2 D_{ij} \tag{11.9}$$

where $D_{ij} = exp(u_{ij}^{(2)^{-1}})$ and $EM_j \in [0, 1]$. According to (11.9), the measure is used to generate a new fuzzy rule and new functional link bases for new incoming data are described as follows. The maximum entropy measure

$$EM_{max} = max_{1 \le j \le R_{(t)}} EM_j \tag{11.10}$$

is determined, where $R_{(t)}$ is the number of existing rules at time t. If $EM_{max} \le \overline{EM}$, then a new rule is generated, where $\overline{EM} \in [0,1]$ is a prespecified threshold that decays during the learning process.

In the structure learning phase, the threshold parameter $\overline{EM}$ is an important parameter. The threshold is set to between zero and one. A low threshold leads to the learning of coarse clusters (i.e., fewer rules are generated), whereas a high threshold leads to the learning of fine clusters (i.e., more rules are generated). If the threshold value equals zero, then all the training data belong to the same cluster in the input space. Therefore, the selection of the threshold value $\overline{EM}$ will critically affect the simulation results. As a result of our extensive experiments and by carefully examining the threshold value $\overline{EM}$, which uses the range [0, 1], we concluded the relationship between threshold value $\overline{EM}$ and the number of input variables. Accordingly, $\overline{EM}$ is defined as 0.26-0.3 times of the number of input variables.

Once a new rule has been generated, the next step is to assign the initial mean and variance to the new membership function and the corresponding link weight for the consequent part. Since the goal is to minimize an objective function, the mean, variance and weight are all adjustable later in the parameter learning phase. Hence, the mean, variance and weight for the new rule are set as follows;

$$m_{ij}^{(R_{(t+1)})} = x_i \tag{11.11}$$

$$\sigma_{ij}^{(R_{(t+1)})} = \sigma_{init} \tag{11.12}$$

$$w_{kj}^{(R_{(t+1)})} = random[-1, 1] \tag{11.13}$$

where x_i is the new input and σ_{init} is a prespecified constant. The whole algorithm for the generation of new fuzzy rules and fuzzy sets in each input variable is as follows. No rule is assumed to exist initially exist:

Step 1: IF x_i is the first incoming pattern THEN do
{ Generate a new rule
with mean $m_{i1} = x_i$, variance $\sigma_{i1} = \sigma_{init}$, weight w_{k1}=random[-1, 1]
where σ_{init} is a perspecified constant.

}

Step 2: ELSE for each newly incoming x_i, do

{ Find $EM_{max} = max_{1 \leq j \leq R_{(t)}} EM_j$

IF $EM_{max} \geq \overline{EM}$

do nothing

ELSE

$R_{(t+1)} = R_{(t)} + 1$ generate a new rule

with mean $m_{ij}^{(R_{(t+1)})} = x_i$, variance $\sigma_{ij}^{(R_{(t+1)})} = \sigma_{init}$,

weight $w_{kj}^{(R_{(t+1)})}$ = random[-1, 1]

where σ_{init} is a perspecified constant.

}

11.3.2 Parameter Learning Phase

After the network structure has been adjusted according to the current training data, the network enters the parameter learning phase to adjust the parameters of the network optimally based on the same training data. The learning process involves determining the minimum of a given cost function. The gradient of the cost function is computed and the parameters are adjusted with the negative gradient. The back-propagation algorithm is adopted for this supervised learning method. When the single output case is considered for clarity, the goal to minimize the cost function E is defined as

$$E(t) = \frac{1}{2}[y(t) - y^d(t)]^2 = \frac{1}{2}e^2(t) \tag{11.14}$$

where $y^d(t)$ is the desired output and $y(t)$ is the model output for each discrete-time t. In each training cycle, starting at the input variables, a forward pass is adopted to calculate the activity of the model output $y(t)$.

When the back-propagation learning algorithm is adopted, the weighting vector of the FLNFN model is adjusted such that the error defined in (11.14) is less than the desired threshold value after a given number of training cycles. The well-known back-propagation learning algorithm may be written briefly as

$$W(t+1) = W(t) + \triangle W(t) = W(t) + \{-\eta \frac{\partial E(t)}{\partial W(t)}\} \tag{11.15}$$

where η and W represent the learning rate and the tuning parameters of the FLNFN model, respectively, in this case. Let $\mathbf{W} = [m, \sigma, w]^T$ be the weighting vector of the FLNFN model. Then, the gradient of error $E(.)$ in (11.14) with respect to an arbitrary weighting vector W is

$$\frac{\partial E(t)}{\partial W} = e(t)\frac{\partial y(t)}{\partial W}. \tag{11.16}$$

Recursive applications of the chain rule yield the error term for each layer. Then the parameters in the corresponding layers are adjusted. With the FLNFN model and the cost function as defined in (11.14), the update rule for w_j can be derived as follows;

$$w_{kj}(t+1) = w_{kj}(t) + \triangle w_{kj}(t) \tag{11.17}$$

where

$$\triangle w_{kj}(t) = -\eta_w \frac{\partial E}{\partial w_{kj}} = -\eta_w \cdot e \cdot [\frac{u_j^{(3)}\phi_k}{\sum_{j=1}^{R} u_j^{(3)}}]. \tag{11.18}$$

Similarly, the update laws for m_{ij}, and σ_{ij} are

$$m_{ij}(t+1) = m_{ij}(t) + \triangle m_{ij}(t) \tag{11.19}$$

and

$$\sigma_{ij}(t+1) = \sigma_{ij}(t) + \triangle \sigma_{ij}(t) \tag{11.20}$$

where

$$\triangle m_{ij}(t) = -\eta_m \frac{\partial E}{\partial m_{ij}} = -\eta_m \cdot e \cdot [\frac{u_j^{(4)}}{\sum_{j=1}^{R} u_j^{(3)}}] \cdot [\frac{2(u_i^{(1)} - m_{ij})}{\sigma_{ij}^2}] \tag{11.21}$$

and

$$\triangle \sigma_{ij}(t) = -\eta_\sigma \frac{\partial E}{\partial \sigma_{ij}} = -\eta_\sigma \cdot e \cdot [\frac{u_j^{(4)}}{\sum_{j=1}^{R} u_j^{(3)}}] \cdot [\frac{2(u_i^{(1)} - m_{ij})^2}{\sigma_{ij}^3}], \tag{11.22}$$

where η_w, η_m and η_σ are the learning rate parameters of the weight, the mean, and the variance, respectively. In this study, both the link weights in the consequent part and the parameters of the membership functions in the precondition part are adjusted by using the back-propagation algorithm. Recently, many researchers [13,20] tuned the consequent parameters using either least mean squares (LMS) or recursive least squares (RLS) algorithms to obtain optimal parameters. However, they still used the back-propagation algorithm to adjust the precondition parameters.

11.4 Simulation Results

This study demonstrated the performance of the FLNFN model for nonlinear system control. This section simulates various control examples and compares the performance of the FLNFN model with that of other models. The FLNFN model is adopted to design controllers in four simulations of nonlinear system control problems - water bath temperature control system [21], control of a bounded-input and bounded-output (BIBO) nonlinear plant [22], control of the ball and beam system [23], and multi-input multi-output plant control [2].

Example 11.1. (Control of Water Bath Temperature System) The goal of this section is to elucidate the control of the temperature of a water bath system according to,

$$\frac{dy(t)}{dt} = \frac{u(t)}{C} + \frac{Y_0 - y(t)}{T_R C} \tag{11.23}$$

where $y(t)$ is the output temperature of the system in ^oC; $u(t)$ is the heat flowing into the system; Y_0 is room temperature; C is the equivalent thermal capacity of the system, and T_R is the equivalent thermal resistance between the borders of the system and the surroundings.

T_R and C are assumed to be essentially constant, and the system in (11.23) is rewritten in discrete-time form to some reasonable approximation. The system

$$y(k+1) = e^{-\alpha T_s} y(k) + \frac{\frac{\delta}{\alpha}(1 - e^{-\alpha T_s})}{1 + e^{0.5y(k)-40}} u(k) + [1 - e^{-\alpha T_s}] y_0 \tag{11.24}$$

is obtained, where α and δ are some constant values of T_R and C. The system parameters used in this example are $\alpha = 1.0015e^{-4}$, $\delta = 8.67973e^{-3}$ and Y_0=25.0(^oC), which were obtained from a real water bath plant considered elsewhere [21]. The input $u(k)$ is limited to 0, and 5V represent voltage unit. The sampling period is T_s=30.

The conventional online training scheme is adopted for online training. Figure 11.4 presents a block diagram for the conventional online training scheme. This scheme has two phases - the training phase and the control phase. In the training phase, the switches S1 and S2 are connected to nodes 1 and 2, respectively, to form a training loop. In this loop, training data with input vector $I(k) = [y_p(k+1) y_p(k)]$ and desired output $u(k)$ can be defined, where the input vector of the FLNFN model is the same as that used in the general inverse modeling [24] training scheme. In the control phase, the switches S1 and S2 are connected to nodes 3 and 4, respectively, forming a control loop. In this loop, the control signal $\hat{u}(k)$ is generated according to the input vector $I'(k) = [y_{ref}(k+1) y_p(k)]$, where y_p is the plant output and y_{ref} is the reference model output.

A sequence of random input signals $u_{rd}(k)$ limited to 0 and 5V is injected directly into the simulated system described in (11.24), using the online training scheme for the FLNFN model. The 120 training patterns are selected based on the input-outputs characteristics to cover the entire reference output. The temperature of the water is initially 25^oC, and rises progressively when random input signals are injected. After 10000 training iterations, four fuzzy rules are generated.

This study compares the FLNFN controller to the proportional integral derivative (PID) controller [25], the manually designed fuzzy controller [4], the FLNN [17] and the TSK-type neural fuzzy network [12]. Each of these controllers is applied to the water bath temperature control system. The performance measures include the set points regulation, the influence of impulse noise, and a large parameter variation in the system, and the tracking capability of the controllers.

The first task is to control the simulated system to follow three set points.

$$y_{ref}(k) = \begin{cases} 35^oC, & \text{for } k \leq 40 \\ 55^oC, & \text{for } 40 < k \leq 80 \\ 75^oC, & \text{for } 80 < k \leq 120. \end{cases} \tag{11.25}$$

Figure 11.5(a) presents the regulation performance of the FLNFN controller. The regulation performance was also tested using the FLNN controller and the TSK-type NFN controller. Figure 11.5(b) plots the error curves of the FLNFN controller, the FLNN controller and the TSK-type NFN controller between k=81 and k=100. In this figure, the FLNFN controller obtains smaller errors than the other two controllers. To test their regulation performance, a performance index, the sum of absolute error (SAE), is defined by

$$SAE = \sum_k \|y_{ref}(k) - y(k)\| \tag{11.26}$$

where $y_{ref}(k)$ and $y(k)$ are the reference output and the actual output of the simulated system, respectively. The SAE values of the FLNFN controller, the PID controller, the fuzzy controller, the FLNN controller and the TKS-type NFN controller are 352.8, 418.5, 401.5, 379.2 and 361.9, which values are given in the second row of Table 11.1. The proposed FLNFN model has a much better SAE value of regulation performance than the other controllers.

The second set of simulations is performed to elucidate the noise-rejection ability of the five controllers when some unknown impulse noise is imposed on the process. One impulse noise value -5^oC is added to the plant output at the 60th sampling

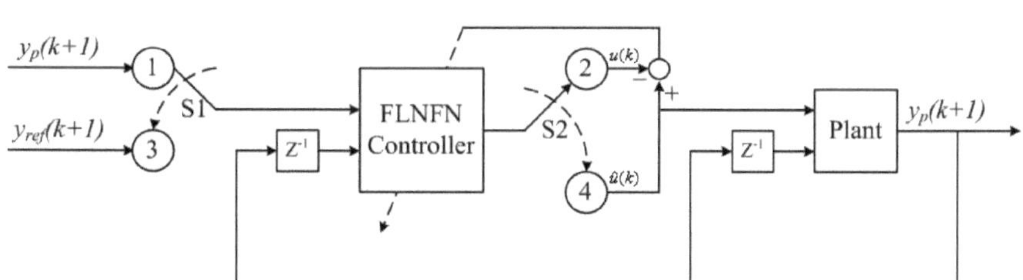

Fig. 11.4 Conventional online training scheme

instant. A set point of 50^oC is adopted in this set of simulations. For the FLNFN controller, the same training scheme, training data and learning parameters as were used in the first set of simulations. Figures 11.6(a) and (b) present the behaviors of the FLNFN controller under the influence of impulse noise, and the corresponding errors, respectively. The SAE values of the FLNFN controller, the PID controller, the fuzzy controller, the FLNN controller and the TSK-type NFN controller are 270.4, 311.5, 275.8, 324.51 and 274.75, which are shown in the third row of Table 11.1. The FLNFN performs quite well. It recovers very quickly and steadily after the occurrence of the impulse noise.

One common characteristic of many industrial-control processes is that their parameters tend to change in an unpredictable way. The value of $0.7 \times u(k-2)$ is added to the plant input after the 60th sample in the third set of simulations to test the robustness of the five controllers. A set point of 50^oC is adopted in this set of

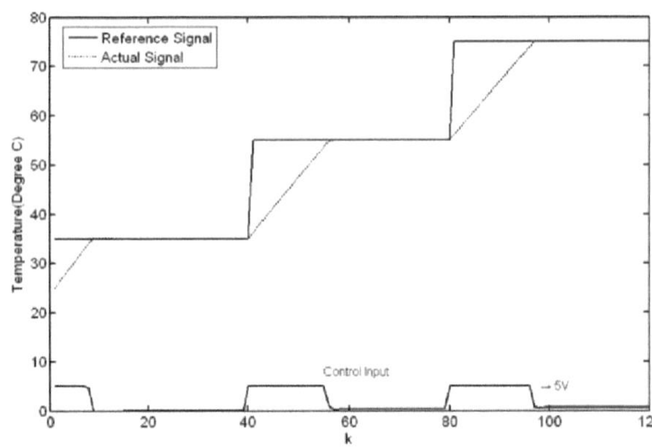

(a)

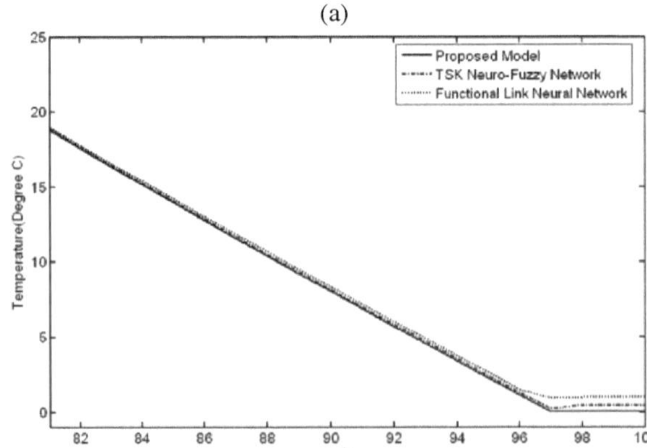

(b)

Fig. 11.5 (a) Final regulation performance of FLNFN controller in water bath system; and (b) error curves of the FLNFN controller, TSK-type NFN controller and FLNN controller between k=81 and k=100 simulations. Figure 11.7(a) presents the behaviors of the FLNFN controller when in the plant dynamics change. Figure 11.7(b) presents the corresponding errors of the FLNFN controller, the FLNN controller and the TSK-type NFN controllers. The SAE values of the FLNFN controller, the PID controller, the fuzzy controller, the FLNN controller and the TSK-type NFN controller are 263.3, 322.2, 273.5, 311.5 and 265.4, which are shown in the fourth row of Table 11.1. The results present the favorable control and disturbance rejection capabilities of the trained FLNFN controller in the water bath system.

In the final set of simulations, the tracking capability of the FLNFN controller with respect to ramp-reference signals is studied. Define

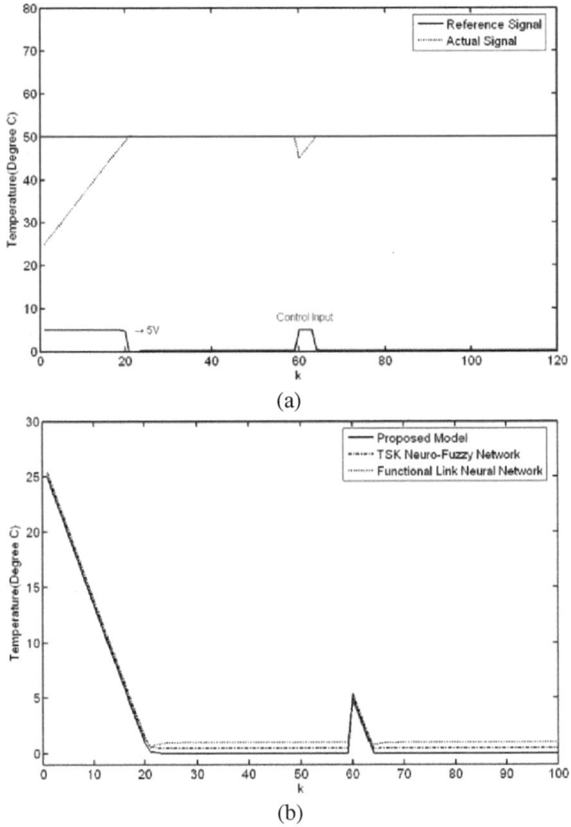

Fig. 11.6 (a) Behavior of FLNFN controller under impulse noise in water bath system; and (b) error curves of FLNFN controller, TSK-type NFN controller and FLNN controller

$$y_{ref}(k) = \begin{cases} 34^oC, & \text{for } k \leq 30 \\ (34+0.5*(k-30))^oC, & \text{for } 30 < k \leq 50 \\ (44+0.8*(k-50))^oC, & \text{for } 50 < k \leq 70 \\ (60+0.5*(k-70))^oC, & \text{for } 70 < k \leq 90 \\ 70^oC, & \text{for } 90 < k \leq 120. \end{cases} \tag{11.27}$$

Figure 11.8(a) presents the tracking performance of the FLNFN controller. Figure 11.8(b) presents the corresponding errors of the FLNFN controller, the FLNN controller, and the TSK-type NFN controller. The SAE values of the FLNFN controller, the PID controller, the fuzzy controller, the FLNN controller, and the TSK-type NFN controller are 44.2, 100.6, 88.1, 98.4 and 54.2, which are shown in the fifth row of Table 11.1. The results present the favorable control and disturbance rejection capabilities of the trained FLNFN controller in the water bath system. The

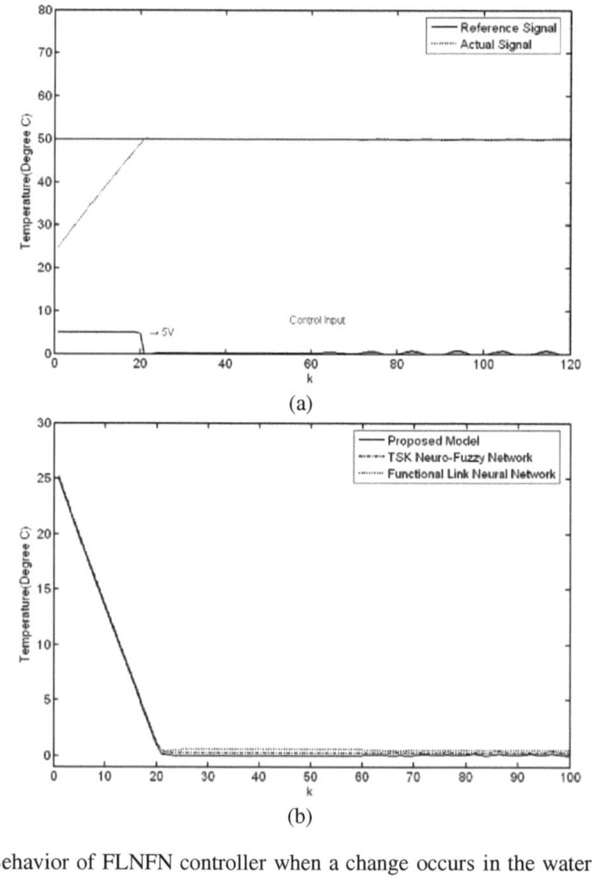

Fig. 11.7 (a) Behavior of FLNFN controller when a change occurs in the water bath system (b) Error curves of FLNFN controller, TSK-type NFN controller and FLNN controller

aforementioned simulation results, presented in Table 11.1, demonstrate that the proposed FLNFN controller outperforms other controllers.

Table 11.1 Comparison of performance of various controllers in Example 11.1

Method	FLNFN Controller	PID Controller [25]	Fuzzy Controller [3]	FLNN Controller [17]	TSK-type NFN Controller [12]
Regulation Performance	352.84	418.5	401.5	379.22	361.96
Influence of Impulse Noise	270.41	311.5	275.8	324.51	274.75
Effect of Change in Plant Dynamics	263.35	322.2	273.5	311.54	265.48
Tracking Performance	44.28	100.6	88.1	98.43	54.28

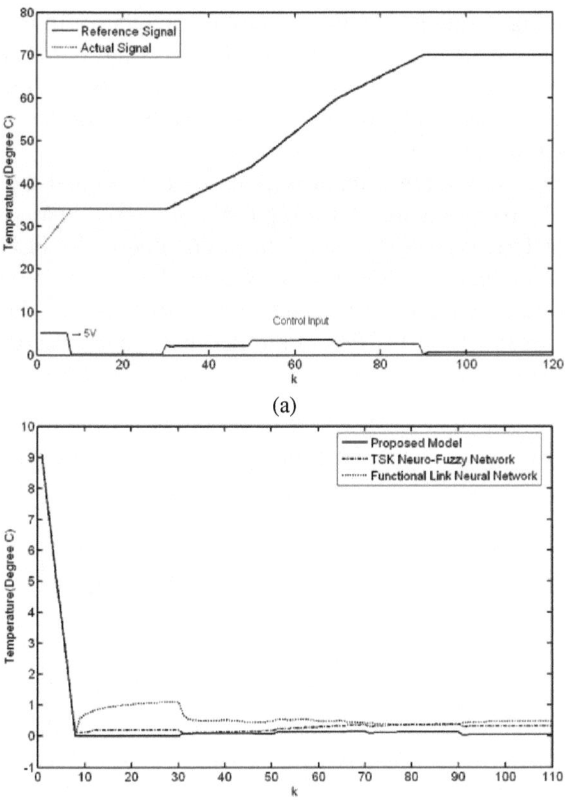

(b)

Fig. 11.8 (a) Tracking of FLNFN controller when a change occurs in the water bath system; and (b) error curves of FLNFN controller, TSK-type NFN controller and FLNN controller

Example 11.2. (Control of BIBO Nonlinear Plant) In this case, the plant is described by the difference equation

$$y(k+1) = \frac{y(k)}{1+y^2(k)} + u^3(k). \tag{11.28}$$

The reference model is described by the difference equation

$$y_r(k+1) = 0.6y_r(k) + r(k) \tag{11.29}$$

where $r(k) = \sin(2\pi k/10) + \sin(2\pi k/25)$. Figure 11.9 present the block diagram of the FLNFN-based control system. The inputs to the FLNFN are the reference input, the previous plant output, and the previous control signal; the output of the FLNFN is the control signal to the plant. The online algorithm developed in this study is adopted to adjust the structure and the parameters of the FLNFN such that the error between the output of the plant and the desired output from a reference model approaches a small value after some train cycles.

After 500 training iterations, six fuzzy rules are generated. In this example, the proposed FLNFN controller is compared to the FLNN controller [17] and the TSK-type NFN controller [12]. Each of the controllers is applied to control the BIBO nonlinear plant. In the following four cases, the FLNFN controller is demonstrated to outperform the other models.

In the first case, the reference input is given by (11.29) and the final result is shown in Figure 11.10(a). Figure 11.10(b) presents the error curves of the FLNFN controller and the TSK-type NFN controller. In this figure, the FLNFN controller yields smaller errors than the TSK-type NFN controller. In the second case, after 100 epochs, the reference input is changed to $r(k) = \sin(2\pi k/25)$. Figures 11.11(a) and (b) plot the result of the FLNFN controller and the corresponding errors of

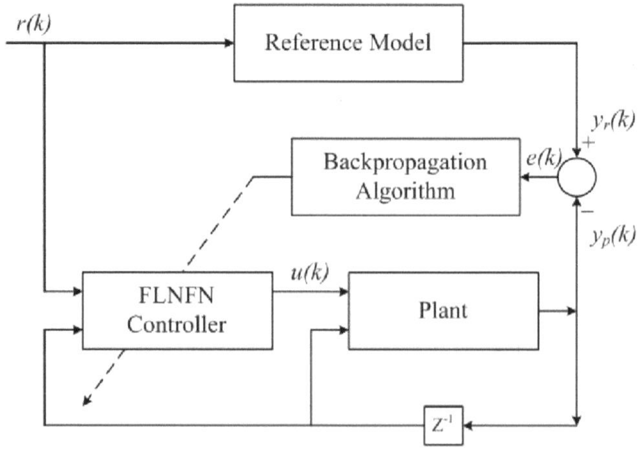

Fig. 11.9 Block diagram of FLNFN-based control system

the FLNFN controller and the TSK-type NFN controller. In the third case, after 100 epochs, the reference input is changed to an impulse signal. Figure 11.12(a) presents the simulation result. Figure 11.12(b) present the corresponding errors of the FLNFN controller, the FLNN controller and the TSK-type NFN controllers. In the fourth case, a disturbance of 2.0 is added to the system between the 100th and the 150th epochs. In this case, the FLNFN-based control system can recover from the disturbance quickly, as shown in Figure 11.13. The root mean square (RMS) error is adopted to evaluate the performance. Table 11.2 presents the RMS errors of the FLNFN controller, the FLNN controller and the TSK-type NFN controller. Table 11.2 shows that, according to the simulation results, the proposed FLNFN outperforms the other models.

Table 11.2 Comparison of the performance of various controllers in Example 11.2

Method	FLNFN Controller	FLNN Controller [17]	TSK-type NFN Controller [12]
Training Steps	500	1000	500
Parameter Numbers	60 parameters (6 rules)	79 parameters	63 parameters (9 rules)
RMS Error of Case1	0.0004	0.0211	0.0084
RMS Error of Case2	0.0006	0.0208	0.0075
RMS Error of Case3	0.0007	0.0303	0.0095

Example 11.3. (Control of Ball and Beam System) Figure 11.14 presents the ball and beam system [23]. The beam is made to rotate in the vertical plane by applying a torque at the center of rotation and the ball is free to roll along the beam. The ball must remain in contact with the beam.

The ball and beam system can be written in state space form as

$$\begin{bmatrix} \dot{x_1} \\ \dot{x_2} \\ \dot{x_3} \\ \dot{x_4} \end{bmatrix} = \begin{bmatrix} x_2 \\ B(x_1x_4^2 - G\sin x_3) \\ x_4 \\ 0 \end{bmatrix} + \begin{bmatrix} 0 \\ 0 \\ 0 \\ 1 \end{bmatrix} u \tag{11.30}$$
$$y = x_1$$

where $x = (x_1, x_2, x_3, x_4)^T \equiv (r, \dot{r}, \theta, \dot{\theta})^T$ is the state of the system and $y = x_1 \equiv r$ is the output of the system. The control u is the angular acceleration $(\ddot{\theta})$ and the parameters $B = 0.7143$ and $G = 9.81$ are chosen in this system. The purpose of control is to determine $u(x)$ such that the closed-loop system output y will converge to zero from different initial conditions.

According to the input/output-linearization algorithm [23], the control law $u(x)$ is determined as follows: for state x, compute $v(x) = -\alpha_3\phi_4(x) - \alpha_2\phi_3(x) - \alpha_1\phi_2(x) - \alpha_0\phi_1(x)$, where $\phi_1(x) = x_1$, $\phi_2(x) = x_2$, $\phi_3(x) = -BG\sin x_3$, $\phi_4(x) = -BGx_4\cos x_3$,

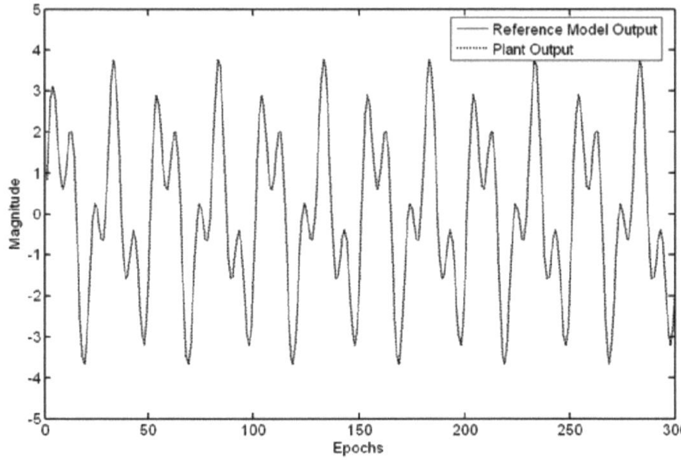

(a)

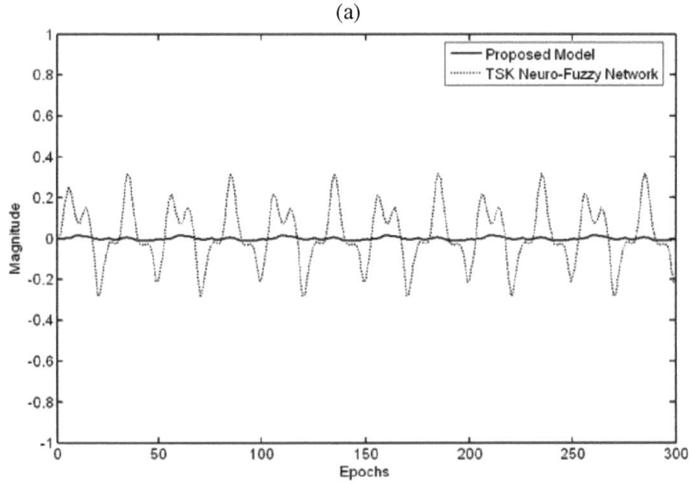

(a)

Fig. 11.10 Final system response in first case (a) the dashed line represents plant output and the solid line represents the reference model; and (b) error curves of FLNFN controller and TSK-type NFN controller

and the α_i are chosen so that $s^4 + \alpha_3 s^3 + \alpha_2 s^2 + \alpha_1 s + \alpha_0$ is Hurwitz polynomial. Compute $a(x) = -BG\cos x_3$ and $b(x) = BGx_4^2 \sin x_3$; then $u(x) = [v(x) - b(x)]/a(x)$.

In this simulation, the differential equations are solved using the second/third-order Runge-Kutta method. The FLNFN model is trained to approximate the aforementioned conventional controller of a ball and beam system. $u(x) = [v(x) - b(x)]/a(x)$ is adopted to generate the input/output train pair with x obtained by randomly sampling 200 points in the region $U = [-5,5] \times [-3,3] \times [-1,1] \times [-2,2]$. After online structure-parameter learning, 14 fuzzy rules are generated. The controller after learning was tested under the following four initial conditions; $x(0) =$

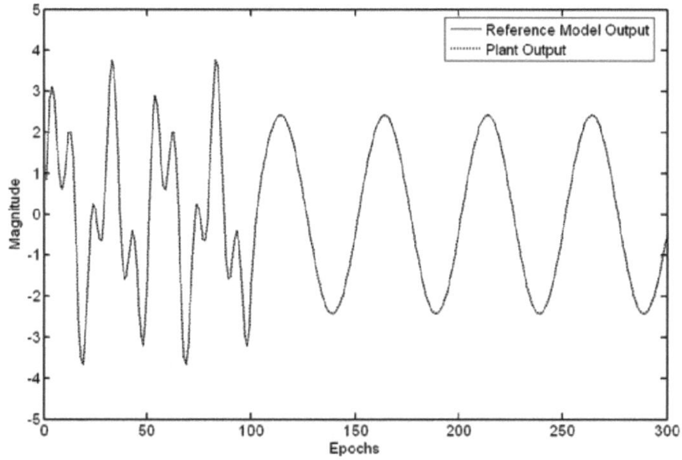

(a)

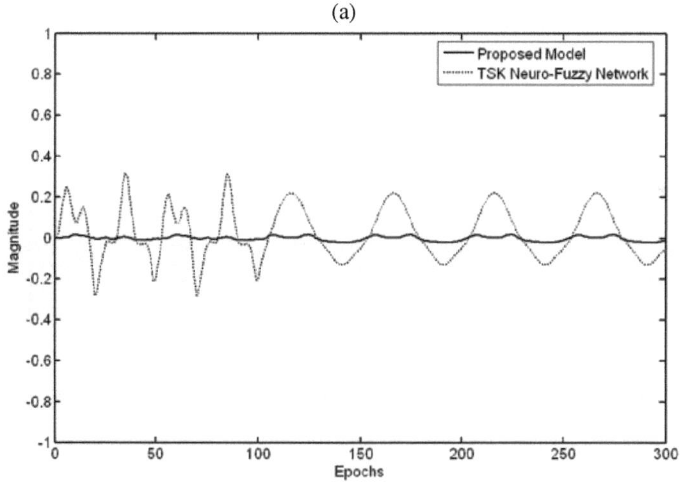

(b)

Fig. 11.11 Final system response in second (a) the dashed line represents plant output and the solid line represents the reference model; and (b) error curves of FLNFN controller and TSK-type NFN controller

$[2.4, -0.1, 0.6, 0.1]^T$, $[1.6, 0.05, -0.5, -0.05]^T$, $[-1.6, -0.05, 0.5, 0.05]^T$ and [-2.4, 0.1, -0.6, -0.1]^T. Figure 11.15 plots the output responses of the closed-loop ball and beam system controlled by the FLNFN model and the TSK-type NFN model. These responses approximate those of the original controller under the four initial conditions. In this figure, the curves of the FLNFN model tend quickly to stabilize. Figure 11.16 also presents the behavior of the four states of the ball and beam system, starting at the initial condition $[-2.4, 0.1, -0.6, -0.1]^T$. In this figure, the four states of the system decay gradually to zero. The results demonstrate the perfect control capability of the trained FLNFN model. The performance of the FLNFN controller

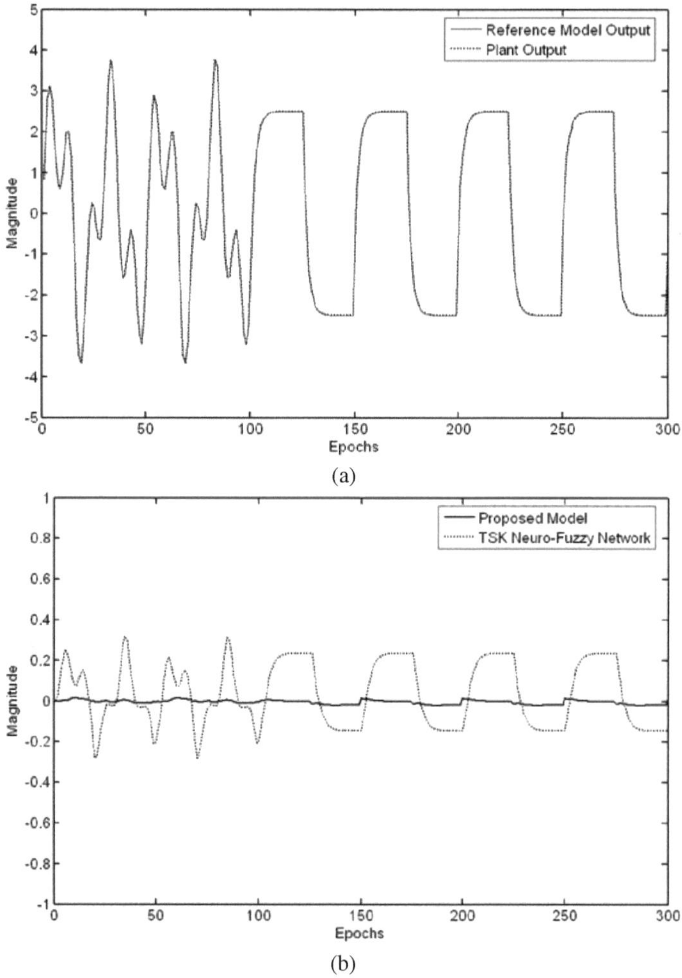

Fig. 11.12 Final system response in third case (a) the dashed line represents plant output and the solid line represents the reference model; and (b) error curves of FLNFN controller and TSK-type NFN controller

is compared with that of the FALCON controller [8], the FLNN controller [17] and the TSK-type NFN controller [12]. Table 11.3 presents the comparison results. The results demonstrate that the proposed FLNFN outperforms other controllers.

Example 11.4. (Control of Multi-input Multi-output (MIMO) Plant) In this example, the MIMO plants [2] to be controlled are described by the equations

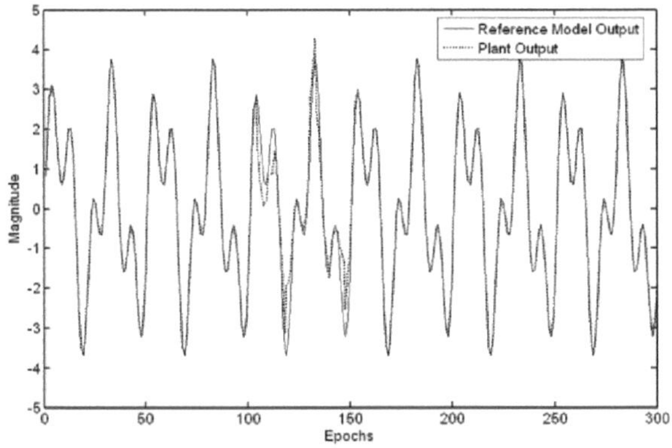

Fig. 11.13 Final system response in fourth case of Example 2. The dashed line represents plant output and the solid line represents the reference model

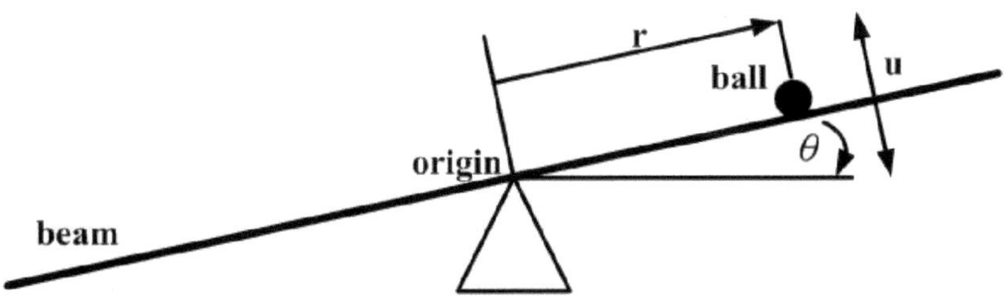

Fig. 11.14 Ball and beam system

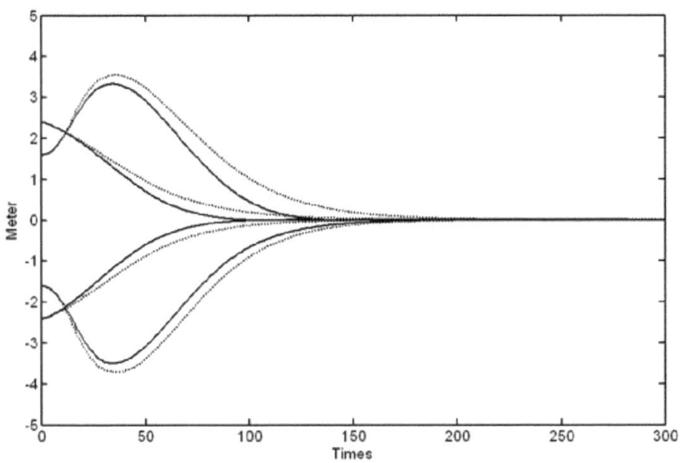

Fig. 11.15 Responses of ball and beam system controlled by FLNFN model (solid curves) and TSK-type NFN model (dotted curves) under four initial conditions

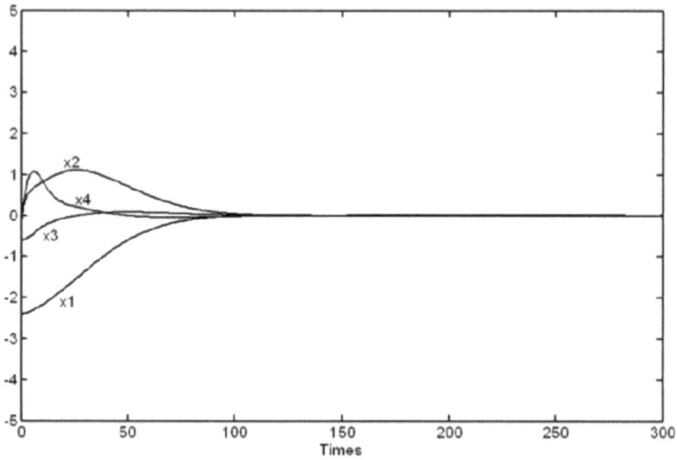

Fig. 11.16 Responses of four states of ball and beam system under the control of the trained FLNFN controller

Table 11.3 Comparison of performance of various controllers in Example 11.3

Method	FLNFN Controller	FALCON Controller [8]	FLNN Controller [17]	TSK-type NFN Controller [12]
Training Steps	500	50000	1000	500
Parameter Numbers	280 parameters (14 rules)	280 parameters (28 rules)	317 parameters	286 parameters (22 rules)
RMS errors	0.056	0.2	0.153	0.079

$$\begin{bmatrix} y_{p1}(k+1) \\ y_{p2}(k+1) \end{bmatrix} = \begin{bmatrix} 0.5\frac{y_{p1}(k)}{1+y_{p2}^2(k)} \\ 0.5\frac{y_{p1}(k)y_{p2}(k)}{1+y_{p2}^2(k)} \end{bmatrix} + \begin{bmatrix} u_1(k) \\ u_2(k) \end{bmatrix}. \tag{11.31}$$

The controlled outputs should follow the desired output y_{r1} and y_{r2} as specified by the following 250 pieces of data;

$$\begin{bmatrix} y_{r1}(k) \\ y_{r2}(k) \end{bmatrix} = \begin{bmatrix} \sin(k\pi/45) \\ \cos(k\pi/45) \end{bmatrix}. \tag{11.32}$$

The inputs of the FLNFN controller are $y_{p1}(k)$, $y_{p2}(k)$, $y_{r1}(k)$ and $y_{r2}(k)$, and the outputs are $u_1(k)$ and $u_2(k)$. After 500 training iterations, four fuzzy rules are generated. In this example, the proposed FLNFN controller is compared to the FLNN controller [17] and the TSK-type NFN controller [12]. Each of the controllers is applied to control the MIMO plant. To demonstrate the performance of the proposed controller, Figures 11.17(a) and (b) plot the control results of the desired output and the model output using FLNFN controller. Figures 11.17(c) and (d) show the error curves of the FLNFN controller and the TSK-type NFN controller. Table 11.4

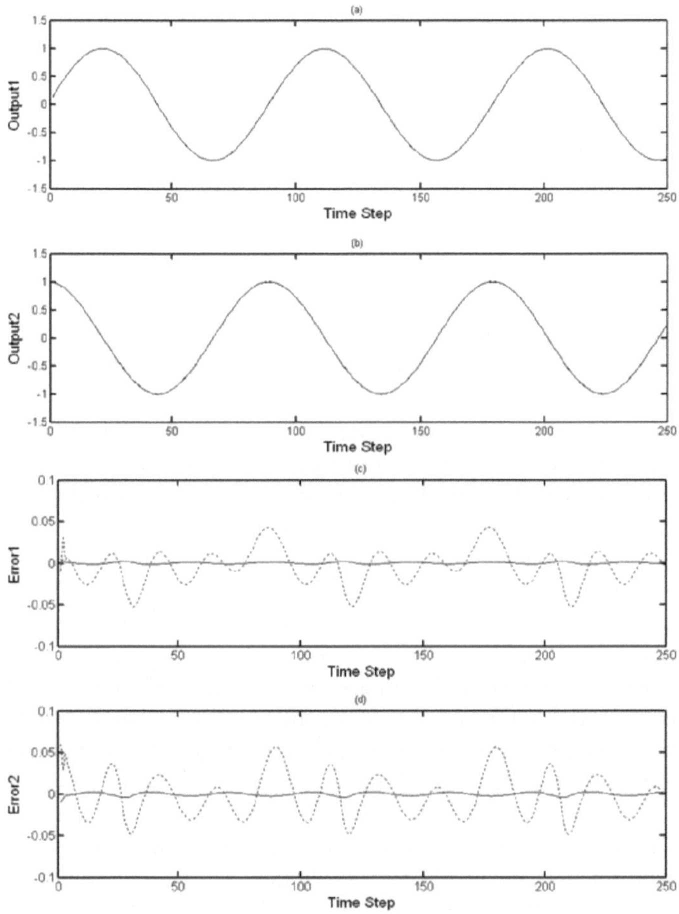

Fig. 11.17 Desired output (solid line) and model output using FLNFN (dotted line) of (a) output 1; (b) output 2 in Example 11.4. Error curves of FLNFN controller (solid line) and TSK-type NFN controller (dotted line) for (c) output 1; and (d) output 2

presents the RMS errors of the FLNFN controller, the FLNN controller and the TSK-type NFN controller. Table 11.4 shows that, according to the simulation results, the proposed FLNFN controller is better than the other controllers.

11.5 Conclusion and Future Works

This study proposes a FLNFN structure for nonlinear system control. The FLNFN model uses a FLNN to the consequent part of the fuzzy rules. The FLNFN model can automatically construct and adjust free parameters by performing online struc-

Table 11.4 Comparison of performance of various controllers in Example 11.4

Method	FLNFN Controller	FLNN Controller [17]	TSK-type NFN Controller [12]
Training Steps	500	1000	500
Parameter Numbers	128 parameters (4 rules)	161 parameters	140 parameters (10 rules)
RMS errors	0.0002	0.0738	0.0084

ture/parameter learning schemes concurrently. Finally, the proposed FLNFN model yields better simulation results than other existing models under some circumstances.

Three advanced topics on the proposed FLNFN model should be addressed in future research. First, the FLNFN model will tend to apply high-order nonlinear or overly complex systems if it can suitably adopt the consequent part of a nonlinear combination of input variables, and a functional expansion of multiple trigonometric polynomials. Therefore, it should be analyzed to see how many trigonometric polynomials are needed for future functional expansion. Secondly, the back-propagation technique is slow for many practical applications. To speed up the convergence and make the algorithm more practical, some variations are necessary. For example, the heuristic techniques include such ideas as varying the learning rate and using momentum [29], and the standard numerical optimization techniques in the BP procedure. In the standard numerical optimization techniques, the conjugate gradient algorithm [30] and the Levenberg-Marquardt algorithm [31] have been applied to the training of neural networks and shown a faster convergence than the basic BP algorithm. On the other hand, since the back-propagation technique is used to minimize the error function, the results may reach the local minima solution. In future work, we will adopt genetic algorithms (GAs) [26, 27] to solve the local minima problem. A GA is a global search technique. Because it can simultaneously evaluate many points in the search space, it is more likely to converge toward the global solution. Thirdly, it would be better if the FLNFN model has the ability to delete unnecessary or redundant rules. The fuzzy similarity measure [28] determines the similarity between two fuzzy sets in order to prevent existing membership functions from being too similar.

References

1. Lee, C. C.: Fuzzy logic in control systems: fuzzy logic controller-Parts I, II. *IEEE Trans. Syst., Man, Cybern.*, 20(2), 404–435, 1990.
2. Narendra, K. S., Parthasarathy, K.: Identification and control of dynamical systems using neural networks. *IEEE Trans. Neural Networks*, 1(1), 4–27, 1990.
3. Astrom, K. J., Wittenmark, B.: Adaptive Control. MA: Addison-Wesley, 1989.
4. Lin, C. T., Lee, C. S. G.: Neural Fuzzy Systems: A Neuro-Fuzzy Synergism to Intelligent System. NJ: Prentice-Hall, 1996.

5. Mitra, S., Hayashi, Y.: Neuro-fuzzy rule generation: survey in soft computing framework. *IEEE Trans. Neural Networks*, 11(3), 748–768, 2000.
6. Sun, F., Sun, Z., Li, L., Li, H. X.: Neuro-fuzzy adaptive control based on dynamic inversion for robotic manipulators. *Fuzzy Sets and Systems*, 134(1), 117–133, 2003.
7. Wang, L. X., Mendel, J. M.: Generating fuzzy rules by learning from examples. *IEEE Trans. on Syst., Man, and Cybern.*, 22(6), 1414–1427, 1992.
8. Lin, C. J., Lin, C. T.: An ART-based fuzzy adaptive learning control network. *IEEE Trans. Fuzzy Systems*, 5(4), 477–496, 1997.
9. Lin, W. S., Tsai, C. H., Liu, J. S.: Robust neuro-fuzzy control of multivariable systems by tuning consequent membership functions. *Fuzzy Sets and Systems*, 124(2), 181–195, 2001.
10. Takagi, T., Sugeno, M.: Fuzzy identification of systems and its applications to modeling and control. *IEEE Trans. on Syst., Man, Cybern.*, 15, 116–132, 1985.
11. Li, C., Lee, C. Y.: Self-organizing neuro-fuzzy system for control of unknown plants. *IEEE Trans. Fuzzy Systems*, 11(1), 135–150, 2003.
12. Jang, J. S. R.: ANFIS: Adaptive-network-based fuzzy inference system. *IEEE Trans. on Syst., Man, and Cybern.*, 23, 665–685, 1993.
13. Juang, C. F., Lin, C. T.: An on-line self-constructing neural fuzzy inference network and its applications. *IEEE Trans. Fuzzy Systems*, 6(1), 12–31, 1998.
14. Takagi, H., Suzuki, N., Koda, T., Kojima, Y.: Neural networks designed on approximate reasoning architecture and their application. *IEEE Trans. Neural Networks*, 3, 752–759, 1992.
15. Mizutani, E., Jang, J. S. R.: Coactive neural fuzzy modeling. in *Proc. Int. Conf. Neural Networks*, 760–765, 1995.
16. Pao, Y. H.: Adaptive Pattern Recognition and Neural Networks. MA: Addison-Wesley, 1989.
17. Patra, J. C., Pal, R. N., Chatterji, B. N., Panda, G.: Identification of nonlinear dynamic systems using functional link artificial neural networks. *IEEE Trans. on Syst., Man, and Cybern.*, 29, 1999.
18. Pao, Y. H., Phillips, S. M., Sobajic, D. J.: Neural-net computing and intelligent control systems. *International Journal of Control*, 56(2), 263–289, 1992.
19. Patra, J. C., Pal, R. N.: A functional link artificial neural network for adaptive channel equalization. *Signal Process.*, 43, 181–195, 1995.
20. Wang, L. X., Mendel, J. M.: Fuzzy adaptive filters, with application to nonlinear channel equalization. *IEEE Trans. Fuzzy Systems*, 1(3), 161–170, 1993.
21. Tanomaru, J., Omatu, S.: Process control by on-line trained neural controllers. *IEEE Trans. on Ind. Electron.*, 39, 511–521, 1992.
22. Ku, C. C., Lee, K. Y.: Diagonal recurrent neural networks for dynamic systems control. *IEEE Trans. Neural Networks*, 6, 144–156, 1995.
23. Hauser, J., Sastry, S., Kokotovic, P.: Nonlinear control via approximate input-output linearization: the ball and beam example. *IEEE Transactions on Automatic Control*, 37, 392–398, 1992.
24. Psaltis, D., Sideris, A., Yamamura, A.: A multilayered neural network controller. *IEEE Contr. Syst.*, 8, 17–21, 1988.
25. Phillips, C. L., Nagle, H. T.: Digital Control System Analysis and Design. Prentice Hall, 1995.
26. Juang, C. F., Lin, J. Y., Lin, C. T.: Genetic reinforcement learning through symbiotic evolution for fuzzy controller design. *IEEE Trans. Syst., Man, Cybern. B*, 30(2), 17–21, 2000.
27. Lin, C. J., Chuang, H. C., Xu, Y. J.: Face detection in color images using efficient genetic algorithms. *Optical Engineering*, 45(4), 2006.
28. Lin, C. J., Ho, W. H.: An asymmetry-similarity-measure-based neural fuzzy inference system. *Fuzzy Sets and Systems*, 152(3), 535–551, 2005.
29. Rumelhart, R., Hinton, G., Williams, J.: Learning internal representations by error propagation. *Parallel Distributed Processing*, 318–362, 1986.
30. Charalambous, C.: Conjugate gradient algorithm for efficient training of artificial neural networks. *IEEE Proceedings*, 139, 301–310, 1992.
31. Hagan, M. T., Menhaj, M.: Training feedforward networks with the Marquardt algorithm. *IEEE Trans. Neural Networks*, 5, 989–993, 1994.

Chapter 12
An Adaptive Neuro-fuzzy Controller for Robot Navigation

Anmin Zhu and Simon X. Yang

Abstract Real-time autonomous navigation in unpredictable environments is an essential issue in robotics and artificial intelligence. In this chapter, an adaptive neuro-fuzzy controller is proposed for mobile robot navigation with local information. A combination of multiple sensors is used to sense the obstacles near the robot, the target location, and the current robot speed. A fuzzy logic system with 48 fuzzy rules is designed. Two learning algorithms are developed to tune the parameters of the membership functions in the proposed neuro-fuzzy model and automatically suppress redundant fuzzy rules from the rule base. The "dead cycle" problem is resolved by a state memory strategy. Under the control of the proposed neuro-fuzzy model, the mobile robot can preferably "see" the surrounding environment, avoid static and moving obstacles automatically, and generate reasonable trajectories toward the target. The effectiveness and efficiency of the proposed approach are demonstrated by simulation and experiment studies.

12.1 Introduction

Mobile robot navigation using only onboard sensors is an essential issue in robotics and artificial intelligence. For real-time autonomous navigation in unknown and changing environments, the robot should be capable of sensing its environment, interpreting the sensed information to obtain the knowledge of its location and the environment, planning a real-time trajectory from an initial position to a target with

Anmin Zhu
The College of Information Engineering, Shenzhen University, Shenzhen 518060, P.R.China, e-mail: azhu@szu.edu.cn

Simon X. Yang
The Advanced Robotics and Intelligent System (ARIS) Laboratory, School of Engineering, University of Guelph, Guelph, ON N1G 2W1, Canada, e-mail: syang@uoguelph.ca

obstacle avoidance, and controlling the robot direction and velocity to reach the target.

There are some conventional approaches to real-time autonomous navigation. Map-based methods [3, 8] combine a graph searching technique with obstacle pruning to generate trajectory for mobile robots. These graph-based methods first need to construct a data structure that is then used to find paths between configurations of the robot. The data structure tends to be very large, and the geometric search computation required is complex and expensive. In addition, these methods are usually used only for robot path planning without considering the robot motion control. Artificial potential field methods [4, 7] assumes that each obstacle in the environment exerts a repulsive force on the mobile robot, and the target exerts an attractive force. These two types of forces are combined at every step and used to determine the next step of the robot movement. These methods are faster than map-based methods, but have to consider robot as a point, being trapped at local minima; unintended stoppage between closely spaced obstacles; and oscillations in the presence of multiple obstacles or in narrow passages. In order to overcome shortcomings of the potential field methods, the vector field method was proposed and extended [6, 19]. These methods are faster then potential field and less trapped in local minima. However the methods ignored the dynamic and kinematic constraints of mobile robots, And may fail to choose the most appropriate direction and get trapped in a "dead cycle" situation. Neural network-based approaches [12, 21] use a neural field approach described by an integro-differential equation. It can be discretized to obtain a nonlinear competitive dynamical system affecting a set of artificial neurons. The next movement step of the robot depends on a neural dynamics mechanism, which actively selects a movement direction from a set of possible directions. However, the robot path is not continuous and the kinematic constraint of the robot is not considered in these algorithms.

Because of the ability of making inference under uncertainty, fuzzy logic approaches are proposed for controlling a mobile robot in dynamic unknown environments. Li and Yang [9] proposed an obstacle avoidance approach using fuzzy logic, but the input sensors are separately inferred, and only a few simple cases are shown in the paper. Yang and Patel [22] developed a navigation algorithm for a mobile robot system by combining a fuzzy logic architecture with a virtual centrifugal effect algorithm (VCEA). In this model, the goal seeking sub-problem and obstacle avoidance sub-problem are solved by two separate fuzzy logic systems. But this algorithm focus on the direction control without considering velocity control. Furthermore, it cannot solve the "dead cycle" problem in environments with an U-shaped obstacle. Aguirre and Gonzalez [1] proposed a perceptual model-based on fuzzy logic in a hybrid deliberative-reactive architecture. This model improved the performance in the two aspects of robot navigation: perception and reasoning. Fuzzy logic is used in different parts of the perceptual model, but the model focuses on map building, and thus is computationally expensive. Park and Zhang [15] proposed a dual fuzzy logic approach. But the design of the two 81 fuzzy rules is not systematized, and redundance is obvious. Fuzzy logic offers a framework for representing imprecise, uncertain knowledge. They make use of human knowledge in the form of linguistic rules. But the disadvantages are that fuzzy logic needs highly abstract heuristics, needs experts for rule discovery with data relationships, and more importantly it lacks self-organizing and self-tuning mechanisms. This makes it difficult to decide the parameters of the membership functions. Another drawback is lack of a systematic procedure to transform expert knowledge into the rule base. This results in many redundant rules in the rule base.

On the other hand, neural networks are model-free systems which are organized in a way to simulate the cells of a human brain. They learn from the underlying relationships of data. Neural networks have a self-learning capability, self-tuning capability, dot no need to know data relationships, and can be used to model various systems. Therefore, fuzzy logic and neural networks can be combined to solve the complex robot navigation control problem and improve the performance. Marichal *et al.* [11] presented a neuro-fuzzy controller by a three-layer neural network with a competitive learning algorithm for a mobile robot. It automatically extracts the fuzzy rules and the membership functions through a set of trajectories obtained from human guidance. Because it is difficult to determine the fuzzy rules for complex environments with obstacles, this model is suitable to very simple environments. Song *et al.* [17] developed a heuristic fuzzy neural network using a pattern-recognition approach. This approach can reduce the number of rules by constructing the environment (e.g. obstacles) using several prototype patterns. It is suitable for simple environments, because the more complex the environment is, the more difficult to construct the patterns are. Hagras *et al.* [5] developed a converging online-learning genetic algorithm mechanism for learning the membership functions of individual behaviors of an autonomous mobile robot. In that approach, a hierarchical fuzzy controller is used to reduce the number of rules, while the genetic algorithm is applied to tune the parameters of the membership functions. The robot needs to be equipped with a short time memory to store the previous 6000 actions and the robot has to go back to some previous positions to evaluate a new solution. Rusu and Petriu [16] proposed a neuro-fuzzy controller for mobile robot navigation in an indoor environment. Infrared and contact sensors are used for detecting targets and avoiding collisions. Two levels with several behaviors are designed for the controller. Fuzzy inference is used in each behavior. A neural network is used to tune the system parameters. A switching coordination technique is used to select the appropriate behavior. Command fusion is used to combine the output of several neuro-fuzzy subsystems. However, the design of the rule base for the controller is not clear. The meanings of system parameters are vague when being trained by neural network. Furthermore, there is no evidence that the controller is able to resolve the "dead cycle" problem. Some methods have been proposed to resolve the "dead cycle" problem, such as [13, 14], However, in these algorithms, the robot is considered as a point, and the robot velocity is not considered at all. Furthermore, these algorithms have to memorize the hit and leave points; this is difficult for real robot systems to realize. Besides the robot navigation, adaptive neuro-fuzzy based approaches have been applied to wide applications, such as robotic manipulator control [18], harmonic-drive robotic actuators [10], and simultaneous localization and mapping of mobile either robots or vehicles [2].

The advantages and disadvantages of previous models for the motion planning and control of a mobile robot using only onboard sensors are summarized in Table 12.1. The theoretical design, suitable to complicated environments, and "dead cycle" problem are the main problems of these existing approaches.

In this chapter, a novel adaptive neuro-fuzzy controller is presented for the reactive navigation of mobile robots in unknown environments. The work presented in this chapter is the summary and extension of our previous work [23,24]. The inputs of the controller are the environment information around the robot, including the obstacle distances obtained from the left, front and right sensor groups, the target direction, and the current robot speed. A set of linguistic fuzzy rules are developed to implement expert knowledge under various situations. The output signals from the fuzzy controller are the accelerations of left and right wheels, respectively. A learning algorithm based on neural network techniques will be developed to tune the parameters of the membership functions. Another learning algorithm is developed to autonomously suppress the redundant fuzzy rules. A state memory strategy is proposed for resolving the "dead cycle" problem. Under the control of the proposed adaptive neuro-fuzzy controller, the mobile robot can generate reasonable trajectories toward the target in various situations.

This chapter is organized as follows. Section 12.2 presents the overall structure of the proposed adaptive neuro-fuzzy controller to mobile robot navigation with limited sensors. The adaptive neuro-fuzzy controller is designed in Section 12.3. Sections 12.3.5 and 12.3.6 describe the algorithms to tune the model parameters, and to suppress redundant rules. The state memorizing strategy is given in Section 12.3.7. Sections 12.4 and 12.5 provide simulation and experiment results. Finally some concluding remarks are given in Section 12.6.

Table 12.1 Summary of the previous methods for mobile robot navigation

Methods (references)	Main advantages	Main disadvantages
Map-based [8]; [3]	Ideal simple and visual	Computation expensive, Not for real-time control
Potential field [7]; [4]	Faster than graph-based	Local minima, "Dead cycle" problem Consider robot to a point
VFH [19]; [6]	Faster than potential field Less trapped in local minima	Ignored dynamic constraints, Ignored dynamic constraints, "Dead cycle" problem
Neural networks-based [21]; [12]	Faster than potential field Without local minima	Discrete paths, Consider constant speed,
Fuzzy logic [22]; [15]	Various environments	No systematic design, Speed jump, "Dead cycle" problem
Neuro-fuzzy [16]; [11]	Various environments Learning capability	Difficult to design, Complicated, "Dead cycle" problem

12.2 The Overall Structure of the Neuro-fuzzy Controller

Navigation is a very easy task for human beings or animals. Like the human thinking process, while a mobile robot is moving in an unknown and changing environment, it is important to the compromise between avoiding collisions with obstacles and moving toward targets, depending on the sensed information about the environment. Without any obstacles, the robot moves straight toward the target, accelerating or decelerating according to the target direction and the distance between the robot and the target. When there are obstacles in the environment, the robot moves according to the distance to the obstacles and the location of the target. When obstacles are very close, the robot should slow down and make a turn to avoid the obstacles. The first part of a robot system are the sensor systems. Sensors must be mounted on the robot to sense the environment and interpret the sensed information. The main sensors of the mobile robot is shown in Figure 12.1. The robot has two front co-axle wheels driven by different motors separately, and a third passive omnidirectional caster. Through adjusting the accelerations of the two driven wheels respectively, the velocity and motion direction of the mobile robot can be controlled. For the reactive navigation to be easily realized, nine ultrasonic sensors are incorporated on the robot, so that the distances between the robot and the obstacles can be measured. These sensors are equipped on the left, the right and in the middle of the front part of the robot to cover a semicircular area around the front half of the robot and to protect the robot from collisions. The nine sensors are divided into three groups (each group has three sensors) to measure the distance from obstacles to the left, right and front of the robot. In order to reach a target, a simple optical range finder with a directing beam and a rotating mirror, a global positioning system (GPS), or a indoor positioning system [20] would be used to obtain the target direction. A speed meter is equipped on the robot to measure the current robot speed.

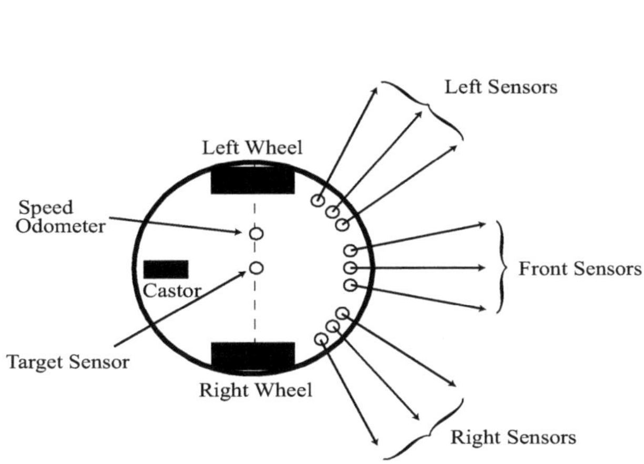

Fig. 12.1 The mobile robot model with onboard sensors

The robot motion is controlled by adjusting the speeds or accelerations of two driven wheels. At first, the robot system has to judge the distance as "far" or "close", the speed as "fast" or "slow" and so on, and then decide the motion commands. These informations ("far", "close", "fast", "slow", etc) are uncertain and imprecise. They are difficult to be represented by conventional logic systems, or mathematical models. However, they are easy for human beings to use, as people do not need precise, numerical information input to make a decision, and they are able to perform highly adaptive control regardless, because there is "fuzzy" concept in human knowledge.

"Fuzzy" concept of human knowledge can be used in logic systems. The logic system then becomes a fuzzy logic system, which is basically an expert system based on a conventional logic system, but the conditions and results in the "IF THEN" rules are described in fuzzy terms. Fuzzy logic is known to be an organized method for dealing with imprecise knowledge. Using linguistic rules, the fuzzy logic system mimics human decision-making to deal with unclearly expressed concepts, to deal with imprecise or imperfect information, and to improve knowledge representation and uncertain reasoning. Therefore, a fuzzy logic method is selected to deal with the sensor-based robot motion control problem. The proposed fuzzy logic method is simple, easy to understand, having the intelligence of human beings, and having a quick reaction capability.

Fuzzy logic offers a framework for representing imprecise, uncertain knowledge. But the disadvantages are that fuzzy logic needs highly abstract heuristics, needs experts for rule discovery with data relationships, and more importantly it lacks self-organizing and self-tuning mechanisms. This makes it difficult to decide the parameters of the membership functions. Another drawback is lack of a systematic procedure to transform expert knowledge into the rule base. This results in many redundant rules in the rule base. Neural networks are unmodel-based systems which are organized in a way to simulate the cells of a human brain. They learn from the underlying relationships of data. Neural networks have a self-learning capability, a self-tuning capability, do not need to know data relationships, and can be used to model various systems. Therefore, fuzzy logic and neural networks can be combined to solve the complex robot navigation control problem and improve the per-

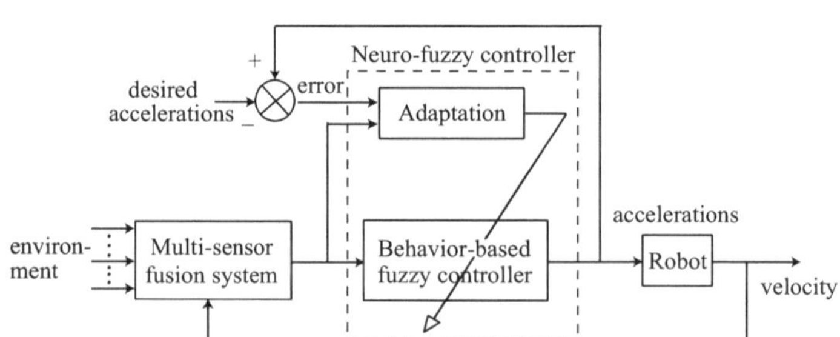

Fig. 12.2 A schematic diagram of the proposed adaptive neuro-fuzzy controller

formance. Figure 12.2 shows a brief structure of the proposed adaptive neuro-fuzzy controller, which consists of a fuzzy controller and a learning adaptation model, and the connection with other parts of the robot system, which consists of a multi-sensor fusion system and a robot acceleration control system.

12.3 Design of the Neuro-fuzzy Controller

The structure of the proposed adaptive neuro-fuzzy controller is shown in Figure 12.3. This is a five-layer neuro-fuzzy network. The inputs of the fuzzy controller are the outputs from the multi-sensor system: the obstacle distances d_l, d_f, d_r obtained from the left, front and right sensor groups, the target direction θ_d (that is the angle between the robot moving direction and the line connecting the robot center with the target), and the current robot speed r_s. The second layer denotes the terms of input membership variables. The third layer denotes the rule base. The fourth layer denotes the terms of output membership variables. In the fifth layer, the output signals from the fuzzy controller are the accelerations of left and right wheels, a_l and a_r, respectively.

There are differences between the proposed approach and most conventional neuro-fuzzy methods (e.g., [5, 11, 16, 17]). First of all, far fewer rules are needed in the proposed approach, while some conventional methods need a large number of rules. This simplifies the structure of the neuro-fuzzy model by reducing the number of the hidden layer neuron, and reduces the computational time. The physical meaning of the parameters remains the same during the training, which is lost in conventional neuro-fuzzy methods. The neural network based methods are developed to improve the performance in the proposed approach, including the parameter tuning and the redundant rule reducing. Lastly, a state memorizing strategy is designed to resolve the "dead cycle" problem in the proposed approach.

Fuzzification, inference mechanism, and defuzzification are considered to create the fuzzy logic controller. After that, neural network approaches are used to improve the performance.

12.3.1 Fuzzification and Membership Function Design

The fuzzification procedure maps the crisp input values to the linguistic fuzzy terms with membership values between 0 and 1. In most fuzzy logic systems, non-fuzzy input data is mapped to fuzzy sets by treating them as Gaussian, triangle, trapezoid, sharp peak membership functions, etc. In this chapter, triangle functions, S-type and Z-type functions are chosen to represent fuzzy membership functions. Because the number of the potential rules of fuzzy logic systems depends on the number of linguistic terms of the input variables, a relatively small number of terms for the input variables are selected. The input variables d_l, d_f and d_r are simply expressed using two linguistic terms: NEAR and FAR. The variable θ_d is expressed by three terms: LEFT, CENTER and RIGHT. The variable r_s is expressed by two terms: FAST and SLOW. The output variables a_l and a_r are expressed by five terms: PB, PS, Z, NS, and NB abbreviated from positive big, positive small, zero, negative small and negative big respectively, and denote a big increment, a small increment, no change, a small decrement, and a big decrement of accelerations of two wheels. It is noted that the real speeds of the two wheels are constrained in the desired upper bounds. The robot will stop whenever it approaches the target.

The membership functions for all the terms of the input and output variables in this controller are shown in Figure 12.4. The outputs of the fuzzification procedure are given as follows. For a triangle function,

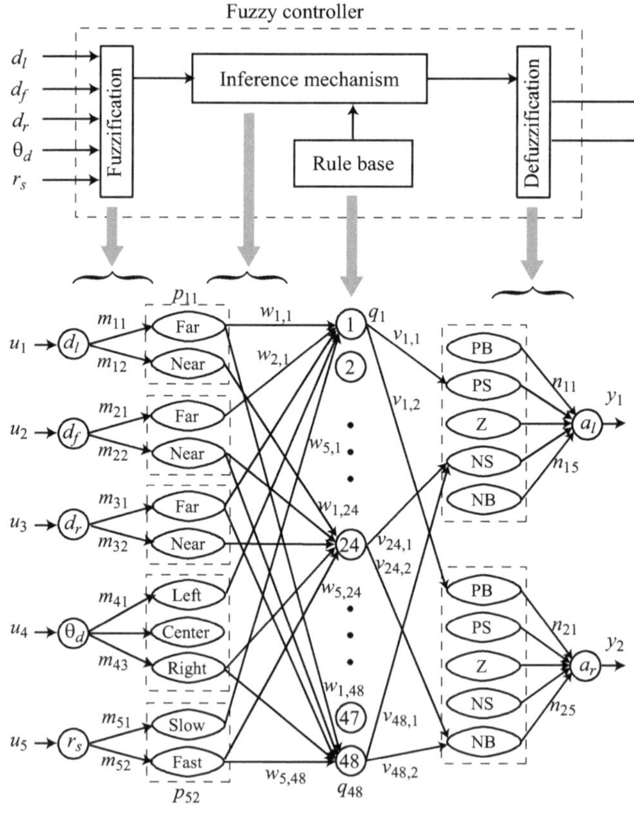

Fig. 12.3 Block diagram of the proposed neuro-fuzzy controller, d_l, d_f, d_r: obstacle distances to the left, front and right of the robot; θ_d: target direction; r_s: current robot speed; a_l, a_r: accelerations of the left and right wheels; $[u_1, u_2, u_3, u_4, u_5] = [d_l, d_f, d_r, \theta_d, r_s]$; m_{ij}: the centers of membership functions for input variables; n_{ls}: the centers of MFs for the output variables; p_{ij}: the degree of memberships; q_k: conjunction degree of IF part of rules; $w_{i,k}$: weights related to m_{ij}; $v_{k,l}$: weights related to n_{ls}; and $[y_1, y_2] = [a_l, a_r]$

$$p_{ij} = \begin{cases} 1 - \frac{2|u_i - m_{ij}|}{\sigma_{ij}}, & \text{if } m_{ij} - \frac{\sigma_{ij}}{2} < u_i < m_{ij} + \frac{\sigma_{ij}}{2}, \\ 0, & \text{otherwise}; \end{cases} \tag{12.1}$$

for a S-type function,

$$p_{ij} = \begin{cases} 0, & \text{if } u_i < m_{ij} - \frac{\sigma_{ij}}{2}, \\ 1, & \text{if } u_i > m_{ij}, \\ 1 - \frac{2|u_i - m_{ij}|}{\sigma_{ij}}, & \text{otherwise}; \end{cases} \tag{12.2}$$

for a Z-type function,

$$p_{ij} = \begin{cases} 0, & \text{if } u_i > m_{ij} + \frac{\sigma_{ij}}{2}, \\ 1, & \text{if } u_i < m_{ij}, \\ 1 - \frac{2|u_i - m_{ij}|}{\sigma_{ij}}, & \text{otherwise}, \end{cases} \tag{12.3}$$

where $i = 1, 2, \cdots, 5$, is the number of input signals; $j = 1, 2, \cdots, 5$, is the number of terms of the input variables; p_{ij} is the degree of membership for the i-th input corresponding to the j-th term of the input variable; u_i is the i-th input signal to the fuzzy controller, $[u_1, u_2, u_3, u_4, u_5] = [d_l, d_f, d_r, \theta_d, r_s]$; m_{ij} is the center of the membership function corresponding to the i-th input and the j-th term of the input variable; and σ_{ij} is the width of the membership function corresponding to the i-th input and the j-th term of the input variable. For example, u_4=θ_d, represents the input value about the target direction; $m_{42} = 0$ means the value of the membership function center related to the 2th term ("center") of the 4th input variable (u_4 or θ_d) is 0; and $\sigma_{42} = 90$ is the width of the membership function.

12.3.2 Inference Mechanism and Rule Base Design

The inference mechanism is responsible for decision-making in the fuzzy controller using approximate reasoning. The rule base is essential for the controller, and stores the rules governing the input and output relationship of the proposed controller. The inference rules are in same form as the example shown in Figure 12.5.

48 rules are formulated for the proposed controller given in Table 12.2. There are three behaviors in the rule base with TS representing the target seeking behavior, OA representing the obstacle avoidance behavior, and BF representing the barrier following behavior.

In general, the target seeking behavior is used to change the direction of the robot toward the target when there are no obstacles blocking the robot. The obstacle avoidance behavior is used to turn away from the obstacles disregarding the direction of the target when obstacles are close to the robot. The barrier following behavior is used to follow the obstacle by moving along or slightly toward the obstacle when the target is behind the obstacle and the obstacle is not too far or too close to the robot.

Table 12.2 The rule base of the proposed fuzzy logic controller. N: near; F: far; L: left; C: center; R: righ; SL: slow; FS: fast; PB: positive big, PS: positive small, Z: zero, NS: negative small, NB: negative big; TS: target seeking; OA: obstacle avoidance; BF: barrier following; ∗: removed rules

Rule	Input					Output			Redundant
No.	d_l	d_f	d_r	θ_d	r_s	a_l	a_r	behavior	rule
1	F	F	F	L	SL	PS	PB	TS	
2	F	F	F	L	FS	NS	Z	TS	
3	F	F	F	C	SL	PB	PB	TS	
4	F	F	F	C	FS	Z	Z	TS	
5	F	F	F	R	SL	PB	PS	TS	
6	F	F	F	R	FS	Z	NS	TS	
7	F	F	N	L	SL	Z	PS	TS	
8	F	F	N	L	FS	NS	Z	TS	
9	F	F	N	C	SL	Z	PS	OA	
10	F	F	N	C	FS	NS	Z	OA	
11	F	F	N	R	SL	NS	Z	BF	
12	F	F	N	R	FS	NB	NS	BF	∗
13	F	N	N	L	SL	NS	Z	OA	
14	F	N	N	L	FS	NB	NS	OA	∗
15	F	N	N	C	SL	NS	Z	BF	
16	F	N	N	C	FS	NB	NS	BF	∗
17	F	N	N	R	SL	NS	Z	BF	
18	F	N	N	R	FS	NB	NS	BF	∗
19	N	N	N	L	SL	NS	Z	BF	
20	N	N	N	L	FS	NB	NS	BF	∗
21	N	N	N	C	SL	NS	Z	BF	
22	N	N	N	C	FS	NB	NS	BF	∗
23	N	N	N	R	SL	Z	NS	BF	
24	N	N	N	R	FS	NS	NB	BF	
25	N	F	N	L	SL	Z	Z	BF	
26	N	F	N	L	FS	NS	NS	BF	
27	N	F	N	C	SL	PS	PS	TS	
28	N	F	N	C	FS	NS	NS	TS	
29	N	F	N	R	SL	Z	Z	BF	
30	N	F	N	R	FS	NS	NS	BF	
31	N	F	F	L	SL	Z	NS	BF	
32	N	F	F	L	FS	NS	NB	BF	
33	N	F	F	C	SL	PS	Z	OA	
34	N	F	F	C	FS	Z	NS	OA	∗
35	N	F	F	R	SL	PS	Z	TS	
36	N	F	F	R	FS	Z	NS	TS	
37	N	N	F	L	SL	Z	NS	BF	
38	N	N	F	L	FS	NS	NB	BF	
39	N	N	F	C	SL	Z	NB	BF	
40	N	N	F	C	FS	NS	NB	BF	∗
41	N	N	F	R	SL	Z	NS	OA	
42	N	N	F	R	FS	NS	NB	OA	
43	F	N	F	L	SL	NS	Z	OA	
44	F	N	F	L	FS	NB	NS	OA	∗
45	F	N	F	C	SL	NS	Z	BF	
46	F	N	F	C	FS	NB	NS	BF	∗
47	F	N	F	R	SL	Z	NS	OA	
48	F	N	F	R	FS	NS	NB	OA	

In every rule, the IF part is defined alternatively: If condition A is true, AND B is true, AND C is true, AND $\cdots$ with five conditions. Using the fuzzy logic operators, the AND can mathematically be represented by the min operator in the aggregation step. The output of the aggregation procedure to get degree of IF part of every rule is given as

$$q_k = \min\{p_{1k_1}, p_{2k_2}, p_{3k_3}, p_{4k_4}, p_{5k_5}\}, \tag{12.4}$$

where q_k is the conjunction degree of the IF part of the k-th rule, k =1, 2, $\cdots$, 48; and p_{ik_i} is the degree of the membership for the i-th input contributing to the k-th rule, $i = 1, 2, \cdots, 5$; $k_i = 1, 2, \cdots, 5$.

12.3.3 Defuzzification

The defuzzification procedure maps the fuzzy output from the inference mechanism to a crisp signal. There are many methods that can be used to convert the conclusion of the inference mechanism into the actual output of the fuzzy controller.

The "center of gravity" method is used in the proposed controller, combining the outputs represented by the implied fuzzy sets from all rules to generate the gravity centroid of the possibility distribution for a control action. The value of the output variables a_l and a_r are given as

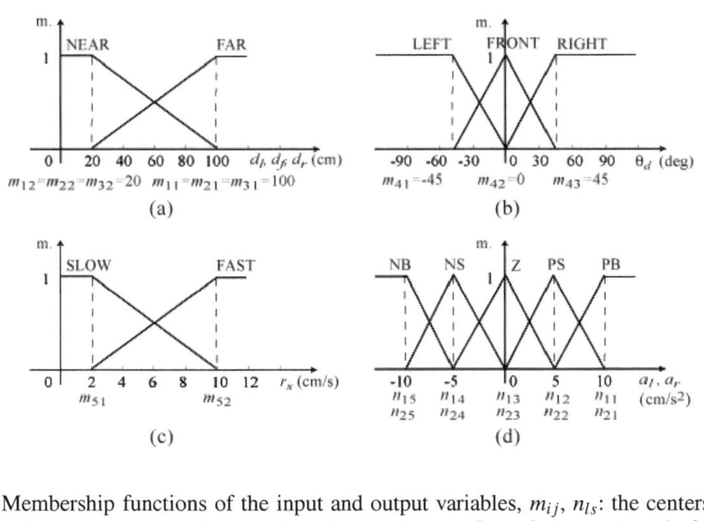

Fig. 12.4 Membership functions of the input and output variables, m_{ij}, n_{ls}: the centers of membership functions; m.: membership; meaning of NB to PB: see foot of (a) membership functions of obstacle distances; (b) membership functions of the target direction; (c) membership functions of the current robot speed; and (d) membership functions of the output variables accelerations of two wheels

IF

obstacle distance on left (d_l) is NEAR and

obstacle distance on front (d_f) is FAR and

obstacle distance on right (d_r) is FAR and

target direction (θ_d) is RIGHT and

current robot speed (r_s) is SLOW

THEN

acceleration of left wheel (a_l) is PB

acceleration of right wheel (a_r) is PS

Fig. 12.5 An example of the inference rules. d_l = {NEAR, FAR}, d_f = {NEAR, FAR}, d_r = {NEAR, FAR}, θ_d = {LEFT, Center, RIGHT}, r_s = {SLOW, FAST}, a_l = {PB, PS, Z, NS, NB}, a_r = {PB, PS, Z, NS, NB}

$$a_l = \frac{\sum_{k=1}^{48} v_{k,1} q_k}{\sum_{k=1}^{48} q_k}, \tag{12.5}$$

$$a_r = \frac{\sum_{k=1}^{48} v_{k,2} q_k}{\sum_{k=1}^{48} q_k}, \tag{12.6}$$

where $v_{k,1}$ and $v_{k,2}$ denote the estimated values of the outputs provided by the k-th rule, which are related to the center of membership functions of the output variables.

12.3.4 The Physical Meanings of Variables and Parameters

The proposed model keeps the physical meanings of the variables and parameters during the processing, while the conventional fuzzy logic based model [11] cannot keep the physical meanings of variables.

The physical meaning of fuzzy variables is explained in Figure 12.3. In this chapter, index $i = 1, \cdots, 5$ represents the number of inputs, $j = 1, \cdots, 5$ represents the number of the terms of the input variables, $k = 1, \cdots, 48$ represents the number of rules, $l = 1, 2$ represents the number of the output, $s = 1, \cdots, 5$ represents number of the terms of the output variables, and $[u_1, u_2, u_3, u_4, u_5] = [d_l, d_f, d_r, \theta_d, r_s]$ is the input vector used in (12.1), (12.2) and (12.3), $[y_1, y_2] = [a_l, a_r]$ is the output vector described in (12.5) and (12.6). The variable p_{ij} is the degree of membership for the i-th input corresponding to the j-th term of the input variable obtained by either (12.1), (12.2) or (12.3) according to different membership functions; q_k is conjunction degree of the IF part of the k-th rule obtained by (12.4); $w_{i,k}$ denotes the center of the membership function corresponding to the i-th input and the k-th rule, and can be assigned to one of the m_{ij} according to the rule base, e.g., $w_{1,1} = m_{11}$, $w_{2,1} = m_{21}$, $w_{5,1} = m_{51}$, $w_{1,24} = m_{,2}$, $w_{5,1} = m_{52}$, $w_{1,48} = m_{11}$, and $w_{5,48} = m_{52}$; and $v_{k,l}$ are the estimated value of the outputs provided by the k-th rule, which are related to one of the center of membership functions of the output variables. Assume n_{ls} denote the centers of the membership functions of variable a_l and a_r. Assume the width of the membership functions are constant (e.g., 1). Then $v_{1,1} = n_{12}$, $v_{1,2} = n_{21}$, $v_{24,1} = n_{14}$, $v_{24,2} = n_{25}$, $v_{48,1} = n_{14}$, and $v_{48,2} = n_{25}$.

12.3.5 Algorithm to Tune the Model Parameters

To smooth the trajectory generated by the fuzzy logic model, a learning algorithm based on a neural networks technique is developed. The effect of the output variables mainly depends on the center of the membership functions when the rule base is designed. The widths of the membership functions can be disregarded, and can usually be set to constant values. The membership function center values of the input and output variables may be improved by the neural networks learning property. The vector of 21 parameters to be tuned in the proposed model is set as

$$Z = \{m_{11}, m_{12}, m_{21}, m_{22}, m_{31}, m_{32}, m_{41}, m_{42}, m_{43}, m_{51}, m_{52}, \\ n_{11}, n_{12}, n_{13}, n_{14}, n_{15}, n_{21}, n_{22}, n_{23}, n_{24}, n_{25}\}. \tag{12.7}$$

One of the most widely used algorithms is the least-mean square (LMS) algorithm based on the idea of stochastic approximation method to adjust parameters of an application system. There are errors between the desired output and the actual output of the system shown in Figure 12.2. Using the errors, the LMS algorithm adjusts the system parameters, thus, altering its response characteristics by minimizing a measure of the error, thereby closing the performance loop. In this chapter, the LMS algorithm is used to minimize the following criterion function.

$$E = \frac{1}{2} \sum_{l=1}^{2} (y_l - \hat{y}_l)^2, \tag{12.8}$$

where $[y_1, y_2] = [a_l, a_r]$ is the output vector described in (12.5) and (12.6); and $[\hat{y}_1, \hat{y}_2]$ is the desired output vector, obtained by a derivation of the desired trajectory that is manually marked on the simulator. Thus, the parameters would be adapted as

$$Z(t+1) = Z(t) - \varepsilon \frac{\partial E}{\partial Z}, \tag{12.9}$$

where Z is the parameter vector to adapt; ε is the learning rate; $Z(t)$ is the parameter vector Z at time t; and t is the number of iterations. Thus, the equations for the adaptation of the parameters are given as

$$m_{ij}(t+1)=m_{ij}(t)-\varepsilon_m\frac{\partial E}{\partial m_{ij}},\begin{cases} i=1,\cdots,5, \\ j=1,2,3; \end{cases} \tag{12.10}$$

$$n_{ls}(t+1)=n_{ls}(t)-\varepsilon_n\frac{\partial E}{\partial n_{ls}},\begin{cases} l=1,2, \\ s=1,\cdots,5, \end{cases} \tag{12.11}$$

where ε_m and ε_n are the learning rates. Therefore, it is only necessary to calculate the partial derivative of the criterion function with respect to the particular parameter to get the expression for each of them. From (12.1)-(12.6), the corresponding expressions for the membership function centers of input and output variables are given as

$$\begin{aligned}\frac{\partial E}{\partial m_{ij}}&=\sum_{l=1}^{2}\left(\frac{\partial E}{\partial y_l}\frac{\partial y_l}{\partial q_k}\frac{\partial q_k}{\partial p_{ij}}\frac{\partial p_{ij}}{\partial m_{ij}}\right)\\ &=-2\sum_{l=1}^{2}\left[(y_l-\hat{y}_l)\frac{v_{k,l}\sum_{k=1}^{48}q_k-\sum_{k=1}^{48}v_{k,l}q_k}{(\sum_{k=1}^{48}q_k)^2}\right]\frac{\text{sign}(u_i-m_{ij})}{\sigma_{ij}};\end{aligned} \tag{12.12}$$

$$\frac{\partial E}{\partial n_{ls}}=\sum_{l=1}^{2}\left(\frac{\partial E}{\partial y_l}\frac{\partial y_l}{\partial v_{k,l}}\frac{\partial v_{k,l}}{\partial n_{ls}}\right)=(y_l-\hat{y}_l)\frac{q_k}{\sum_{k=1}^{48}q_k}. \tag{12.13}$$

The iterative procedure for the adaptation of the parameters and for the minimization of the criterion function are summarized in Figure 12.6. At first, design the center values of the membership function by experience, then get the weight vectors, set the learning rates, and set the tolerant rates. After that, read the sensor information and the desired output accelerations. Then get the output by fuzzy logic algorithm (including fuzzification, fuzzy inference, and defuzzification three steps). Then modify the parameters. Finally, evaluate the criterion function. If it is suitable to the criterion, the procedure is stopped, otherwise, repeat from reading sensor information to evaluating the criterion.

12.3.6 Algorithm to Suppress Redundant Rules

48 rules are defined in the fuzzy logic based model. However, it is difficult to define the controller rules accurately and without redundant rules, if the number of input variables increases, or the number of the terms of variables increases to fit a more complex environment. To solve the problem, a selection algorithm is added to the fuzzy controller model to suppress redundant fuzzy rules automatically.

After the training to tune the model parameters, the learning algorithm to suppress redundant rules is considered. It is clear from the Figure 12.3 that the variables $w_{i,k}$ and $v_{k,l}$, which can be obtained from the model parameters m_{ij} and n_{ls} described above in Section 12.3, determine the response of the fuzzy rules to the input signals. Every rule is related to a weight vector

$$W_k=\{w_{1,k},\cdots,w_{5,k},v_{k,1},v_{k,2}\},\quad k=1,2,\cdots,48. \tag{12.14}$$

BEGIN

set m_{ij}, n_{ls} and σ_{ij} by experience;

get the weight vectors $w_{i,k}$ and $v_{k,l}$ from m_{ij} and n_{ls};

set learning rates ε_m, ε_n;

set tolerance δ;

DO (Training procedure)

Data input $\{ u_1, u_2, u_3, u_4, u_5 \}$ and $\{ \hat{y}_1, \hat{y}_2 \}$;

Fuzzification;

Fuzzy inference;

Defuzzification;

Data output $\{ y_1, y_2 \}$;

Adaptation of parameters m_{ij};

Adaptation of parameters n_{ls};

Evaluation of the criterion function E;

UNTIL $(E < \delta)$;

END

Fig. 12.6 The algorithm to tune the parameters

If the Euclidean distance between two weight vectors is small enough, both vectors will generate similar rules in the sense that a similar result is obtained for the same input. So, by calculating the Euclidean distances between the weight vectors, the redundant rules can be reduced. Based on this idea, the proposed algorithm is summarized in Figure 12.7. At first, design the center values of the membership function by experience, then get the weight vectors, normalize the weights, and set the tolerant rates. After that, compare the Euclidean distance of every two vectors. If the distance is less than the tolerant value, remove one of the rules, which relate to the two vectors.

After applying the algorithm, a minimum number of rules is obtained. Thus, the minimum number of nodes in the second layer of the structure in Figure 12.3 is obtained. For example, if the environment is simple, the tolerance can be chosen to be relatively large. Consequently, some of the rules will be suppressed and the number of useful rules will be smaller than 48. This algorithm has obvious benefits over rule bases with hundreds or even thousands of fuzzy rules.

12.3.7 State Memorizing Strategy

Usually, the fuzzy behavior-based robot, like other reactive robots, suffers from the "symmetric indecision" and the "dead cycle" problems. The proposed model resolves the "symmetric indecision" problem by designing several mandatory-turn rules (e.g. Rules 21, 22, 45, 46). When an obstacle is near the front of the robot, and no obstacles exist on either side or the same situated obstacles exist on either side, the robot will turn left (or right) without hesitation. But the "dead cycle" problem may still occur in some cases, even if the barrier following behavior has been added to the fuzzy inference rule base. For example, a robot wanders indefinitely in a loop inside a U-shaped obstacle, because it does not remember the place visited before, and its navigation is only based on the local sensed environment.

```
BEGIN
    set m_ij, n_ls and σ_ij by experience;
    get the weight vectors w_i,k and v_k,l from m_ij and n_ls;
    normalize the values of weights to [0, 1];
    assign the number of rules to variable N;
    set tolerance δ;
    initialize index N_1 = 1; N_2 = 1;
    WHILE ( N_1 < N ) DO
        N_2 = N_1 + 1;
        WHILE ( N_2 < N + 1) DO
            d_N1N2 = ||W_N1 - W_N2||; //distance;
            IF ( d_N1N2 < δ ) THEN
                N = N - 1; //decrease number of rules;
                FOR ( N_3 = N_2 to N ) DO
                    W_N3 = W_N3+1; //reduce the vector W_N2;
                ENDDO
            ENDIF
            N_2 = N_2 + 1;
        ENDDO
        N_1 = N_1 + 1;
    ENDDO
END
```

Fig. 12.7 The algorithm to suppress redundant rules

A typical "dead cycle" situation is shown in Figure 12.8(a). First, the robot moves directly toward the target according to Rules 3 and 4, using the target seeking behavior, because there are no obstacles sensed in the front of the robot and the robot thinks this is the ideal shortest path to the target. When the robot detects obstacles directly in front, it makes a left turn according to Rules 21 and 22, using the barrier following behavior. After that the robot goes to the left along the obstacles according to Rules 11, 12, 17, and 18, using the barrier following behavior, because the obstacles are on the right side of the robot and the target is behind the obstacles. As a result, the robot moves along the curved path from Position 1 to 3 via 2. At Position 3, the robot turns to the left according to the barrier following behavior, since the obstacles exist on the right side and in the front side of the robot while the left side of the robot is empty and the target is on the left behind the robot. After Position 3, the robot is attracted to the target according to the target seeking behavior because there are either no obstacles or far away obstacles detected by the robot. Similarly, the robot follows the path from Position 4 to 6 via 5, and then to Position 1. A "dead cycle" occurs.

By careful examination of the above "dead cycle" path under the control of the fuzzy rules, it can be found that Positions 3 and 6 are the critical points resulting in the "dead cycle" path. At Position 3, if the robot can go straight instead of a left turn, the problem may be resolved. Without changing the designed control rule base, if the robot can assume the target just on the right front side of the robot and behind the obstacles at Position 3, the robot will go straight and try to pass around the obstacles. Based on this idea, an assistant state memorizing strategy is developed for the system. There are three states designed for the robot shown in Figure 12.9. State 0 represents the normal state or other situations except State 1 and 2; State 1 for both the target and obstacles being on the left side of the robot; and State 2 for both the target and obstacles being on the right side of the robot.

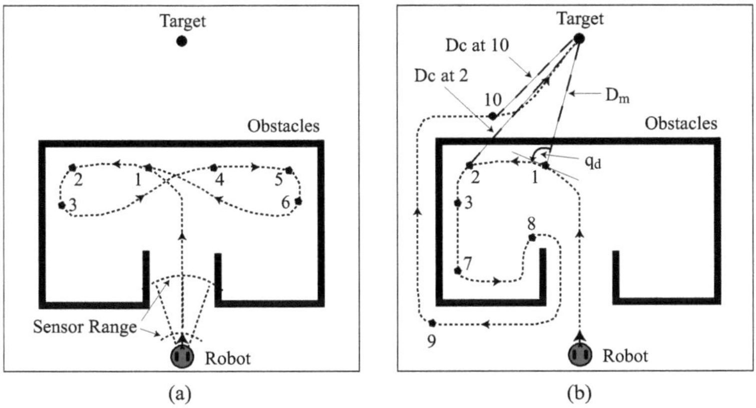

Fig. 12.8 Robot navigation in a "dead cycle" situation with limited sensor range. (a) without the proposed state memory strategy; and (b) with the proposed state memory strategy

The flow diagram of the algorithm is shown in Figure 12.10. Initially, the robot is in State 0, which means it is following the control rules. The robot changes to State 2 when the target and the obstacles on the right side and the target direction θ_d shown in Figure 12.8(b), which is the angle between the robot moving direction and the line connecting the robot center with the target angle, is increasing at Position 1, while the robot changes to State 1 if at Position 4. When the robot changes to State 1 or 2, the robot memorizes the state and the distance between the target and the robot denoted by D_m shown in Figure 12.8(b). During State 1 or 2, the robot assumes the target just on the front left or right side of the robot and behind the obstacles. Under this assumption and when the distance between the target and the current position D_c at Positions $1, 2, 3, 7, 8$ and 9 being longer than D_m memorized at Position 1, the robot goes along the curve through Positions $1, 2, 3, 7, 8$ and 9 in the Figure 12.8(b). The robot changes back to State 0 when D_c at Position 10, shown in Figure 12.8(b), is shorter than D_m. This means that the robot has passed round the obstacles. Finally, the robot will reach the target and stop there.

12.4 Simulation Studies

To demonstrate the effectiveness of the proposed fuzzy-logic-based controller, simulations using a mobile robot simulator (MobotSim Version 1.0.03 by Gonzalo Rodriquez Mir) are performed. The robot is designed as shown in Figure 12.1. The diameter of the robot plate is set to 0.25 m, distance between wheels is set as 0.18 m, wheel diameter is 0.07 m, and wheel width is 0.02 m. In addition to the target sensor and the speed odometer, there are nine ultrasonic sensors mounted on the front part of the robot. The angle between sensors is 20°. The sensor ring radius is 0.1 m. The radiation cone of the sensors is 25°. The sensing range of the ultrasonic sensors is from 0.04 m to 2.55 m. The upper bound of the wheel speed is 0.15 m/s. In every case, the environment is assumed to be completely unknown for the robot, except the target location; and the sensing range of the on-board robot sensors are limited.

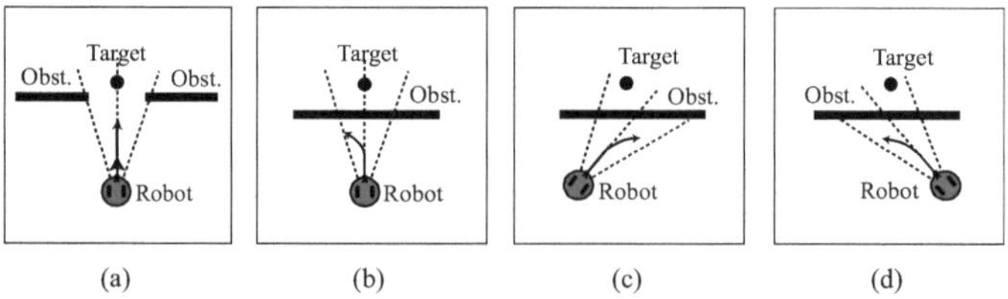

Fig. 12.9 (a) An illustration of robot states; (b) state 0; (c) state 1; and (d) state 2

12.4.1 Off-line Training Phase to Tune Model Parameters

Initially in the training phase, the robot moves under the control of the supervisor. The goal of the learning step is to adjust the parameters of the membership functions, and smooth the trajectory. To suit any situations for the mobile robot, the training environment should be designed in a relatively complicated way, where, there are left turns, right turns, straight stretches, and different obstacles for the robot. The workspace with a mobile robot, a target, and several obstacles is designed as in Figure 12.11(a).

According to the experience, the model parameters are initialized as

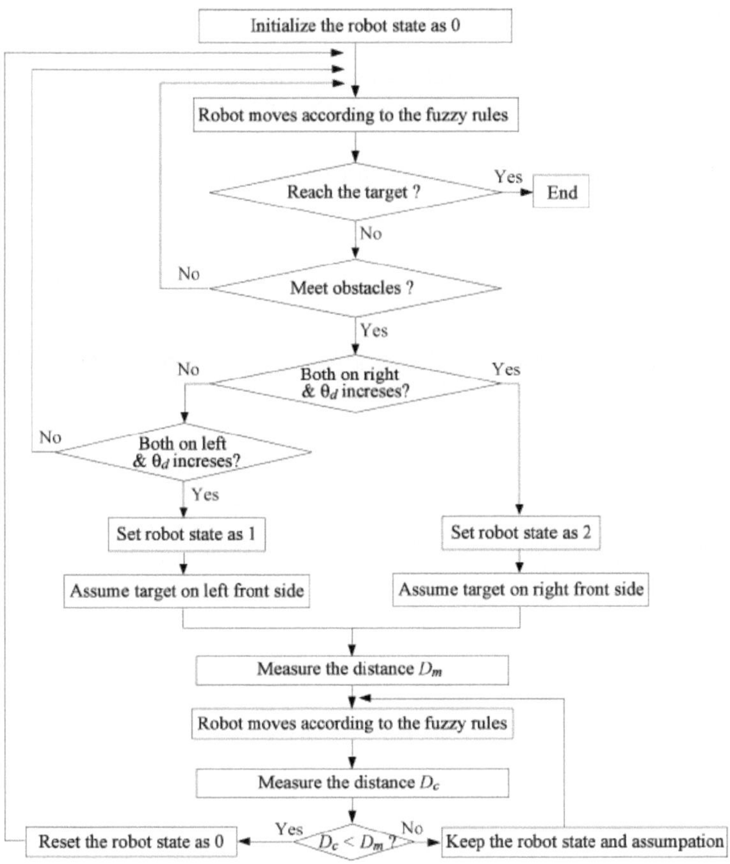

Fig. 12.10 Flow diagram of the developed state memorizing strategy

$$\{m_{11}, m_{12}, m_{21}, m_{22}, m_{31}, m_{32}, m_{41}, m_{42}, m_{43}, m_{51}, m_{52}\} =$$
$$\{100, 20, 100, 20, 100, 20, -45, 0, 45, 2, 10\}; \quad (12.15)$$
$$\{n_{11}, n_{12}, n_{13}, n_{14}, n_{15}, n_{21}, n_{22}, n_{23}, n_{24}, n_{25}\} =$$
$$\{10, 5, 0, -5, -10, 10, 5, 0, -5, -10\}; \quad (12.16)$$
$$\{\sigma_{11}, \sigma_{12}, \sigma_{21}, \sigma_{22}, \sigma_{31}, \sigma_{32}, \sigma_{41}, \sigma_{42}, \sigma_{43}, \sigma_{51}, \sigma_{52}\} =$$
$$\{160, 160, 160, 160, 160, 160, 90, 90, 90, 16, 16\}. \quad (12.17)$$

To get the desired outputs $[\hat{y}_1, \hat{y}_2]$, a reasonable path is drawn by an expert first as shown in Figure 12.11(b). Then the robot follows the path from the start position to the target position. The accelerations of the robot are then recorded at every time interval as the desired accelerations in the learning algorithm. During the training phrase, the adaptation is done at every time interval. The trajectory during the training phase is shown in Figure 12.11(c), where the trajectory is winding at the beginning period, but smoother at the end period. As mentioned before, the effect of the controller outputs mainly depends on the center of the membership functions, while the widths can be disregarded. The parameters are tuned as set in (12.7). In this simulation, after the learning, the better parameters of the membership functions are

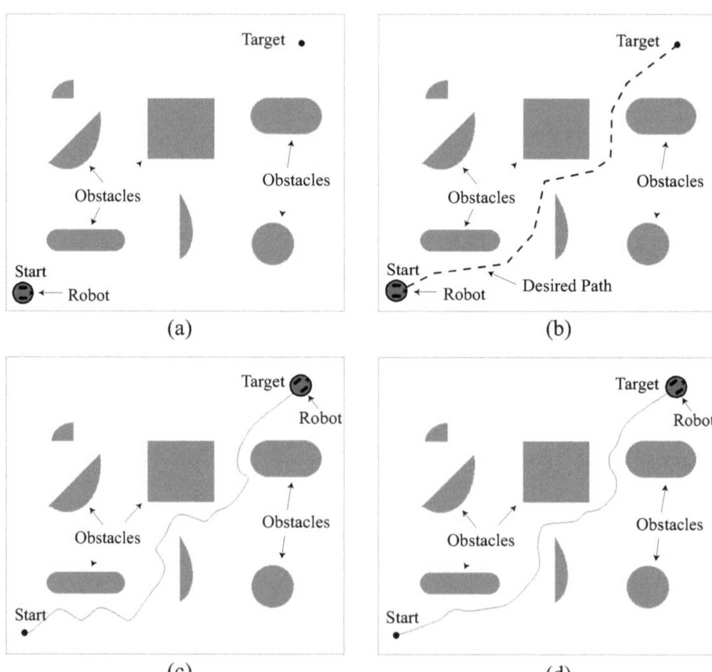

Fig. 12.11 The process of the training phrase. (a) the workspace; (b) the desired trajectory; (c) the trajectory using initial model parameters and during the training; (d) the trajectory using new model parameters got from the training

obtained as

$$\{m_{11}, m_{12}, m_{21}, m_{22}, m_{31}, m_{32}, m_{41}, m_{42}, m_{43}, m_{51}, m_{52}\} = \{94, 18, 105, 22, 95, 19, -48, 0, 47, 1.2, 9.4\}; \quad (12.18)$$

$$\{n_{11}, n_{12}, n_{13}, n_{14}, n_{15}, n_{21}, n_{22}, n_{23}, n_{24}, n_{25}\} = \{9.5, 5.4, 0, -5.5, -9.3, 9.4, 5.6, 0, -5.3, -9.4\}. \quad (12.19)$$

$$\{\sigma_{11}, \sigma_{12}, \sigma_{21}, \sigma_{22}, \sigma_{31}, \sigma_{32}, \sigma_{41}, \sigma_{42}, \sigma_{43}, \sigma_{51}, \sigma_{52}\} = \{160, 160, 160, 160, 160, 160, 90, 90, 90, 16, 16\}. \quad (12.20)$$

By using the new parameters, the mobile robot can work very well. The trajectory is shown in Figure 12.11(d). Figure 12.12(a) shows the smooth trajectories generated in other case by new model parameters in (12.18 and 12.19) after the training. But the trajectories are winding in Figure 12.12(b), if without the training phrase.

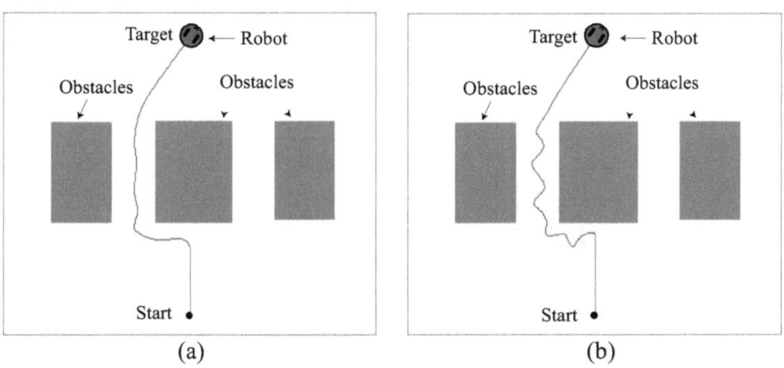

Fig. 12.12 Mobile robot trajectories (a) generating a smooth trajectory with the proposed learning algorithm; and (b) generating a winding trajectory without the learning algorithm

12.4.2 Off-line Training to Remove Redundant Rules

As mentioned above, every rule is related to a weight vector in (12.14). And the weight vectors depend on (12.18) and (12.19). Table 12.3 shows the weights related to the rules.

After the applying of the selection algorithm, the number of the rules will be less than 48, depending on the tolerance δ. Table 12.4 shows the relationship between the number of the useful rules and the tolerance δ. The bigger δ, the less the number of rules, but the poorer the performance.

The robot trajectories are shown in Figure 12.13 when δ = (1)0, (2)0.001, (3)0.005, (4)0.01. It is obvious that the trajectories in Figures 12.13(a), (b) and (c) are almost the same, except in Figures 12.13(d). This means that, with only 38 rules,

Table 12.3 The weights related to the rules after the tuning

Rule No.	w_{1k}	w_{2k}	w_{3k}	w_{4k}	w_{5k}	v_{k1}	v_{k2}
1	94	105	95	-48	1.2	5.4	9.4
2	94	105	95	-48	9.4	-5.5	0
3	94	105	95	0	1.2	9.5	9.4
4	94	105	95	0	9.4	0	0
5	94	105	95	47	1.2	9.5	5.6
6	94	105	95	47	9.4	0	-5.3
7	94	105	19	-48	1.2	0	5.6
8	94	105	19	-48	9.4	-5.5	0
9	94	105	19	0	1.2	0	5.6
10	94	105	19	0	9.4	-5.5	0
11	94	105	19	47	1.2	-5.5	0
12	94	105	19	47	9.4	-9.3	-5.3
13	94	22	19	-48	1.2	-5.5	0
14	94	22	19	-48	9.4	-9.3	-5.3
15	94	22	19	0	1.2	-5.5	0
16	94	22	19	0	9.4	-9.3	-5.3
17	94	22	19	47	1.2	-5.5	0
18	94	22	19	47	9.4	-9.3	-5.3
19	18	22	19	-48	1.2	-5.5	0
20	18	22	19	-48	9.4	-9.3	-5.3
21	18	22	19	0	1.2	-5.5	0
22	18	22	19	0	9.4	-9.3	-5.3
23	18	22	19	47	1.2	0	-5.3
24	18	22	19	47	9.4	-5.5	-9.4
25	18	105	19	-48	1.2	0	0
26	18	105	19	-48	9.4	-5.5	-5.3
27	18	105	19	0	1.2	5.4	5.6
28	18	105	19	0	9.4	-5.5	-5.3
29	18	105	19	47	1.2	0	0
30	18	105	19	47	9.4	-5.5	-5.3
31	18	105	95	-48	1.2	0	-5.3
32	18	105	95	-48	9.4	-5.5	-9.4
33	18	105	95	0	1.2	5.4	0
34	18	105	95	0	9.4	0	-5.3
35	18	105	95	47	1.2	5.4	0
36	18	105	95	47	9.4	0	-5.3
37	18	22	95	-48	1.2	0	-5.3
38	18	22	95	-48	9.4	-5.5	-9.4
39	18	22	95	0	1.2	0	-9.4
40	18	22	95	0	9.4	-5.5	-9.4
41	18	22	95	47	1.2	0	-5.3
42	18	22	95	47	9.4	-5.5	-9.4
43	94	22	95	-48	1.2	-5.5	0
44	94	22	95	-48	9.4	-9.3	-5.3
45	94	22	95	0	1.2	-5.5	0
46	94	22	95	0	9.4	-9.3	-5.3
47	94	22	95	47	1.2	0	-5.3
48	94	22	95	47	9.4	-5.5	-9.4

Table 12.4 The relationship between the number of the useful rules and the tolerance δ

tolerance δ	0	0.0005	0.001	0.005	0.01	0.05
numberofrules	48	48	47	38	26	18

the system can get a reasonable result in making the robot navigate. There are 10 redundant rules that the system automatically removes. They are marked in Table 12.2. According to the experience, the redundant rules are Rules 12, 14, 16, 18, 20, 22, 34, 40, 44 and 46 in Table 12.2. This can be explained from the rule based directly. For example, Rule 40 is similar to Rule 39, and is therefore redundant.

12.4.3 Effectiveness of the State Memory Strategy

In this section, after suppressing the redundant rules, the proposed controller is applied to the typical complicated situation that causes the "dead cycle" problem existing in many conventional approaches.

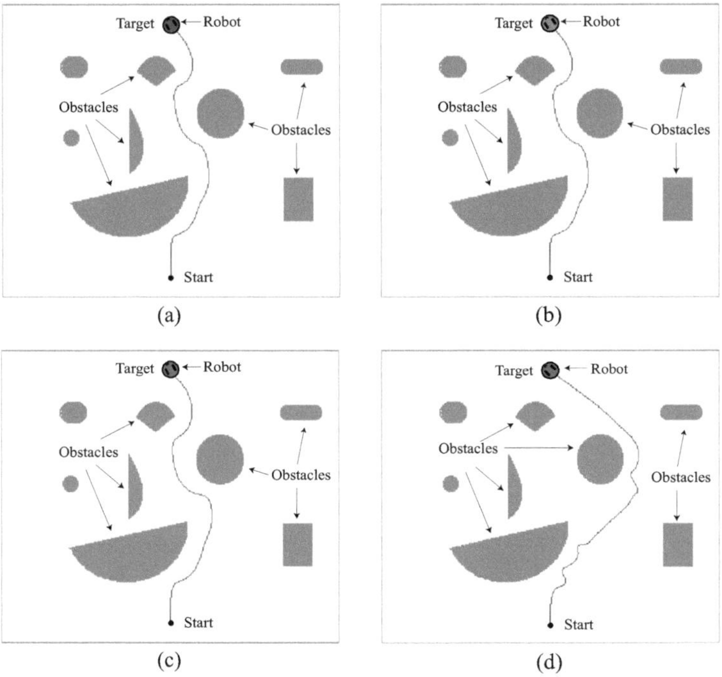

Fig. 12.13 Robot trajectories with different number of rules when the tolerance δ is selected (a) 0 with 48 rules; (b) 0.001 with 47 rules; (c) 0.005 with 38 rules; and (d) 0.01 with 26 rules

A robot moving in a U-shaped obstacles situation is shown in Figure 12.14. Because of the limited sensors, the robot will get in the U-shaped area first, but by the state memory strategy, the robot will escape from the trap and eventually reaches the target. It shows that the robot trajectory from the start position to the target does not suffer from the "dead cycle" problem.

12.4.4 Dynamic Environments and Velocity Analysis

Figure 12.15(a) shows the trajectory of the robot in a dynamic environment that the target moves in a cycle. Assume a target is moving in a cycle from the position (14, 6) around the point (10, 6) with 4 radius and clockwise direction, while the robot starts from position (10, 18) in the workspace and is moving to the right. At the beginning, the robot turns left, and then follows the target with a smooth trajectory.

The recorded velocity profile of both wheels in this simulation is presented in Figure 12.15(b). It can be seen from this figure that at the beginning robot turns left, the velocity of the right wheel increases much more than the velocity of the left wheel. The velocity of the left wheel increases a little and the velocity of the right wheel decreases a little when the robot turns right during its navigation procedure.

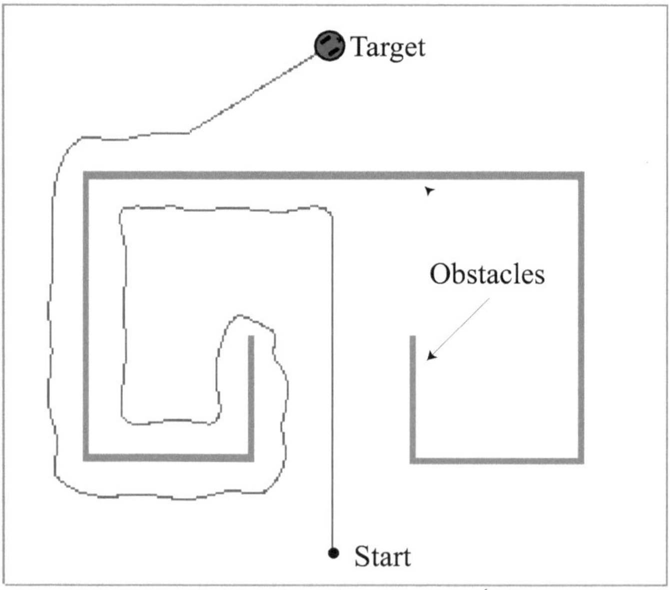

Fig. 12.14 Robot navigation in a complicated situation without suffering from the "dead cycle" problem

In the second case, the robot navigation in a complicated environment is demonstrated in Figure 12.16(a), where the robot meets some static or movable obstacles and the target is also movable. Assume that the target moves in a line from the position (5, 5) to (16, 5) and then back to (5, 5); besides some static obstacles, one robot considering an obstacle is moving in a line from the position (3, 12) to (16, 12) and back to (3, 12); one robot is moving in a cycle from the position (15, 9) around the point (11.7, 9) with a 3.3 radius and an anticlockwise direction; another robot is moving randomly; and the controlled robot with a direction to the right starts from position (10, 18) in the workspace. A smooth trajectory is executed in obstacle avoidance and travelling to the target.

The recorded velocity profile of both wheels in this simulation is presented in Figure 12.16(b). It can be seen from this figure that at the beginning, the velocities of both wheels of the controlled robot increases; and the velocity of the right wheel increases and the velocity of the left wheel decreases when the robot turns to the left to avoid obstacles. The velocity increases or decreases a little when the robot makes a small turn, and changes much more for large turns.

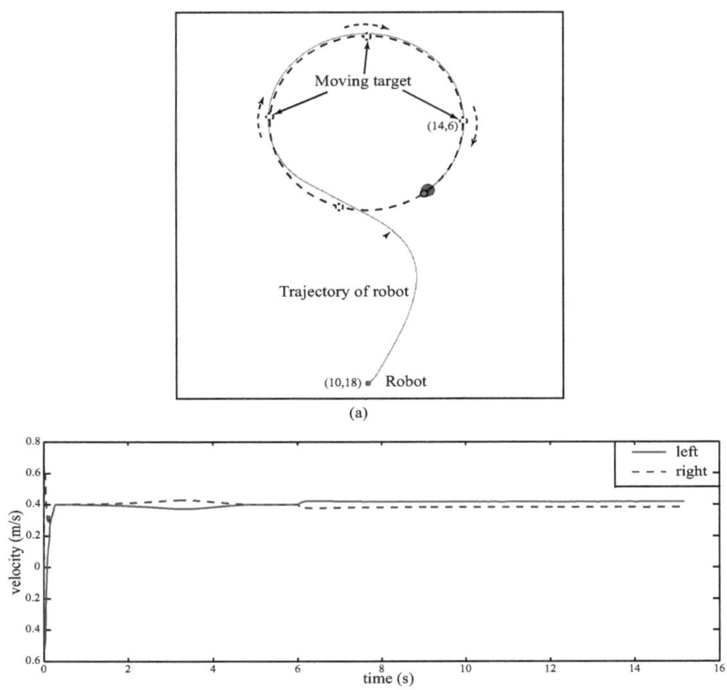

Fig. 12.15 Robot navigation in a dynamic environment with a moving target along a cycle. (a) the generated trajectory; (b) the velocities of the robot

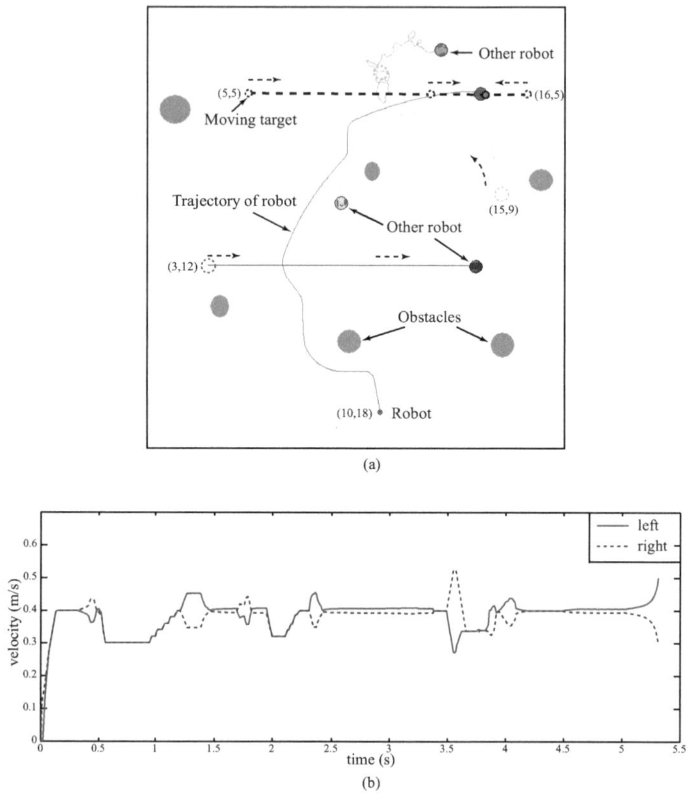

Fig. 12.16 Robot navigation in a dynamic environment with a moving target and moving obstacles. (a) the generated trajectory; (b) the velocities of the robot

12.5 Experiment of Studies

As a test bed, real robots were employed to test the performance of the proposed neuro-fuzzy approach, which navigated autonomously in an unknown environment using onboard sensors.

The shape of robots used as a test bed is shown in Figure 12.17. The left one includes the respective mechanical structure, electronic modules as well as the software development kit (SDK). The mechanical structure is already pre-built, including two 70mm diameter separated driving wheels with the maximum speed 9 meter per minute, animated head system, five servos, two direct current (DC) motors and so on. The electronic system is setup with a multimedia controller (PMB5010), a sensing-motion controller (PMS5005) and various peripheral electronic modules, includes one color image module with camera, four ultrasonic range sensor modules, two pyroelectric human motion sensor modules, one tilt/acceleration sensor model, one ambient temperature sensor module, one sharp infrared distance measuring sensor module, four rotary sensor modules, one wireless communication module, one microphone, one speaker, one battery pack, and so on. The software component is installed on a PC, being responsible to establish a wireless connection and exchange data with the robot. The users can develop their own applications in VC++ or VB using application programming interface (API), which can access the sensor information, send control commands, and configure the system settings. The right robot is the upgraded model of the left one.

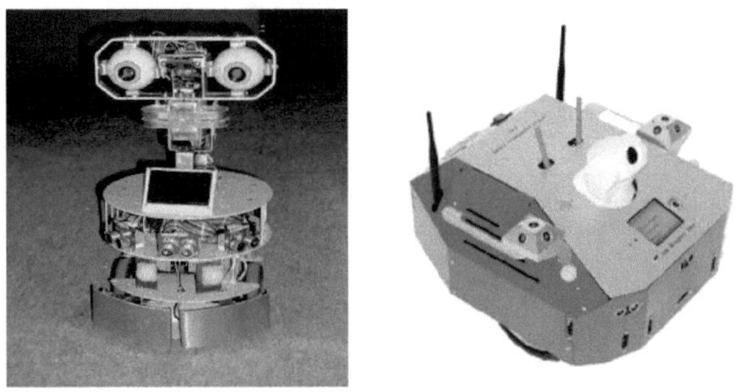

Fig. 12.17 Real robots used as a test bed

Two issues are worth mentioning here. First, because of the limitation of the equipments of the robot, only three ultrasonic sensors with the range between 0.04 m to 2.55 m are used to get the obstacle distances from the robot at left, in front, and at right. The direction of the target, and the distance between the robot and the target, are ignored. In this situation, the robot can wander on a level floor with obstacle avoidance using the proposed neuro-fuzzy approach. Secondly, because of the sensor noise and the misreading of the sensors, a simple sensor fusion algorithm is used to filter the noise of sensors. Based on these situations, two experiments in a variety of environments are given as follows.

A situation of the robot avoiding the obstacle when it is encountered on the left side of the robot is shown in Figure 12.18. Figure 12.18(a) shows the simulation trajectory of the robot in this situation, where the robot turns right to avoid the obstacle on the left side, then goes straight and turns left when it meets obstacles on the right side. Figures 12.18(b), (c), (d), and (e) show the pictures of the real robot at position 1, 2, 3, and 4 during its navigation. Figure 12.18(f) presents the recorded sensor profiles of left, front and right sensors of the robot in this experiment. Figure 12.18(g) presents the recorded velocity profiles of left and right wheels of the robot in this experiment.

A situation in which the robot avoids the obstacle when the obstacle is encountered in front of the robot is shown in Figure 12.19. Figure 12.19(a) shows the simulation trajectory of the robot when it turns left to avoid the obstacle in front of robot, then goes straight and turns left again when it meets obstacles on the right side. Figures 12.19(b), (c), (d), and (e) show the pictures of the real robot at position

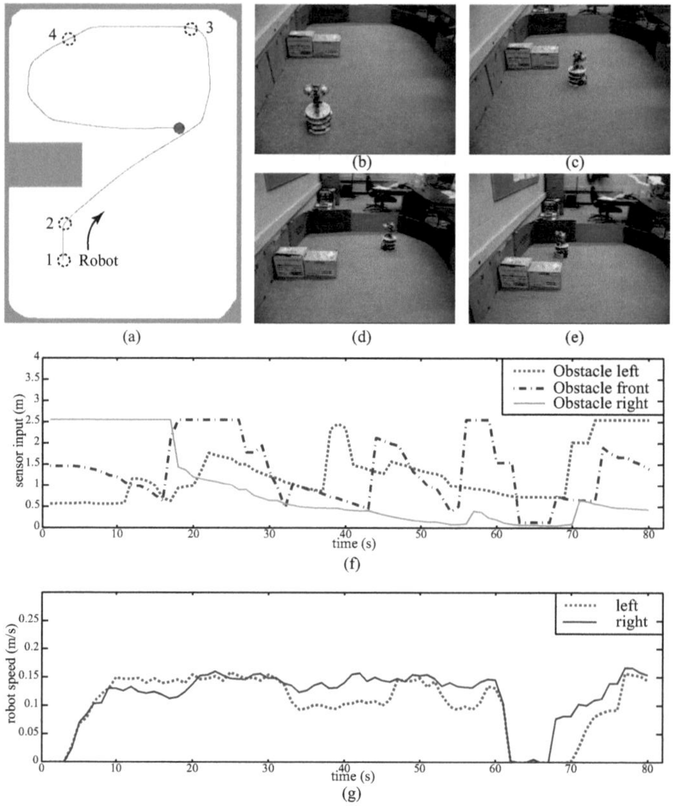

Fig. 12.18 Robot moves when obstacles on its left (a) trajectory; (b)-(e) snapshots; (f) sensor input; and (g) robot speed

1, 2, 3, and 4 during its navigation. Figure 12.19(f) presents the recorded sensor profiles of left, front and right sensors of the robot in this experiment. Figure 12.19(g) presents the recorded velocity profiles of left and right wheels of the robot in this experiment.

From the movies of the experiments, we can see that the robot can go smoothly without obvious oscillation in a workspace with different situations. From the robot speed analysis of the experiments, we can see that the speeds of the left and right wheels of the robot are smooth. In general, "smoothness" is synonymous to "having small high-order derivatives". Smoothness is relative. It can be seen obviously from a picture, or defined as the change less than a value. In this study, the "smoothness" is defined as without obvious oscillation. It can be defined as that the speed change is less than 0.05 m/s^2 and the angle change is less than 30 $degree/s$.

Comparing the performance between the experiment results and and the simulation results, the velocity of the simulation is smoother than the experiment. That is reasonable, because the input information, and the velocity control of the simulation

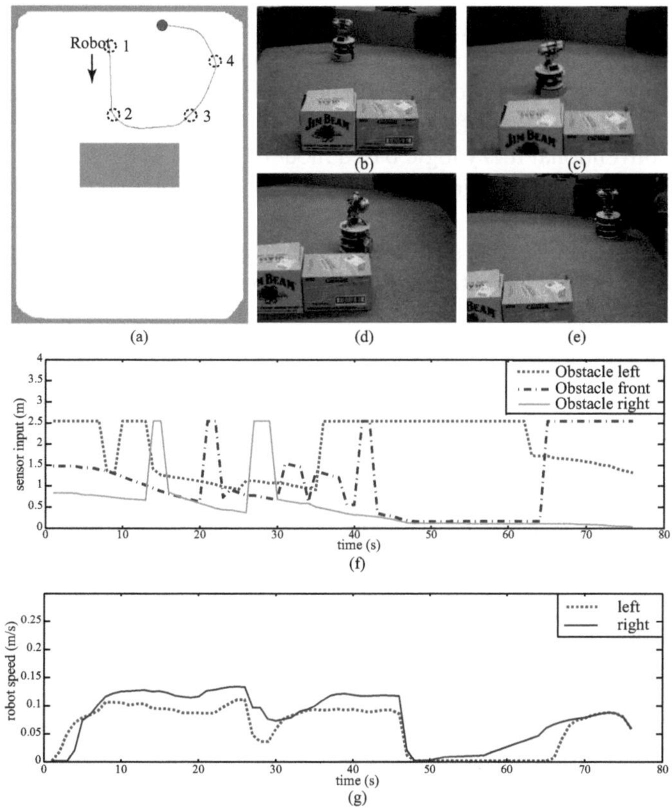

Fig. 12.19 Robot moves when obstacles are in front of it (a) trajectory; (b)-(e) snapshots; (f) sensor input; and (g) robot speed

are perfect, but in a practical robot, the sensor noise, and the physical moving of the real robot will infect the performance.

12.6 Summary

In this chapter, a novel fuzzy logic based control system, combining sensing and a state memory strategy, was proposed for real-time reactive navigation of a mobile robot. Under the control of the proposed fuzzy logic based model, the mobile robot can autonomously reach the target along a smooth trajectory with obstacle avoidance. Several features of the proposed approach are summarized as follows.

- The proposed model keeps the physical meanings of the variables and parameters during the processing, while some conventional models cannot.

- 48 fuzzy rules and three behaviors are designed in the proposed model, much fewer than conventional approaches that use hundreds of rules.
- The structure of the proposed fuzzy logic based model is very simple with only 11 nodes in the first layer of the structure, while hundreds of nodes are necessary in some conventional fuzzy logic based models.
- The proposed selection algorithm can automatically suppress redundant fuzzy rules when there is a need.
- A very simple yet effective state memory strategy is designed. It can resolve the "dead cycle" problem existing in some previous approaches without changing the fuzzy control rule base.

References

1. Aguirre E, Gonzalez A (2003) A fuzzy perceptual model for ultrasound sensors applied to intelligent navigation of mobile robots. Applied Intelligence, 19:171-187.
2. Chatterjee A, Matsuno F (2007) A neuro-fuzzy assisted extended kalman filter-based approach for simultaneous localization and mapping (slam) problems. IEEE Transactions on Fuzzy Systems, 15(5):984-997.
3. Filliat D, Meyer JA (2003) Map-based navigation in mobile robots: a review of localization strategies I. Cognitive Systems Research, 4:243-282.
4. Gao J, Xu D, Zhao N, Yan W (2008) A potential field method for bottom navigation of autonomous underwater vehicles. Intelligent Control and Automation, 7th World Congress on WCICA, 7466-7470.
5. Hagras H, Callaghan V, Colley M (2000) Online learning of the sensors fuzzy membership functions in autonomous mobile robots. Proceedings of the 2000 IEEE International Conference on Robotics and Automation, 3233-3238, San Francisco.
6. Hong J, Choi Y, Park K (2007) Mobile robot navigation using modified flexible vector field approach with laser range finder and ir sensor. International Conference on Control, Automation and Systems, 721-726.
7. Kim CT, Lee JJ. (2005) Mobile robot navigation using multi-resolution electrostatic potential field. 31st Annual Conference of IEEEIndustrial Electronics Society, 1774-1778.
8. Kim MY, Cho H (2004) Three-dimensional map building for mobile robot navigation environments using a self-organizing neural network. Journal of Robotic Systems, 21(6):323-343.
9. Li H, Yang SX (2003) A behavior-based mobile robot with a visual landmark recognition system. IEEE Transactions on Mechatronics, 8(3):390-400.
10. Machado C, Gomes S, Bortoli A, *et al.* (2007) Adaptive neuro-fuzzy friction compensation mechanism to robotic actuators. Seventh International Conference on Intelligent Systems Design and Applications, 581-586.
11. Marichal GN, Acosta L, Moreno L, *et al.* (2001) Obstacle avoidance for a mobile robot: A neuro-fuzzy approach. Fuzzy Sets and Systems, 124(2):171-179.
12. Na YK, Oh SY (2003) Hybrid control for autonomous mobile robot navigation using neural network based behavior modules and environment classification. Autonomous Robots, 15:193-206.
13. Noborio H, Nogami R, Hirao S (2004) A new sensor-based path-planning algorithm whose path length is shorter on the average. IEEE International Conference on Robotics and Automation, 2832-2839, New Orleans, LA.
14. Nogami R, Hirao S, Noborio H (2003) On the average path lengths of typical sensor-based path-planning algorithms by uncertain random mazes. IEEE International Symposium on Computational Intelligence in Robotics and Automation, 471-478, Kobe, Japan.

15. Park K, Zhang N (2007) Behavior-based autonomous robot navigation on challenging terrain: A dual fuzzy logic approach. IEEE Symposium on Foundations of Computational Intelligence, 239-244.
16. Rusu P, Petriu EM, Whalen TE, et al (2003) Behavior-based neuro-fuzzy controller for mobile robot navigation. IEEE Transactions on Instrumentation and Measurement, 52(4):1335-1340.
17. Song KT, Sheen LH (2000) Heuristic fuzzy-neuro network and its application to reactive navigation of a mobile robot. Fuzzy Sets and Systems, 110(3):331-340.
18. Sun F, Li L, Li HX, et al (2007) Neuro-fuzzy dynamic-inversion-based adaptive control for robotic manipulators- discrete time case. IEEE Transactions on Industrial Electronics, 54(3):1342-1351.
19. Ulrich I, Borenstein J (2000) Vfh*: Local obstacle avoidance with look-ahead verification. Proceedings of the 2000 IEEE International Conference on Robotics and Automation, 2505-2511, San Francisco, CA.
20. HaiPeng Xie (2008) Wireless networked autonomous mobile robot-i90 sentinel2 user guide. Available at http://www.drrobot.com/products/item_downloads/Sentinel2_1.pdf.
21. Yang SX, Meng QH (2003) Real-time collision-free motion planning of mobile robots using neural dynamics based approaches. IEEE Transactions on Neural Networks, 14(6):1541-1552.
22. Yang SX, Moallem M, Patel RV (2003) A novel intelligent technique for mobile robot navigation. IEEE Conference on Control Applications, 674-679.
23. Zhu A, Yang SX (2004) A fuzzy logic approach to reactive navigation of behavior-based mobile robots. IEEE International Conference on Robotics and Automation, 5:5045-5050.
24. Zhu A, Yang SX (2007) Neurofuzzy-based approach to mobile robot navigation in unknown environments. IEEE Transactions on Systems, Man, and Cybernetics, Part C: Applications and Reviews, 37(4):610–621.

Part IV
Intelligent Control

Chapter 13
Flow Control of Real-time Multimedia Applications in Best-effort Networks

Aninda Bhattacharya and Alexander G. Parlos

Abstract Real-time multimedia applications use the user datagram protocol (UDP) because of the inherent conservative nature of the congestion avoidance schemes of transmission control protocol (TCP). The effects of such uncontrolled flows on the Internet have not yet been felt because UDP traffic constitutes at most only $\sim 20\%$ of the total Internet traffic. However, it is pertinent that real-time multimedia applications become better citizens of the Internet, while at the same time delivering acceptable quality of service (QoS). Network flow accumulation is proposed for use as the basis for active flow control. Novel end-to-end active flow control schemes are proposed for unicast real-time multimedia applications transmitting over best-effort networks. The performance of the proposed control schemes is evaluated using the ns-2 simulation environment. The research concludes that active control of hard, real-time flows results in the same or marginally better QoS as no flow control. However, the bandwidth usage of actively controlled flows is significantly lower than that of uncontrolled flows. The scalability of the proposed active control schemes is acceptable.

13.1 Introduction

Multimedia applications can be broadly classified into two categories: real-time and non real-time. Non real-time multimedia applications include certain types of streaming. Streaming applications deliver audio or video streams from a server to a client. Commercial products like RealAudio™and Windows Media® Player are

Aninda Bhattacharya
Remote Prognostics Lab, GE Global Research, Bangalore, India, e-mail: aninda.bhattacharya1@ge.com

Alexander G. Parlos
Department of Mechanical Engineering, Texas A&M University, TX 77843, USA, e-mail: a-parlos@tamu.edu

examples of streaming applications. These applications require no human interactivity during the playtime sessions.

Real-time multimedia applications can be subdivided further into two categories: interactive (hard) and non-interactive (soft). All types of conferencing applications fall under the category of interactive real-time multimedia applications. Examples of popular conferencing applications are Skype™, Google Talk™, Windows® NetMeeting®, and Yahoo® Chat. Conferencing multimedia applications require human interactivity.

Conferencing applications delivering information over best-effort networks must overcome various challenges because of the limitations of human sensory perceptions. If two people are conversing with each other using phones that send information over best-effort networks, then the maximum one-way delay the packets can tolerate without loss of interactivity is 200 ms [1]. This upper value of the end-to-end delay has been shown to be commercially acceptable. If the end-to-end delay reaches 800 ms, then adverse psychological factors impede normal telephonic conversation. A flexible end-to-end delay range between 200 ms and 800 ms is conditionally acceptable for a short portion of a conversation. However, occurrences of delays in the abovementioned range should be rare and far apart. Yensen, Goubran, and Lambadaris [2] report that high end-to-end delays are unsuitable for interactive conversations since they impair normal conversation between participants and may intensify the effect of acoustic or network echoes.

One of the earliest important works regarding adaptive flow control has been done by Ramjee, Kurose, Towsley, and Schulzrinne [4]. They investigate the performance of four different algorithms for adaptively adjusting the playout delay of audio packets in an interactive packet-audio terminal application, in the face of varying network delays. Pinto and Christensen [5] extend the work proposed in [4].

Ranganathan and Kilmartin [8] use novel neural networks and fuzzy systems as estimators of network delay characteristics. The estimated delay is used for developing adaptive playout algorithms for flow control. The performance of their proposed scheme is analyzed in comparison with a number of traditional techniques for both inter- and intra-talkspurt adaptation paradigms. The design of a novel fuzzy trend analyzer system (FTAS) for network delay trend analysis and its usage in intra-talkspurt playout delay adaptation are presented in the paper.

Sisalem and Schulzrinne [9] present a new scheme called the *loss-delay based adjustment algorithm* (LDA) for adapting the transmission rate of multimedia applications to the congestion level of the network. The LDA algorithm is designed to reduce losses and improve utilization in a TCP-friendly way that avoids starving competing TCP connections. The paper by Casetti, DeMartin, and Meo [11], discusses most of the issues of interest in the current area. In practice, the authors implement a very simple empirical control algorithm to control the flow of audio on the network. This algorithm is similar to the often-used additive increase multiplicative decrease (AIMD) algorithm.

Beritteli, Ruggeri, and Schembra [13] state that although several TCP friendly algorithms have been introduced to support real-time applications on the Internet, the only target in optimizing them is to achieve fairness with TCP flows in the network. No attention has been paid to the QoS perceived by their users. Their paper analyzes the problem of transmitting voice over internet protocol (VoIP) when voice sources use one of the most promising TCP-friendly algorithms, route access protocol (RAP) and a transferring receptor (TFRC). They also propose a modification of both RAP and TFRC, in order to take care of the QoS in real-time multimedia flows.

A survey on transport adaptation techniques has been provided in the article written by Homayounfar [14]. This article deals with voice encoders in mobile networks. An adaptive multi rate (AMR) encoder has been considered for VoIP applications. For VoIP, the purpose of rate adaptation is to reduce network congestion, and this can be thought of as an indirect form of implementing TCP-friendliness. Abreu-Sernandez and Garcia-Mateo [15] propose a voice coder with three different bit rates adapted for the transmission of VoIP packets. The encoder has a rate control device that analyzes the traffic congestion of the network and commands the voice coder to switch among five operating modes, if necessary. These modes include mitigation techniques for packet losses.

Bai and Ito [16] have written a good survey paper on different approaches to provide QoS control for video and audio communication in conventional and active networks. Recent work regarding adaptive end-to-end flow control schemes for real-time audio applications has been done by Roychoudhury and Al-Shaer [17]. This paper presents an adaptive rate control framework that performs proactive rate control based on packet loss prediction and on-line audio quality assessment for real-time audio. The proposed framework determines the optimal combination of various audio codecs for transmission, utilizing a thorough analysis of audio quality based on individual codec characteristics, as opposed to ad-hoc codec redundancy used by other approaches.

The current research discusses flow control of unicast multimedia flows from the point of view of VoIP applications. Nevertheless, the flow control algorithms developed for unicast VoIP flows transported over best-effort networks are also applicable for other multimedia applications, such as video. The emphasis is not to develop new audio or video codecs but to develop flow control algorithms that could be used with existing or newly developed codecs.

The present research makes the following contributions:

1. Proposes the use of flow accumulation signal for development of flow control schemes for hard real-time multimedia applications.
2. Proposes two simple flow control schemes: a linear control law (LCL) and a nonlinear control law (NCL).
3. Demonstrates the feasibility of using active flow control schemes to deliver acceptable QoS while utilizing less bandwidth than flows utilizing no flow control.

13.2 Modeling End-to-end Single Flow Dynamics in Best-effort Networks

Traffic on the Internet is self-similar and long-range dependent in nature. Kihong and Willinger [28] state that scale-invariant burstiness is a persistent trait existing across a range of network environments. In the present research a single end-to-end real-time multimedia flow is considered as the system to be modeled. The total sum of the cross traffic in all of the intermediate nodes of the path traveled by the flow, not under the control of the applications at either end, is considered as the disturbance. The rate at which packet flow is generated by the application is considered to be the input to the system.

Output signals are important aspects in the development of a dynamic model. A aspect for the development of a flow model over best-effort networks is the choice of the output signal that reflects the state(s) of the system under observation. Among a host of signals that can be considered as the output signal of the system, three signals, the packet loss rate, the end-to-end forward delay, and the flow accumulation, hold the most promise. Each of these signals will be discussed in the next subsections.

13.2.1 Definitions

Let us assume that a packet i is sent into the network by the source at send time s_i. It arrives at the destination at the arrival time a_i. The destination calculates the state(s) of the network through the associated one way delay. The state(s) of the network is sent back to the source at the departure time f_i. The feedback information about the network state(s), as determined by the passage of the packet i in the forward path, arrives at the source at the receive time r_i. The forward and backward path delays can be defined as

$$T_i^f = a_i - s_i, \tag{13.1}$$

$$T_i^b = r_i - f_i. \tag{13.2}$$

The total round trip time (RTT), T_i^R, of the packet is the sum of the forward and the backward delays. Mathematically, this can be expressed as

$$T_i^R = T_i^f + T_i^b \tag{13.3}$$

$$\Rightarrow T_i^R = (r_i + a_i) - (f_i - s_i). \tag{13.4}$$

The end-to-end delay of a packet includes the amount of time the packet spends in the playout buffer b_i along with the propagation delay D_{prop}, transmission delay D_i^t, and the queuing delay v_i. Mouth-to-ear delay of a talkspurt includes the coding and the decoding delays associated with the specific codec in use along with the end-to-end delay.

13.2.2 Packet Loss Signal

The packet loss signal of a real-time multimedia application is comprised of not only the packets that are dropped in the queues of the intermediate nodes while traveling through the network, but also the packets that arrive at the destination after the expiration of their arrival deadline. A more suitable name for this signal would be comprehensive loss rate (CLR). The loss rate signal is correlated with the existence of congestion in the network. An increase in the magnitude of the loss rate signal signifies congestion in the network. However, this assertion is only valid for wired networks where the bit error rate (BER) is low. In the case of wireless networks, a high loss rate might not be an indication of congestion in the network.

It is very difficult to judge whether or not a packet has been lost while traversing through the network in the presence of flow reversal. Flow reversal is the phenomenon when the sequence at which packets arrive at the destination does not increase monotonically. Flow reversal happens when the packets take different paths while traversing through the network to reach the same destination. In this chapter it is assumed that flows do not undergo any significant flow reversal. This is a valid assumption as real world experiments have shown that the Internet shows very few instances of flow reversal. This simplifies the calculation of the packet loss signal in real-time. In the absence of flow reversal, if a packet with a sequence number j reaches the destination after the packet with a sequence number i such that $j-i>1$, all the packets with sequence numbers in the range of $[i+1, j-1]$ are assumed lost. This type of packet loss is categorized as *pure loss*.

The packets that fail to meet their playout time deadline need to be accounted for in the loss signal too. Let T_i be the time taken by packet i to reach the playout buffer of the destination. Any packet that satisfies equation

$$T_i^f > T_{Th}, \tag{13.5}$$

where T_{Th} is a threshold time that is determined by the needs of the real-time multimedia application, is considered lost. This type of loss is termed as *delay threshold loss*. The larger the T_{Th}, the poorer is the achievable interactivity between the two end users.

Comprehensive loss of packets L_{CL} is defined as

$$L_{CL} = L_T + L_P \tag{13.6}$$

where L_T and L_P are delay threshold loss and pure loss of the packets of a multimedia flow, respectively. The CLR of packets in a flow can be defined as the ratio of the number of comprehensive loss of the packets and the total number of packets sent N_T from source to destination in a time interval $[t_1, t_2)$, as follows:

$$CLR = \frac{L_{CL}}{N_T} \times 100. \tag{13.7}$$

This quantity is expressed in percentage.

CLR is tried as an output signal in order to model the end-to-end dynamics of a single flow. However, the developed linear and nonlinear models using flow bit-rate as the input and the loss signal as the output do not turn out to be very accurate. The delay threshold, T_{Th}, used for calculating the CLR is 230 ms. This is the approximate upper bound of the end-to-end delay for interactive real-time multimedia flows exchanging information over best-effort networks.

13.2.3 Packet Delay Signal

The packet delay signal seems to hold more promise for modeling a single flow as compared to the CLR signal described in the previous section. The packet delay signal seems to be smoother and content rich than CLR. However, there is a major problem in measuring this signal in real-time; dealing with packet losses.

The packet delay signal can only be defined if the packets reach the destination at the other end. If the packets are lost during their travel, the delay signal cannot be defined or measured. Researchers have used various techniques to determine the value of the packet delay signal in the presence of packet losses. The simplest way to deal with this issue is to ignore the packet delay values whenever packet losses occur. However, this approach leads to many mathematical complications. The most rational solution to this issue is to allocate an artificial delay value to the signal in the case of packet losses. This artificial packet delay value determines the shape of the signal in the time domain.

If it is decided to allocate an artificial delay value to the signal in the case of packet losses, the issue at hand is what artificial value should be allocated to the packet delay signal to substitute packet losses. The answer to the above question determines the difficulty with which the flow can be modeled based on the flow bit-rate as the input signal and packet delay as the output signal.

13.2.4 Accumulation Signal

The most suitable contender for the role of the output signal in the present modeling effort is the accumulation signal. The accumulation signal is the difference between the number of bytes sent into the network by the source of a flow and the number of bytes received by the destination of the same flow at time t. Both the cross-traffic flow as well as the input bit-rate from the source affects the accumulation signal. The accumulation signal reflects both packet losses and the effects of delays in the network. Moreover, the accumulation signal is a continuous-time signal. It has a definite value at each instant of time irrespective of packet losses.

Figure 13.1 shows a comparison of the accumulation signal with respect to the delay and comprehensive loss signals for a single UDP flow in a simulation of 120 sec. It can be noticed that the delay signal has the same trend as the accumulation signal. Whenever the delay goes up, accumulation increases as well. This shows a positive correlation between delay and the accumulation. The packet losses increase when the accumulation goes up. Losses also occur intermittently in areas where the accumulation appears normal.

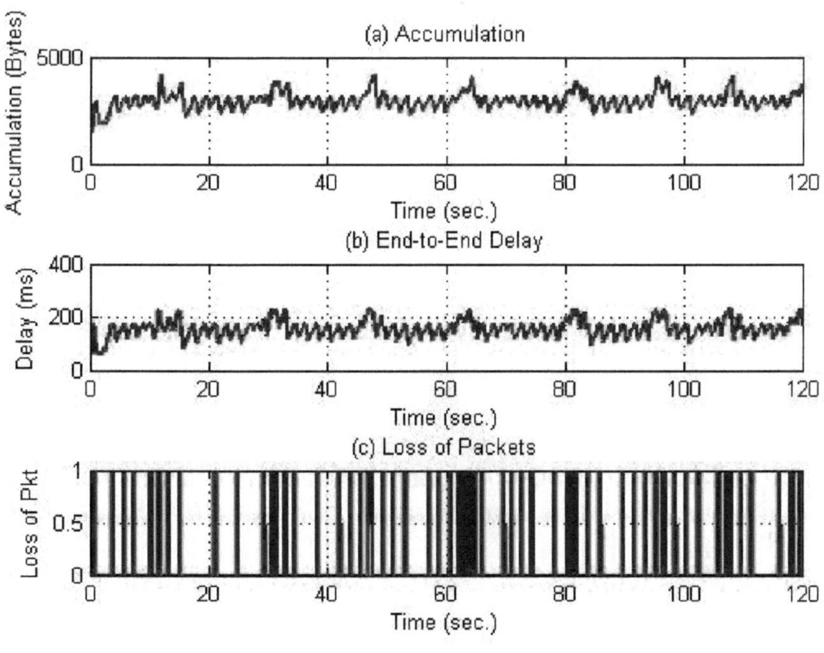

Fig. 13.1 Comparison of the accumulation signal with respect to the delay and comprehensive loss signals of a single flow

13.2.5 Linear State Space Modeling and End-to-end Flows

In this research, a single flow within a network is modeled using fluid flow rather than packet flow. This means that differential equations in the continuous-time domain are used to describe the flow dynamics. A single-input single-output (SISO) model with the bit-rate as the input signal and the accumulation as the output signal is developed. Although this model is derived from conservation principles, it is not used for the development of different control laws. The main purpose for this model is to provide physical insights into the structure of the empirical models that could be used in an adaptive controller framework.

While deriving the flow equations to characterize a single flow in a best-effort network as observed from the application, it is important to describe the different definitions of time and relate them to each other. Let $\tau(t)$ be the time-delay present in the network for the flow traversing from one end to the other under observation at time t. In order to maintain consistency with the notations already used in this research, the time associated with the event of sending packets of a flow into the network from the source is called the send time and is represented by s. Similarly, the time associated with the event of arrival at the destination is called the arrival time and can be represented by a. The two definitions of time are related to each other by the following equation:

$$a = s + \tau(s) \tag{13.8}$$

where $\tau(s)$ is the delay suffered by the flow while traversing through the network. The delay is a function of the send time s and is time varying.

If the state equations are defined with respect to the send time s, the resultant system incorporates the effect of flow reversal. Assuming no flow reversals allows one to write the state equations of the system with respect to the arrival time a at the destination, in the continuous-time domain. However, the output or the measurement equations of the system are defined with respect to the send time, s.

13.2.6 Deriving the State Equations of a Conservative Single Flow System

Let $A(a)$ be the flow accumulation signal at arrival time a and $u(a)$ be the rate at which information is sent into the multimedia flow under observation. Let $z(a)$ be the arrival rate of the information at the destination. It is important to note that all these observations are being made at the destination end of a conservative system. A network flow is considered to be conservative when there are no packet losses within the flow as it traverses from the source to the destination. The continuous-time state space equations are derived first. Later on, these equations are discretized to obtain discrete-time state space equations.

Applying flow conservation, the rate of change of the accumulation signal of a single source can be described by the following expression with respect to the arrival time a:

$$\frac{dA(a)}{da} = u(a - \tau(s)) - z(a), \tag{13.9}$$

where $\tau(s)$ is the variable delay suffered by the flow. Multiplying both sides of (13.9) by da and integrating in the time interval t_1 to t_2 such that $t_2 > t_1$, we obtain

$$\int_{t_1}^{t_2} dA(a) = \int_{t_1}^{t_2} u(a-\tau(s))\,da - \int_{t_1}^{t_2} z(a)da. \tag{13.10}$$

For the sake of simplicity, let us assume that the delay $\tau(s)$ remains constant. Therefore, $\tau(s) = \tau$ and (13.10) can be simplified as

$$\int_{t_1}^{t_2} dA(a) = \int_{t_1}^{t_2} u(a-\tau)\,da - \int_{t_1}^{t_2} z(a)da,$$
$$A(t_2) - A(t_1) = \int_{t_1}^{t_2} u(a-\tau)\,da - \int_{t_1}^{t_2} z(a)da. \tag{13.11}$$

Let $x = a - \tau$. This implies $dx = da$ and the transformed (13.11) can be written as

$$A(t_2) = A(t_1) + \int_{t_1-\tau}^{t_2-\tau} u(x)dx - \int_{t_1}^{t_2} z(a)da,$$
$$A(t_2) = A(t_1) + \int_{t_1-\tau}^{t_1} u(x)dx + \int_{t_1}^{t_2-\tau} u(x)dx$$
$$- \int_{t_1}^{t_2} z(a)da. \tag{13.12}$$

Let $t_1 = kT$ and $t_2 = kT + T$, where T is the sampling period of the system and $k \in \mathbb{N}$. Therefore, the above equation can be written as

$$A(kT+T) = A(kT) + \int_{kT-\tau}^{kT} u(x)dx +$$
$$\int_{kT}^{kT+T-\tau} u(x)dx - \int_{kT}^{kT+T} z(a)da. \tag{13.13}$$

(13.13) must be expressed for three different cases, each pertaining to the value of the constant time-delay τ. The time-delay can have three different values, i.e., $\tau < T$, $\tau = T$, and $\tau > T$.

As it is common in sampled data systems, it is assumed that the input signal passes through a zero order hold (ZOH) before being applied to the system. This implies that the input rate $u(t)$ does not change between time instants kT and $kT + T$.

13.2.6.1 Case I: $\tau < T$

(13.13) can be simplified as

$$\begin{aligned} A(kT+T) &= A(kT) + u(kT-T)\int_{kT-\tau}^{kT} dx + \\ &\quad u(kT)\int_{kT}^{kT+T-\tau} dx - \int_{kT}^{kT+T} z(a)da, \\ \Rightarrow A(kT+T) &= A(kT) + u(kT-T)\tau + u(kT)(T-\tau) \\ &\quad - z(kT)T, \\ \therefore A((k+1)T) &= A(kT) + u((k-1)T)\tau + u(kT)(T-\tau) \\ &\quad - z(kT)T. \end{aligned} \tag{13.14}$$

The final form of (13.14) can be expressed as

$$\begin{aligned} A(k+1) &= A(k) + u(k-1)\tau + u(k)(T-\tau) \\ &\quad - z(k)T. \end{aligned} \tag{13.15}$$

13.2.6.2 Case II: $\tau = T$

When $\tau = T$, (13.13) takes the form

$$\begin{aligned} A(kT+T) &= A(kT) + u(kT-T)\int_{kT-T}^{kT} dx \\ &\quad - z(kT)T, \\ \Rightarrow A(kT+T) &= A(kT) + u(kT-T)T - z(kT)T, \\ \therefore A((k+1)T) &= A(kT) + u((k-1)T)T - z(kT)T. \end{aligned} \tag{13.16}$$

The final form of (13.16) can be expressed as

$$A(k+1) = A(k) + u(k-1)T - z(k)T. \tag{13.17}$$

13.2.6.3 Case III: $\tau > T$

The final equation depends on how large the value of τ is. Without a bound on the value of τ, it is impossible to determine the exact expression for $A(k+1)$. In order to complete the model, we assume that $T < \tau < 2T$ and derive the final form for this case. In this special case, we can represent the time-delay τ as $T + \delta$ with range of δ being defined as $0 < \delta < T$. (13.13) takes the form

$$\begin{aligned}
A(kT+T) &= A(kT) + \int_{kT-T-\delta}^{kT-T} u(x)dx + \\
&\quad \int_{kT-T}^{kT} u(x)dx + \int_{kT}^{kT-\delta} u(x)dx \\
&\quad - \int_{kT}^{kT+T} z(a)da, \\
\Rightarrow A(kT+T) &= A(kT) + u((k-2)T)\delta + u((k-1)T)T \\
&\quad - \int_{kT-\delta}^{kT} u(x)dx - z(k)T, \\
\Rightarrow A(kT+T) &= A(kT) + u((k-2)T)\delta + u((k-1)T)T \\
&\quad - u((k-1)T)\delta - z(k)T.
\end{aligned} \tag{13.18}$$

The final form of (13.18) can also be written as

$$\begin{aligned}
A(k+1) &= A(k) + u(k-2)\delta + u(k-1)(T-\delta) \\
&\quad - z(k)T.
\end{aligned} \tag{13.19}$$

13.2.6.4 Generalization of Flow State Equations with Constant Network Time-delay

If we take a closer look at (13.15) and (13.19), we can arrive at a generalized equation for single flow accumulation in a best-effort network for delay τ that lies between the values $(N-1)T$ and NT, such that $N \in \mathbb{N}$. The delay can be expressed as

$$\tau = (N-1)T + \delta, \tag{13.20}$$

where $0 < \delta < 1$. The accumulation at time instant $(k+1)T$ can be expressed in the following form:

$$\begin{aligned}
A(k+1) &= A(k) + u(k-N)\delta + u(k-N+1)(T-\delta) \\
&\quad - z(k)T.
\end{aligned} \tag{13.21}$$

A simple exercise is performed to check the generalized state (13.21). Let the network delay τ be equal to the sampling period T. From (13.20), it can be seen that δ reduces to 0 and $N = 2$. The accumulation at time instant $(k+1)T$ can be expressed as:

$$A(k+1) = A(k) + u(k-1)T - z(k)T. \tag{13.22}$$

This expression is the same as (13.17).

13.2.7 Deriving the Output/Measurement Equation of a Conservative Single Flow System

The measurement of the accumulation signal (output) is accomplished by the source after receiving the feedback from the destination. Therefore, all the continuous-time equations regarding the measurement of accumulation (output), $y(s)$, are written with respect to send time s. Accumulation measured at time s is the delayed feedback provided by the destination at the other end. This implies

$$y(s) = A(s - \tau(s)). \tag{13.23}$$

For the sake of simplicity, let us assume that the delay $\tau(s)$ remains constant. This implies $\tau(s) = \tau$. Equation 13.23 can be written as

$$y(kT) = A(kT - \tau). \tag{13.24}$$

Like in the previous section, we must express (13.24) for the three cases of τ, i.e., $\tau < T$, $\tau = T$, and $\tau > T$.

13.2.7.1 Case I: $\tau < T$

(13.24) takes the form

$$\begin{aligned} y(kT) &= A((k-1)T) + \Delta A(\tau), \\ \Rightarrow y(kT) &= A((k-1)T) + (u((k-1)T) \\ &\quad - z((k-1)T))\tau. \end{aligned} \tag{13.25}$$

The final form of the measurement equation in the case of $\tau < T$ can be expressed as

$$y(k) = A(k-1) + (u(k-1) - z(k-1))\tau. \tag{13.26}$$

13.2.7.2 Case II: $\tau = T$

(13.24) takes the form

$$\begin{aligned} y(kT) &= A((k-1)T) \\ \Rightarrow y(k) &= A(k-1). \end{aligned} \tag{13.27}$$

This is the final form of the measurement equation when $\tau = T$.

13.2.7.3 Case III: $\tau > T$

The final measurement equation, when $\tau > T$, depends on the upper bound of τ. In order to complete the set of equations, we assume that $T < \tau < 2T$ and derive the final form for this case. In this special case, we can the represent the time-delay τ as $T + \delta$ with range of δ being defined as $0 < \delta < T$. (13.24) takes the form

$$\begin{aligned} y(kT) &= A((k-2)T) + \Delta A(\tau), \\ \Rightarrow y(kT) &= A((k-2)T) + (u((k-2)T) \\ &\quad - z((k-2)T)))\delta. \end{aligned} \tag{13.28}$$

Therefore, the final form of the measurement equation in the case of $T < \tau < 2T$ can be written as

$$y(k) = A(k-2) + (u(k-2) - z(k-2))\delta. \tag{13.29}$$

13.2.7.4 Generalization of Measurement/Output Equations with Constant Flow Time-delay

Let τ be expressed as $\tau = (N-1)T + \delta$ such that $0 < \delta < 1$. Referring to (13.26) and (13.29), the generalized equation of measured accumulation $y(kT)$ in the network with the flow delay τ between time instant $(N-1)T$ and NT, such that $N \in \mathbb{N}$, is:

$$y(k) = A(k-N) + (u(k-N) - z(k-N))\delta. \tag{13.30}$$

13.2.8 Summary of the State Space Equations for a Conservative Flow in a Best-effort Network

If the constant delay, τ, for a conservative flow falls between time instant $(N-1)T$ and NT, such that $N \in \mathbb{N}$, then τ can be written as

$$\tau = (N-1)T + \delta, \tag{13.31}$$

with the range of δ being $0 < \delta < T$. Then the linear state space equations describing the multimedia flow in terms of accumulation signal, A(k), can be written as

$$\begin{aligned} A(k+1) = A(k) &+ u(k-N)\delta + u(k-N+1)(T-\delta) \\ &- z(k)T, \end{aligned} \tag{13.32}$$

and,

$$y(k) = A(k-N) + (u(k-N) - z(k-N))\delta. \tag{13.33}$$

13.3 Proposed Flow Control Strategies

The conservation model implied by (13.32) and (13.33) indicates the presence of delays in both the state dynamics and output. Developing adaptive forms of flow controllers requires the availability of predictors which utilize available measurements. Typical forms of such predictors perform single-step predictions. However, the presence of delays in the flow dynamics implies the need for multi-step predictors for use in adaptive control. Such predictors are quite complex to develop and even more complex to validate for prediction accuracy. An alternative is to utilize simpler predictor forms in the formulation of the flow controllers.

13.3.1 Simple Predictor

For the sake of simplicity a simple form of a predictor is used to design the flow control regimes. A very simple p-step ahead predictor that can be used to design flow control schemes is defined by the following equation:

$$\widehat{A}(k+p|k) = A(k), \tag{13.34}$$

where $p \in \mathbb{N}$. (13.34) states that the value of the accumulation signal at p step ahead in the future is equal to the latest measured value of the signal. Control strategies based on p-step ahead "simple" predictors are essentially reactive control strategies. Two such control laws, one linear and the other nonlinear, are developed based on the "simple" predictor.

13.3.2 Flow Controllers

Two simple control laws based on the simple predictor (SP), the LCL and NCL are proposed. In the current context, accumulation $A(k)$ is the output signal i.e., $y(k) = A(k)$, and the size of the packets in the real-time multimedia flows is the input signal $u(k)$. The disturbance $d(k)$ comprises of the cross flow traffic in the best-effort network that affects the QoS of the flow under observation. It is very difficult to model the disturbance $d(k)$ because of the distributed nature of the Internet.

A reference accumulation level, $r(k)$, must be determined to enable regulation of the flow accumulation level around this reference. This is the most difficult aspect of the problem at hand. If the reference accumulation selected is very small, then the controller will always act conservatively and the controlled real-time multimedia flow can never take advantage of the extra bandwidth available when the network is not congested. If the reference accumulation selected is very large, then the controller will never predict the congestion present in the network and take appropriate actions. Assuming that the interdeparture time of the packets is fixed at

20 ms, the reference accumulation is dependent on the selection of the packet size of the multimedia flow. For example, if the maximum end-to-end delay that can be tolerated while transferring multimedia information over the best-effort network is 230 ms and the bit-rate of the flow is 36 kbps, then the accumulation level signifying a normal network is $\frac{36000\times0.23}{8} = 1035$ bytes. The control action needs to be taken only if the accumulation level goes above or below 1035 bytes. If the bit-rate of the flow changes to 96 kbps, then the accumulation level signifying a normal network is $\frac{36000\times0.23}{8} = 2760$ bytes. This makes the mathematical formulation of the control strategy difficult for both design and implementation.

In order to overcome the difficulties of designing simple flow control strategies based on the error signal as typically done in traditional control system design, an alternative framework is used, as follows:

$$u(k) = f(y(k)). \tag{13.35}$$

(13.35) describes the flow control law. Both the LCL as well as the NCL, described in the next two subsections, are based on this framework that does not have any reference signal and the control law is a function of the measurement of the output signal. A block diagram of this control strategy is depicted in Figure 13.2.

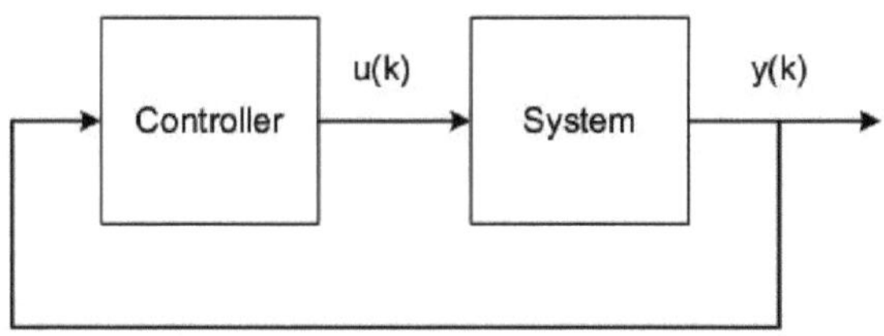

Fig. 13.2 Block diagram of a single flow in a best-effort network modeled as system controlled with the help of a controller employing either LCL or NCL. The controller adjusts the input signal $u(k)$ based on the output signal $y(k)$

13.3.2.1 Linear Control Law with Simple Predictor

The LCL has been designed using the following logic: if the accumulation in the network goes up, then the bit-rate of the controlled real-time multimedia flow should go down and *vice versa*. The control law is not really linear in the right sense of the word. It incorporates saturation of the actuator, i.e. the inability of the codec to generate bit-rates beyond a certain level. The law is linear in a certain specific range of the accumulation level. In spite of this, this control law will be referred to as the LCL in this chapter.

The codec considered in this research has six different bit-rates, varying from 36 kbps to 96 kbps. The end-to-end delay of the VoIP application that has acceptable level of interactivity between the two users at the ends is considered to be 230 ms. The interdeparture time of the packets is 20 ms. In an ideal case, the maximum amount of accumulation, measured in bytes, that a flow can afford to have and preserve interactivity is

$$A_s = \frac{p_s \times 230}{20}, \tag{13.36}$$

where p_s is the size of the packets in the VoIP flow.

This means that the maximum accumulation that can be tolerated by the application is equivalent to the total number of packets in transport when the end-to-end delay of the packets is equal to 230 ms. (13.36) generates six different levels of accumulation for six different bit-rates. These are shown in Table 13.1.

Table 13.1 Equilibrium accumulation corresponding to each of the six bit rate levels in LCL

	Packet Size (Bytes)	Bit-rate (kbps)	Accumulation (Bytes)
1	90	36	1035
2	120	48	1380
3	150	60	1725
4	180	72	2070
5	210	84	2415
6	240	96	2760

Let us assume that at time instant kT, the accumulation signal value $A\left((k-1)T\right)$ is known because of the feedback delay from the destination to the source. This measurement, $A\left((k-1)T\right)$, is utilized to predict the value of accumulation at time instant $(k+1)T$, using a two-step-ahead SP. The two-step-ahead predictor is described by the following equation:

$$\widehat{A}(k+1|k-1) = A(k-1). \tag{13.37}$$

Then the LCL is described by

$$u(k+1) = \begin{cases} 90 & , \text{if } \widehat{A}(k+1|k-1) > 2760 \\ -0.087\widehat{A}(k+1|k-1) + 330 & , \text{if } 1035 \le \widehat{A}(k+1|k-1) \le 2760, \\ 240 & , \text{if } \widehat{A}(k+1|k-1) < 1035 \end{cases} \tag{13.38}$$

where $u(k+1)$ is the size of the packets that must be sent every 20 ms.

The LCL and its implementation in the quantized form is shown together in Figure 13.3.

The quantized implementation of the LCL can be expressed as

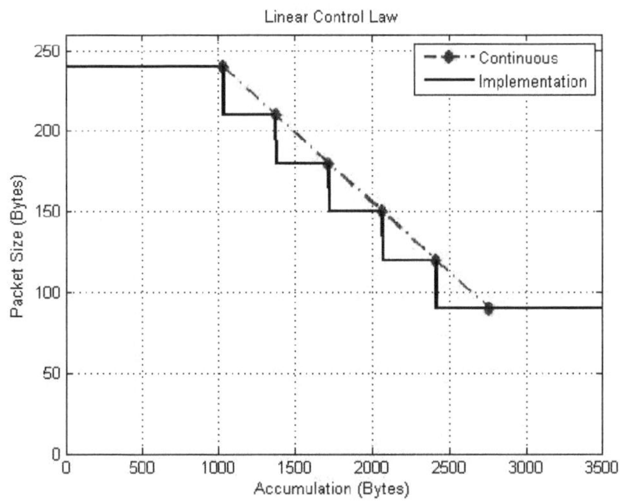

Fig. 13.3 Linear control law

$$u(k+1) = \begin{cases} 90 \text{ , if } \widehat{A}(k+1|k-1) \geq 2415. \\ 120 \text{ , if } 2070 \geq \widehat{A}(k+1|k-1) < 2415. \\ 150 \text{ , if } 1725 \geq \widehat{A}(k+1|k-1) < 2070. \\ 180 \text{ , if } 1380 \geq \widehat{A}(k+1|k-1) < 1725. \\ 210 \text{ , if } 1035 \geq \widehat{A}(k+1|k-1) < 1380. \\ 240 \text{ , if } \widehat{A}(k+1|k-1) < 1035. \end{cases} \tag{13.39}$$

This quantization is necessary as any realistic codec cannot generate the bit-rates specified by (13.38).

13.3.2.2 Nonlinear Control Law with the Simple Predictor

A slight variation to the LCL is used to design another simple flow control law. The LCL is designed on the basis of the concept that the higher the accumulation signal, the lower the bit-rate that must be sent into the network. If the network is congested the need for congestion avoidance is far more urgent. Therefore, the drop in packet sizes during congestion should happen at a faster rate as compared to the increase in packet sizes when the network is not congested. This is inspired from the AIMD control employed by the TCP for congestion avoidance.

The LCL based on the accumulation does not take care of the above mentioned concern. Therefore, a separate control law is designed, the NCL. This control law operates nonlinearly in the specified range of the accumulation. A quantized version of this law is employed for flow control. The nonlinear flow control law is described by the following equation:

$$u(k+1) = \begin{cases} 90 \\ \quad , \text{if } \widehat{A}(k+1|k-1) > 2760 \\ \left(65700 - \frac{65700}{8688825}\widehat{A}(k+1|k-1)^2\right)^{\frac{1}{2}} \\ \quad , \text{if } 1035 \le \widehat{A}(k+1|k-1) \le 2760. \\ 240 \\ \quad , \text{if } \widehat{A}(k+1|k-1) < 1035 \end{cases} \tag{13.40}$$

The NCL and its implementation in the quantized form are shown together in Figure 13.4.

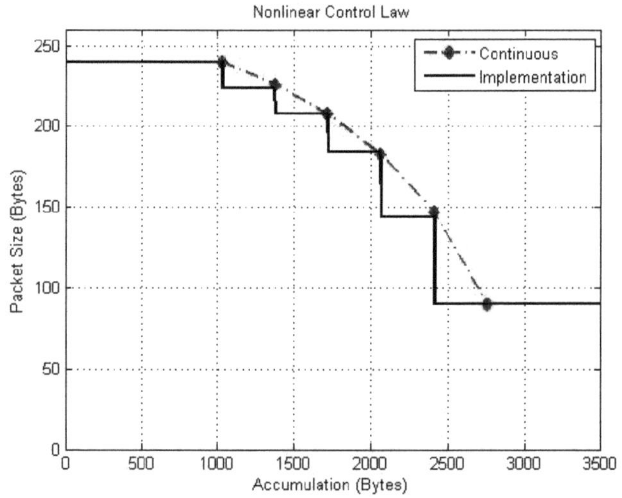

Fig. 13.4 Nonlinear control law

The quantized implementation of the NCL can be described by the following equation

$$u(k+1) = \begin{cases} 90 \;\; , \text{if } \widehat{A}(k+1|k-1) \ge 2415 \\ 144 \; , \text{if } 2070 \ge \widehat{A}(k+1|k-1) < 2415 \\ 184 \; , \text{if } 1725 \ge \widehat{A}(k+1|k-1) < 2070. \\ 208 \; , \text{if } 1380 \ge \widehat{A}(k+1|k-1) < 1725 \\ 224 \; , \text{if } 1035 \ge \widehat{A}(k+1|k-1) < 1380 \\ 240 \; , \text{if } \widehat{A}(k+1|k-1) < 1035 \end{cases} \tag{13.41}$$

13.4 Description of the Network Simulation Scenarios

One of the important premises of the current work is that in a congested network, a flow occupying lesser bandwidth will be able to provide a better end-to-end QoS as compared to a flow that demands more bandwidth. High quality multimedia flows require larger bandwidth as compared to low quality flows. During congestion, the probability of the loss of packets belonging to a high bandwidth flow is higher than that of packets belonging to a low bandwidth flow. The loss of packets of the higher bandwidth flow deteriorates the quality of the flow to make it inferior to that of the quality achieved by the lower bandwidth flow during congestion. The delays and the losses suffered by a UDP flow in a congested network can be reduced by reducing the send-rate of the flow at minimal expense to voice quality.

The bit-rate of a multimedia flow has two degrees of freedom. It can be manipulated by either varying the interdeparture times between the packets or by varying the sizes of the packets comprising the flow. The packets of a flow sent during a connection have dual roles. The first role is related to the transmission of the information from one end to the other. This facilitates the interaction between the two end users. The second role is related to providing the feedback signal about the the state of the network through which the packets pass in order to deliver information.

Since packets also provide the information about the state of the network, varying interdeparture time between the packets means that the information about the network is sampled at different frequencies. This complicates the formulation of control strategies for the system. While traversing through the network, the packets undergo time-variant queuing and processing delays leading to variation in the interarrival times and even losses. This implies that depending on the state of the network, the sampling rate of the state of the network also varies considerably. These time-varying time-delays introduce a lot of complexities in the subsequent mathematical analysis. If it is decided to vary the interdeparture time of the packets then the control algorithms become too complicated. It is prudent that the send-rate of the flow to the network be varied by varying the size of the packets generated.

A simulation environment provided by ns-2: an open source simulator, is selected to demonstrate the ideas of adaptive flow control of real-time multimedia flows traversing through best-effort IP networks. The simulation scenarios developed in ns-2 environment provide the experimental framework for the current research. The application level flow control algorithms for improving the QoS of VoIP applications for this work have been implemented as separate modules in the ns-2 framework and integrated, later on.

Creating a network topology that is sufficiently saturated to serve the purpose of the current research is easy in a simulator framework. The user can decide on the level of congestion by varying the parameters that determine the final topology suitable for demonstrating the efficacy of the ideas. Some important aspects that were taken care of while designing the simulation topologies are:

- The percentage of UDP flows in the network as compared to the TCP flows should reflect what is currently prevalent on the Internet.

- The queue capacity of the routers that carry the main traffic should not be too large so that there is no drop of packets. The capacities should not be too small so that they get saturated by a small number of flows transferring packets at moderate levels.
- The propagation delay between the source and the destination should be large enough to reflect real world conditions. The propagation delay across the mainland United States varies from 60 ms to 90 ms.

For the purpose of the experiments two different network topologies are designed. The first topology is used to design and validate the flow control strategies. The second topology is used for studying the scalability of the flow control strategies.

13.4.1 Simulated Network Topology for Validating the Performance of the Flow Control Strategies

The first topology for validating the designed control strategies is depicted in Figure 13.5. This topology has two links that resemble the backbone links found in networks like Internet2. The backbone links are formed by the three routers named Router 0 (R0), Router 1 (R1), and Router 2 (R2).

The link between R0 and R1 (R0-R1) has a propagation delay of 30.87 ms. The link between R1 and R2 (R1-R2) also has the same propagation delay. Therefore, the total propagation delay in the backbone links adds up to 61.74 ms. The propagation delay in the backbone links is never changed during the experiments. The link between routers R1 and R2 (R1-R2) has been fixed to 73.26 Mbps.

The two backbone links in the current topology, R0-R1 and R1-R2, serve unique roles. The purpose of the link R0-R1 is to cause losses in the flows. Meanwhile, the R1-R2 link causes variation in the end-to-end delay of the flows under observation. The term "flows under observation" refers to UDP flows that we are observing and trying to control. By changing the bandwidth of the link R0-R1 in a suitable manner, the loss rates of the flows under observation are varied. The bandwidth capacity of the link R0-R1 varies between 28.92 Mbps and 50.92 Mbps. Subsequently, the average loss rate of the five UDP flows under observation vary from 15% to 3%.

Besides the bandwidth capacity and the propagation delay of the queues, the queue sizes of the links R0-R1 and R1-R2 are important. All the queues in the current topology have First In First Out (FIFO) buffer management scheme. In ns-2, the output queue of a node is implemented as part of a link. The duplex link R0-R1 has two queues with a capacity of 460800 bytes. The next backbone duplex link R1-R2 has two queues of capacity equal to 1024000 bytes.

If we consider left to right as the forward direction of the flow of traffic in Figure 13.5, the forward cross-flow traffic at the link R0-R1 is comprised of 150 HTTP (TCP) flows, 5 CBR (UDP) flows, and 87 file transfer protocol (FTP) flows. Also, 100 HTTP (TCP), 5 CBR (UDP) flows, and 87 FTP (TCP) flows make up the cross-

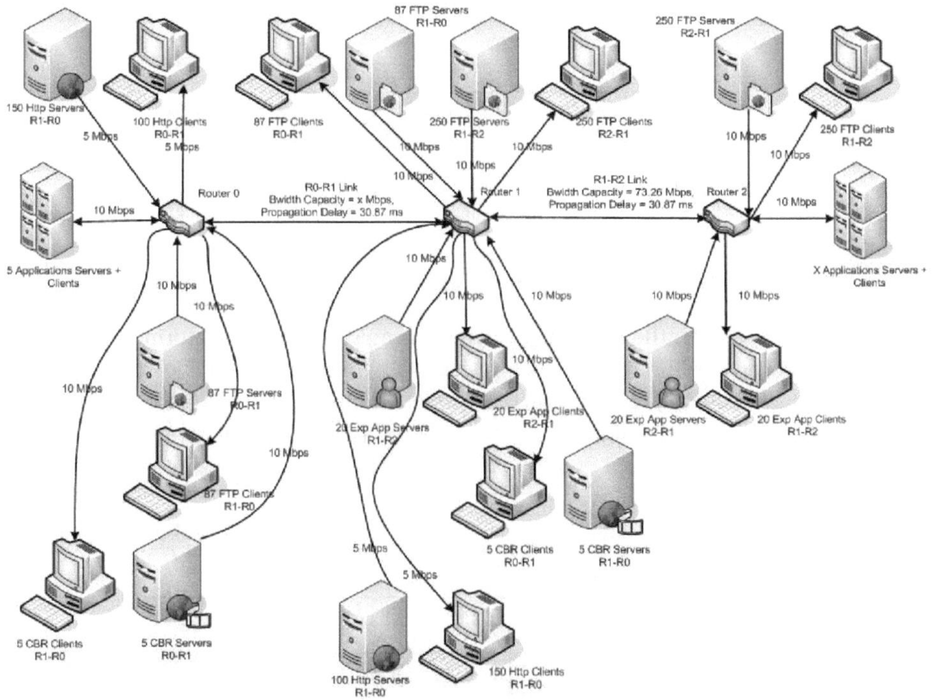

Fig. 13.5 Topology created in the ns-2 simulation environment to design and study the adaptive end-to-end multimedia flow control schemes

flow traffic in the backward direction in the same link. The HTTP nodes are connected to Router 0 and Router 1 using 5 Mbps capacity links. The rest of the nodes are connected using 10 Mbps capacity links.

There is no cross-traffic from HTTP flows in the link R1-R2. The cross-traffic of the link R1-R2 is composed of 250 FTP (TCP) and 20 Exponential (UDP) flows in the forward as well as the backward direction. All the application nodes are connected to the routers R1 and R2 using 10 Mbps links. Table 13.2 summarizes the information about cross-traffic in the topology in terms of the flows. The cross-traffic flows constitute 98.98 % of the total flows in the current simulation scenario. The last column of the table provides the information about the contribution of a type of flow as a percentage of total flows in the topology.

The remaining flows in the topology (1.02%) are comprised of the nodes that run multimedia applications using UDP in the transport layer. These applications use the designed flow control schemes. Henceforth, these flows will be referred to as *controlled UDP* flows. The 10 nodes that provide controlled flows are connected to Router 0 (R0) and Router 2 (R2) using 10 Mbps bandwidth capacity links. The controlled UDP flows traverse both the backbone links.

Describing the topology in terms of the number of flows fails to provide a realistic summary of the contribution of the flows to the cross-traffic. Table 13.3 provides

Table 13.2 Details of the composition of cross-traffic in terms of flows for the network topology used to design and validate the control schemes

Flow Type	R0-R1 (F)	R0-R1 (B)	R1-R2 (F)	R1-R2 (B)	%
HTTP	150	100	0	0	25.41
FTP	87	87	250	250	68.50
CBR	5	5	0	0	1.02
EXP	0	0	20	20	4.06
Total					98.98

the composition of traffic in terms of bytes during simulations performed using the current topology. Although the loss rate of the controlled UDP flows vary from 3% to 15% because of the variation of the bandwidth of the link R0-R1, the relative contribution of each type of traffic to the simulation does not undergo much changes. The variations are within a bound of $\pm 1\%$.

Table 13.3 Details of the composition of cross-traffic in terms of bytes for the network topology used to design and validate the control schemes

Type	% of Total TCP	% of Total UDP	% of Total Traffic
TCP (HTTP)	~ 15.69	N.A.	~ 12.64
TCP (FTP)	~ 84.31	N.A.	~ 67.92
Total % (TCP/Total Traffic)			~ 80.56
Controlled UDP	N.A.	~ 2.98	~ 0.58
UDP (Exp)	N.A.	~ 84.30	~ 16.39
UDP (CBR)	N.A.	~ 12.72	~ 2.47
Total % (UDP/Total Traffic)			~ 19.44

13.4.2 Simulated Network Topology for Studying the Scalability of the Flow Control Strategies

The second topology is designed to study the scalability performance of the adaptive flow control algorithms. There are two objectives of the scalability studies:

- Testing the performance of the flow control algorithms as the share of controlled UDP flows to the total UDP traffic in the network goes up. The ratio of the UDP flows and the TCP flows to the total traffic generated during this set of simulations remains constant. This test is designed to find out whether increasing the percentage of controlled UDP flows as compared to the uncontrolled UDP flows while keeping the contribution of the total UDP flows to the total traffic in the network constant affects the performance of the TCP flows.

- Testing the performance of the flow control algorithms as the share of the UDP flows (both the controlled as well as the uncontrolled UDP flows) to the total traffic in the network goes up. This scalability test is aimed at investigating the advantages of using flow controlled real-time multimedia applications as the share of UDP traffic in the overall traffic over the Internet goes up in the future.

These experiments, using the designed network topology, will help to demonstrate the advantages of using real-time multimedia applications deploying adaptive flow control algorithms over the Internet. The difference between the current topology and the topology designed earlier is with respect to the location of the nodes generating the exponential (UDP) traffic. In the earlier topology, the nodes running exponential applications on top of UDP were sending traffic between the link R1-R2 in both directions. In the current topology, the traffic from these nodes traverses the same path as that of the packets from the controlled UDP flows. This means that the nodes having exponential UDP applications are connected either to Router 0 (R0) or to Router 2 (R2). Figure 13.6 shows the topology generated using ns-2 to study concept scalability. The changed position of the nodes running an exponential

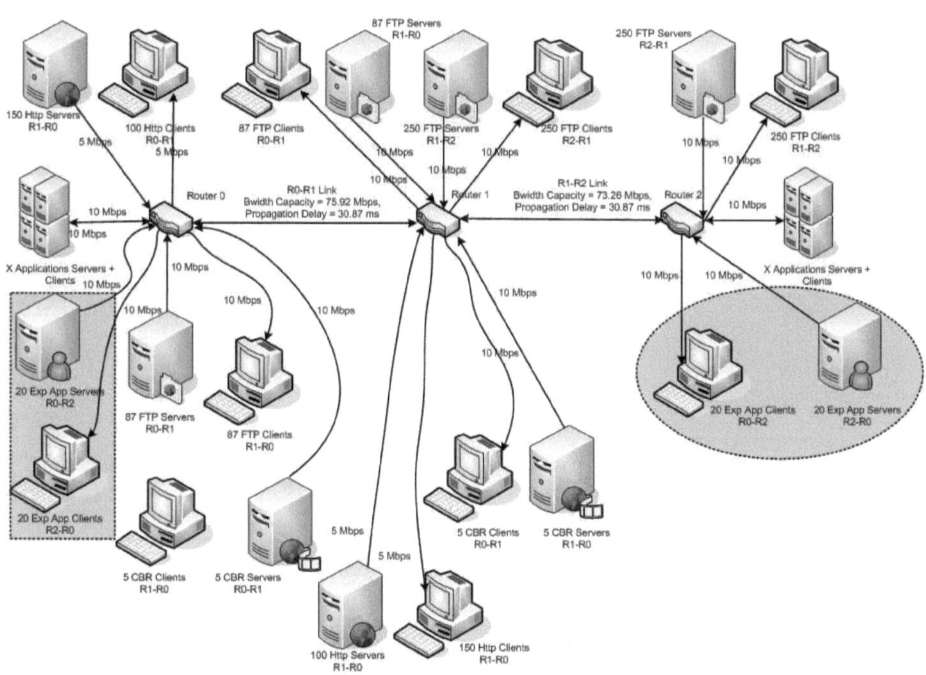

Fig. 13.6 Topology created in the ns-2 simulation environment to study scalability of the adaptive end-to-end multimedia flow control schemes

application using UDP have been shown using shaded areas in Figure 13.6.

Link R0-R1 has to tolerate extra traffic because of the shifting of the locations of the nodes running exponential applications. Simultaneously, the link also has to maintain the average loss rate of the controlled UDP flows around 3% to 4%. Therefore, the sizes of the queues present in the duplex link R0-R1 are increased to 768000 bytes from 460800 bytes. Unlike the previous topology, the bandwidth of the R0-R1 link is fixed at 75.92 Mbps. This is because this set of experiments do not require the loss rates of the controlled UDP flows to be varied by changing the bandwidth or the queuing parameters of the network.

The topology for studying scalability is modified in many ways during the experiments as per the requirements. In order to study the effect of increasing the ratio of controlled UDP flows with respect to the uncontrolled UDP flows, the ratio of total UDP to total TCP traffic is kept constant. The number of nodes using flow controlled multimedia UDP applications, connected to routers R0 and R2, is steadily increased. The bit rates generated by the nodes running exponential applications using UDP, connected to the same routers, is suitably decreased. For studying the second objective of the scalability studies, the number of nodes using flow controlled multimedia UDP applications is steadily increased without making any further changes in the network topology. This leads to an increase in the percentage of UDP traffic contribution in the total traffic during the simulations. Figures 13.7(a), (b), and (c) show the objectives of the scalability experiments.

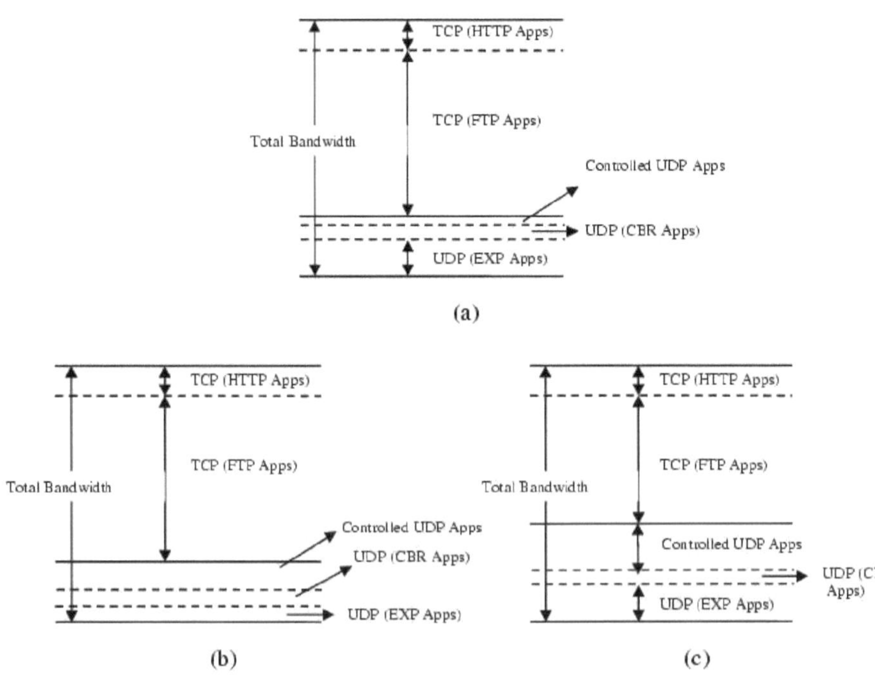

Fig. 13.7 Objectives of the scalability experiments

Figure 13.7(a) represents the original topology. The bandwidth capacity in the network is consumed by five different categories of flows. The two TCP categories (HTTP and FTP) have the largest share. This is followed by the shares of the three types of UDP flows. Figure 13.7(b) shows the change in percentage of the bandwidth used among the UDP flows. The percentage of bandwidth consumed by the controlled UDP flows goes up as compared to the other two categories of the UDP flows. The proportion of the total UDP contribution to the total traffic does not change. Finally, Figure 13.7.(c) depicts the second objective of the scalability tests visually. The proportion of the UDP traffic to the total traffic goes up.

Let $\mathbb{T}_1$ be the total traffic in the network measured in bytes. $\mathbb{T}_1$ is the sum of the UDP traffic $\mathbb{U}_1$ and TCP traffic $\mathbb{P}_1$ in the network. Both of them are also measured in bytes. Therefore,

$$\frac{\mathbb{P}_1}{\mathbb{T}_1} + \frac{\mathbb{U}_1}{\mathbb{T}_1} = 1. \tag{13.42}$$

$\mathbb{U}_1$ comprises of controlled UDP traffic U_{c_1} and uncontrolled UDP traffic U_{u_1}. This implies that

$$\frac{U_{c_1}}{\mathbb{U}_1} + \frac{U_{u_1}}{\mathbb{U}_1} = 1. \tag{13.43}$$

In the first case of scalability experiments, overall contribution of the UDP flows in the total traffic is kept constant. The traffic composition is changed by increasing the share of the controlled UDP flows and decreasing the share of the uncontrolled UDP flows by the same magnitude.

$$\frac{\mathbb{P}_2}{\mathbb{T}_2} + \frac{\mathbb{U}_2}{\mathbb{T}_2} = 1, \tag{13.44}$$

$$\frac{U_{c_2}}{\mathbb{U}_2} + \frac{U_{u_2}}{\mathbb{U}_2} = 1, \tag{13.45}$$

such that $\frac{U_{c_2}}{\mathbb{U}_2} > \frac{U_{c_1}}{\mathbb{U}_1}$ and $\frac{U_{u_2}}{\mathbb{U}_2} < \frac{U_{u_1}}{\mathbb{U}_1}$. Moreover, $\frac{\mathbb{U}_2}{\mathbb{T}_2} = \frac{\mathbb{U}_1}{\mathbb{T}_1}$. This means that $\frac{\mathbb{P}_2}{\mathbb{T}_2} = \frac{\mathbb{P}_1}{\mathbb{T}_1}$.

In the second case of scalability experiments, overall contribution of the UDP flows in the total traffic is increased. This increase is achieved by increasing the share of the controlled UDP flows without changing any other aspect of the traffic as compared to the original case.

$$\frac{\mathbb{P}_3}{\mathbb{T}_3} + \frac{\mathbb{U}_3}{\mathbb{T}_3} = 1, \tag{13.46}$$

$$\frac{U_{c_3}}{\mathbb{U}_3} + \frac{U_{u_3}}{\mathbb{U}_3} = 1, \tag{13.47}$$

such that $\frac{U_{c_3}}{\mathbb{U}_3} > \frac{U_{c_1}}{\mathbb{U}_1}$, $\frac{U_{u_3}}{\mathbb{U}_3} < \frac{U_{u_1}}{\mathbb{U}_1}$, and $\frac{\mathbb{U}_3}{\mathbb{T}_3} > \frac{\mathbb{U}_1}{\mathbb{T}_1}$. This implies that $\frac{\mathbb{P}_3}{\mathbb{T}_3} < \frac{\mathbb{P}_1}{\mathbb{T}_1}$.

Table 13.4 provides information about the composition of traffic in the network topology with five controlled UDP sources to study scalability of the control algorithms measured in terms of the bytes contributed by each type of flow. This traffic composition undergoes changes as the topology is modified for achieving the objectives of the scalability tests.

Table 13.4 Details of the composition of cross-traffic in terms of bytes for the network topology used to study scalability of the control schemes

Type	% of Total TCP	% of Total UDP	% of Total Traffic
TCP (HTTP)	~ 15.19	N.A.	~ 12.35
TCP (FTP)	~ 84.81	N.A.	~ 68.95
Total % (TCP/Total Traffic)			~ 81.30
Controlled UDP	N.A.	~ 3.0	~ 0.56
UDP (Exp)	N.A.	~ 84.15	~ 15.73
UDP (CBR)	N.A.	~ 12.85	~ 2.41
Total % (UDP/Total Traffic)			~ 19.45

13.5 Voice Quality Measurement Test: E-Model

Quality of voice and metrics to determine the quality of voice is an integral part of showing the change in the quality of voice due to the application of the designed adaptive flow control algorithms. Subjective tests like absolute category rating (ACR) and objective tests like perceptual evaluation of speech quality (PESQ) provide effective frameworks to assess voice quality. However, these frameworks fail to provide reliable results in the scenario of best-effort packet-switched IP networks. These frameworks have not been designed to take care of the effects arising out of variable time-dependent end-to-end delay, packet losses, and delay jitter. A suitable framework for estimating call quality from the measured network performance such as delay and loss characteristics of a path is the ITU-T E-Model [19]. E-Model is used by the VoIP industry to measure the the quality of the voice signals.

E-Model defines an R-factor that combines different aspects of voice quality impairments. (13.48) provides the mathematical definition of the R-factor.

$$R = R_0 - I_s - I_e - I_d + A. \tag{13.48}$$

In (13.48), R_0 refers to the effects of various noises, I_s represents the effects of the impairments that occur simultaneously with the original signal, I_e is the effect of impairments caused by packet losses in the network, and I_d shows the effect of impairments because of the delay suffered by the packets. A compensates for the above impairments under various user conditions.

I_d and I_e are the two parameters among all the others that are important for the VoIP systems. (13.48) can be modified after substituting the default values of the other parameters [20]. (13.49) is the modified and final equation to determine the R-factor that determines voice quality in a VoIP application using best-effort networks to transmit information.

$$R = 94.5 - I_e - I_d. \tag{13.49}$$

The R-factor ranges from 0 to 100 and can be mapped to mobile operating system (MOS) using a nonlinear mapping [19].

$$MOS = 1 + 0.035R + 7 * 10^{-6} R(R-60)(100-R). \tag{13.50}$$

(13.49) and (13.50) provide the facility of measuring the quality of voice in terms of MOS in a VoIP application based on the measurement of delay and packet loss in the network. The two variables that needs to be calculated in order to calculate the R-factor and eventually the MOS are I_e and I_d.

I_d is dependent on the mouth-to-ear delay, d. Mouth-to-ear delay is the sum of the end-to-end delay of the packet containing the encoded voice data and the delay due to coding and decoding of the signal. (13.51) shows how the mouth-to-ear delay affects I_d.

$$I_d = 0.024d + 0.11(d - 177.3)\mathbf{I}(d - 177.3). \qquad (13.51)$$

$\mathbf{I}(x)$ is an indicator function that implies

$$\mathbf{I}(x) = \begin{cases} 0 \text{ if } x < 0 \\ 1 \text{ otherwise} \end{cases}$$

As seen from (13.51), I_d does not depend on the type of codec used for encoding voice. It depends on the mouth-to-ear delay. However, calculating the impact of packet losses on quality of voice within the E-Model framework is not easy. I_e depends on parameters that are determined by the properties of the encoder. The relation between I_e and overall packet loss rate e can be expressed by the following equation:

$$I_e = \gamma_1 + \gamma_2 ln(1 + \gamma_3 e) \qquad (13.52)$$

γ_1 is a constant that determines voice quality impairment caused by encoding, γ_2 and γ_3 describe the impact of loss on perceived voice quality for a given codec. e includes both network losses and playout buffer losses.

γ_1, γ_2, and γ_3 are determined by using voice quality testing results under different loss conditions. To read more about E-Model and its applications in determining quality of voice in VoIP applications, the papers by Cole and Rosenbluth [20], Ding and Goubran [21], Tao *et al.* [22], and Sun and Ifeachor [23] need to be referred to. The abovementioned papers also provide the experimentally determined values of the three parameters: γ_1, γ_2, and γ_3, for various existing encoders.

Ye [24] provides the values of γ_1, γ_2, and γ_3 for the codec - Speex [25], at different bit-rates operating without any error-concealment algorithms and in CBR mode in his research work. Speex is a multi mode codec. Ye also describes the experimental procedure of obtaining the parameter values for Speex in his dissertation. Table 13.5 summarizes the results for several codecs as mentioned in [20] and [22]. Table 13.5 provides the γ values for Speex encoder without any error-concealment, encoded at CBR as determined by Ye [24]. In Table 13.5, the column "Quality" refers to the Speex encoding that is controlled by a quality parameter that ranges from 0 to 10. In CBR operation, the quality parameter is an integer.

Table 13.5 The values of γ_1, γ_2 and γ_3 for various codecs

Codec	PLC	γ_1	γ_2	γ_3
G.711	[26]	0	30	15
G.723.1.B-5.3	silence	19	71.38	6
G.723.1.B-6.3	silence	15	90.00	5
G.729	silence	10	47.82	18
G.723.1.A+VAD-6.3	none	15	30.50	17
G.729A+VAD	none	11	30.00	16

Table 13.6 The values of γ_1, γ_2 and γ_3 for Speex

Quality	γ_1	γ_2	γ_3
3	31.01	36.99	10.29
4	31.01	36.99	10.29
5	23.17	29.36	20.03
6	23.17	29.36	20.03
7	16.19	24.91	36.17
8	16.19	24.91	36.17
9	9.78	22.81	58.76
10	6.89	21.99	72.75

13.6 Validation of Proposed Flow Control Strategies

The efficacy of the control strategies will be decided on the basis of the following criteria:

- QoS: The QoS of the real-time audio session is decided by the MOS calculated by using E-Model. The E-Model uses the end-to-end packet delays and the packet loss rates to calculate the R-factor that is mapped to the MOS scale using a non-linear relationship.
- Bandwidth occupancy of the controlled multimedia flow: This factor is related to being a "good citizen of the Internet". Let us assume that a controlled UDP flow employing one of the control strategies devised in the current research has the same end-to-end QoS as compared to an uncontrolled UDP flow. However, if the controlled UDP flow occupied less bandwidth in the network as compared to the uncontrolled UDP flow in order to provide the same level of end-to-end QoS, then using the adaptive flow control strategy is considered to be better than employing no control at all.

13.6.1 Determining I_e Curves for Different Modes of the Ideal Codec

In the current research, MOS of the voice transmitted over the best-effort network is calculated by assuming a desirable codec with six different bit-rates. Therefore, I_e curves for this multimode codec with six different bit-rate modes has to be determined in order to calculate the MOS of the real-time VoIP streams in the ns-2 simulations.

Ye [24] determined the values of the parameters, γ_1, γ_2, and γ_3, for eight bit-rate modes of Speex. The values of the three parameters for each mode of the Speex codec have already been provided in Table 13.5. The I_e curves of the Speex codec corresponding to the "Quality" modes of 4, 8, and 10 are used to emulate the I_e curves of the desirable codec sending information at the bit-rates of 36 kbps, 48 kbps, and 60 kbps, respectively. This is a valid choice as the bit-rates of these three modes at constant bit-rate without the packet headers are very similar to the bit-rates generated by the desirable codec.

The remaining three I_e curves of the codec for the bit-rates of 72 kbps, 84 kbps, and 96 kbps are determined by extrapolating from the three I_e curves for the bit-rates of 36 kbps, 48 kbps, and 60 kbps. The polynomials representing the I_e curves for three highest bit-rates (72 kbps, 84 kbps, 96 kbps) of the desirable codec used for determining the QoS of the VoIP flows are:

$$\begin{aligned} I_{e_{72}} &= 2.187 \times 10^{-12} e^9 - 6.447 \times 10^{-10} e^8 + 8.129 \times 10^{-8} e^7 \\ &\quad -5.736 \times 10{-}6e^6 + 0.002491 e^5 - 0.006914 e^4 + 0.124 \\ &\quad e^3 - 1.444 e^2 + 11.69 e + 5.162, \end{aligned} \tag{13.53}$$

$$\begin{aligned} I_{e_{84}} &= 1.704 \times 10^{-9} e^7 - 3.963 \times 10^{-7} e^6 + 3.778 \times 10^{-5} e^5 \\ &\quad - 0.001913 e^4 + 0.05589 e^3 - 0.9706 e^2 + 10.52 e \\ &\quad + 3.443, \end{aligned} \tag{13.54}$$

$$\begin{aligned} I_{e_{96}} &= 1.3 \times 10^{-9} e^7 - 3.118 \times 10^{-7} e^6 + 3.086 \times 10^{-5} e^5 \\ &\quad - 0.001633 e^4 + 0.05025 e^3 - 0.9234 e^2 + 10.47 e \\ &\quad + 1.722. \end{aligned} \tag{13.55}$$

In (13.53), (13.54), and (13.55) the loss rate e is measured in percentage. These equations have been derived by using the curve-fitting toolbox of MATLAB®.

13.6.2 Calculation of the MOS to Determine QoS in the Simulations

Each of the simulated experiments used to determine the efficacy of the developed adaptive flow control strategies is executed for 120 seconds (two minutes). The MOS of the controlled UDP flows in the simulations are calculated every 10 seconds using the CLR and the mean end-to-end packet delays in the specific time slot. The E-Model framework based on parameters R, I_d, and I_e is used to accomplish this. The final MOS of the controlled UDP flows is determined by taking a mean of all the 12 MOS of the observed flow, calculated every 10 seconds.

13.6.3 Performance of the Flow Control Schemes in the Network Topology Used for Validation

Earlier, two different ns-2 network topologies are discussed. The first topology is used to design and validate the flow control strategies. Six sets of experiments are conducted in order to determine the performance of the four designed control strategies. Each set of experiments is conducted by varying the comprehensive loss percentage of the packets in the network.

Figure 13.8 has two plots. The first plot shows the relative performance of the control schemes in terms of the mean MOS of the five "Controlled UDP" flows with a CLR of $\sim 3\%$. The second part of the plot shows the mean bandwidth utilized by the five flows.

High bit rate (HBR) depicted in Figure 13.8 stands for the experiment in which all the five UDP flows under observation send information at a constant bit rate of 96 kbps. HBR is performed in order to investigate what will happen if the codec sending information at 96kbps (best quality) does not respond to congestion in the network. Low bit rate (LBR) stands for the experiment in which all the five UDP flows under observation send information at a constant bit rate of 36 kbps. LBR investigates the effect on the QoS of the VoIP flows if the codec sends flow at the lowest bit rate of 36 kbps (worst quality). NCL refers to the performance of the VoIP flows employing NCL. LCL shows the performance of the VoIP flows employing LCL.

The plots in Figure 13.8 clearly shows the advantage of adapting the bit-rate of the codec based on the feedback about the congestion level in the network. VoIP flows employing NCL flow control schemes perform better as compared to the cases when the flows do not adapt to the congestion. The flows employing control schemes not only show either similar or better end-to-end QoS but also utilize less bandwidth.

In the set of experiments done with the network topology having an average CLR of around 5%, the average MOS of the five flows employing NCL flow control is almost same as that of the average MOS of the same flows when their bit-rate is

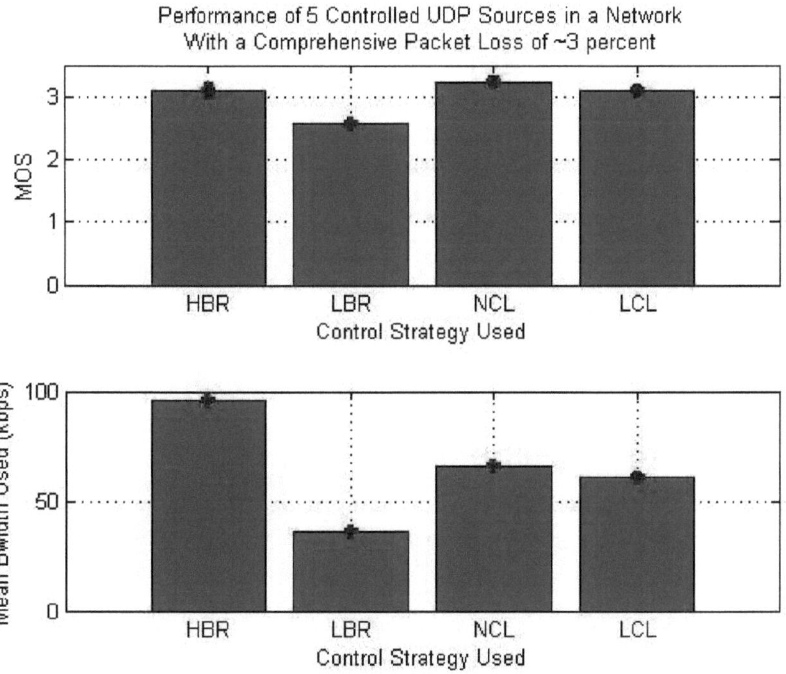

Fig. 13.8 Relative performance of the control schemes in terms of the mean MOS of the five "Controlled UDP" flows in the network topology #1 with a CLR of $\sim 3\%$

equal to 96 kbps. This is shown in the Figure 13.9. However, the average bandwidth capacity utilized by each controlled flow is less than 96 kbps.

The CLR of the packets of the flows under observation is changed to around 7% in the next set of experiments, as shown in Figure 13.10. It can be seen that QoS of the five UDP flows employing LCL with a SP is better than when these flows did not employ any control and sent information at a constant bit-rate of 96 kbps. The average bandwidth used up by these controlled flows is 35.43 kbps less than 96 kbps. The other case in which flows utilize NCL control scheme also performs well while consuming far less bandwidth than the HBR flows.

Figures 13.11, 13.12, and 13.13 show the relative performance of the control schemes in network topologies with average CLR of the observed flows equal to 9%, 11%, and 15%. One common trend that emerges from these figures is that the "controlled UDP" flows perform as good as, or better than the high constant bit-rate flows without employing any kind of control. However, these "controlled UDP" flows utilize far less bandwidth.

Summary of the Performance of the Control Schemes The ideal result while deploying end-to-end flow control schemes for real-time multimedia flows in best-effort networks is that the "controlled UDP" flows use as little bandwidth as possible while delivering highest QoS measured in terms of MOS. However, in practice,

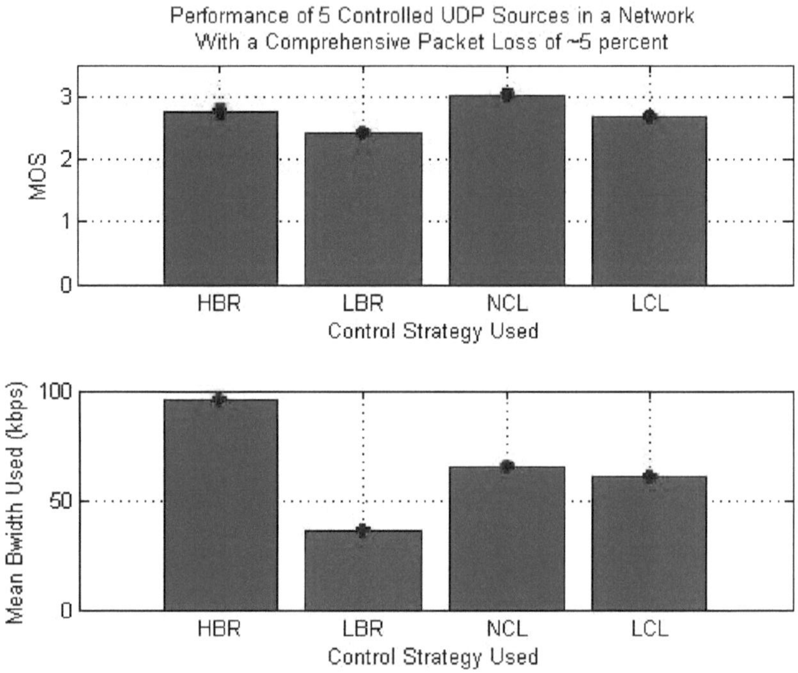

Fig. 13.9 Relative performance of the control schemes in terms of the mean MOS of the five "Controlled UDP" flows in the network topology #1 with a CLR of $\sim 5\%$

there is a limit to which this tradeoff between using less bandwidth while delivering high QoS can be achieved. This is because of the limitations of the voice encoding algorithms. If two metrics, MOS^{-1} and the bandwidth utilized by the real-time multimedia flows are plotted with respect to each other, the plot would look somewhat similar to that shown in Figure 13.14. This figure shows that MOS^{-1} decreases as the bandwidth increases and the function showing the relationship between them is convex and nonlinear.

Assuming that the cost of bandwidth used by a real-time multimedia flow in a congested network and the end-to-end QoS are equally important, then the most optimal result of the problem of tradeoff between bandwidth utilized while delivering good end-to-end QoS can be achieved at the point where the straight line intersects the convex curve. If we go above this point while following the convex curve, bandwidth utilization by the flows will decrease. However, the MOS of the flows will decrease simultaneously. Similarly, if we go below this point, the MOS of the flows will increase but the bandwidth utilized by them will increase.

Figure 13.15 summarizes the results of all the six sets of experiments involving five "controlled UDP" flows using network topologies with different CLRs. There are three distinct regions in the figure. The first region on the left corresponds to the average performance of the five UDP sources sending traffic at the lowest bit rate

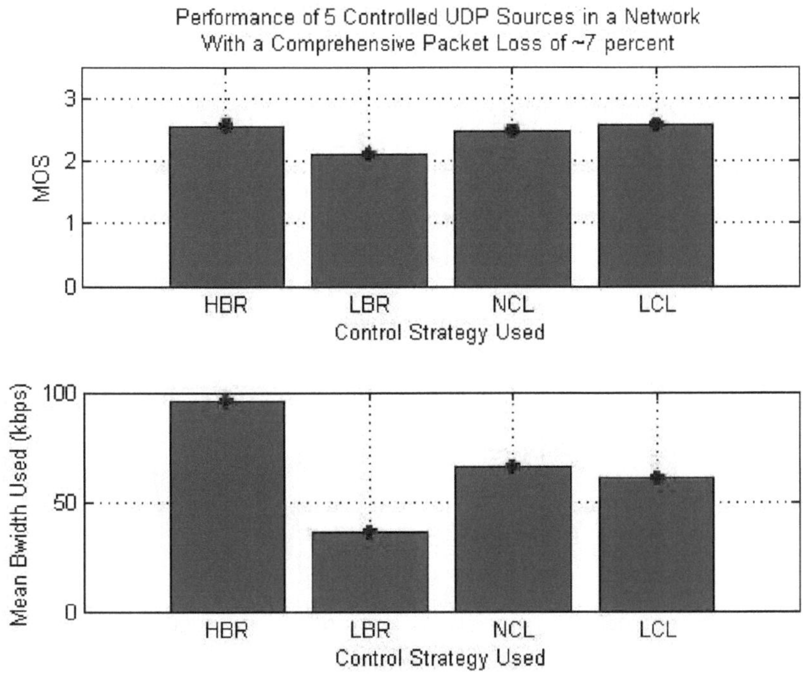

Fig. 13.10 Relative performance of the control schemes in terms of the mean MOS of the five "Controlled UDP" flows in the network topology #1 with a CLR of $\sim 7\%$

of 36 kbps in different network scenarios. The last region on the right shows the performance of the five UDP sources when they sent traffic at the highest bit rate of 96 kbps. The region in the center depicts the performance of the control schemes devised during this research. The gray band shows the range of bandwidth used by flows employing the NCL control scheme. NCL performs a bit worse than the LCL. The red band shows the range of bandwidth used by flows employing LCL control scheme.

Table 13.7 shows the best performances of the control schemes for different CLR networks. LCL and NCL perform well in comparison to the uncontrolled flows. The bandwidth saved because of the control laws while delivering equal, if not better, performance in terms of MOS varies from 33.52% to 43.95%. This shows that the adaptive flow control helps to maintain the QoS of real-time multimedia flows and make them better citizens of the Internet.

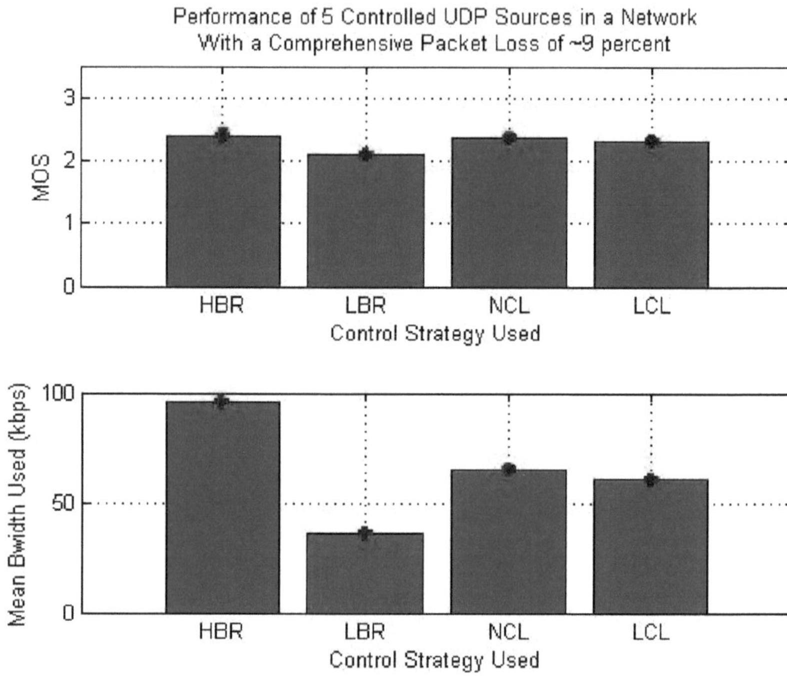

Fig. 13.11 Relative performance of the control schemes in terms of the mean MOS of the five "Controlled UDP" flows in the network topology #1 with a CLR of $\sim 9\%$

Table 13.7 Summary of the best performance of the control laws

CLR (%)	MOS (HBR)	MOS (LBR)	Best	MOS (Best)	BW (kbps)
3	3.09	2.56	NCL	3.22	65.83
5	2.74	2.42	NCL	3.03	53.80
7	2.56	2.09	LCL	2.57	60.57
9	2.39	2.09	NCL	2.37	65.60
11	2.24	1.98	LCL	2.35	60.48
15	2.12	1.67	LCL	1.97	59.99

13.6.4 Scalability of the Flow Control Strategies

One of the important aspects of the study of flow control schemes for real-time multimedia flows is to investigate the scalability issues. The questions that need to be answered are: 1. what happens when a higher number of multimedia flows in the network start employing the proposed flow control schemes? 2. how friendly are these flow control schemes to flows using TCP?

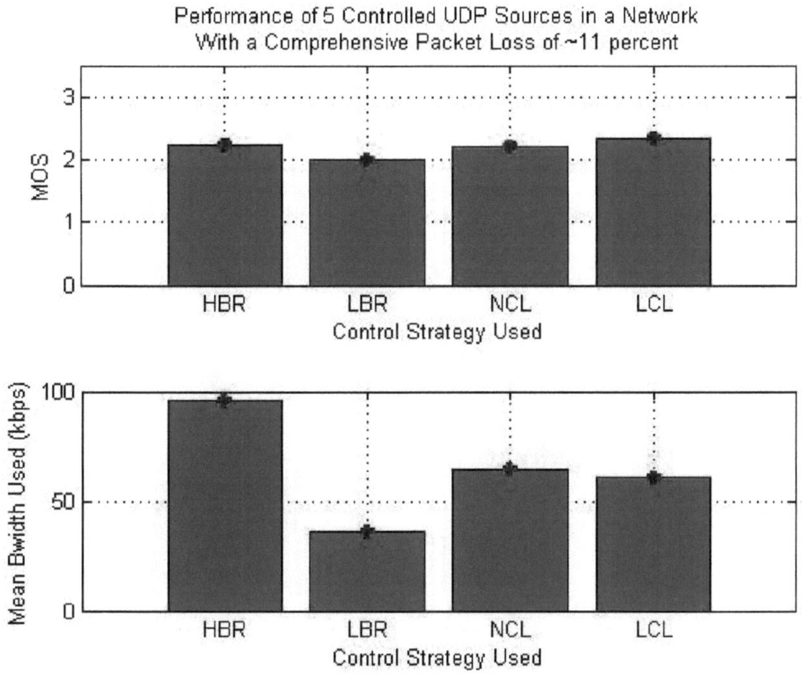

Fig. 13.12 Relative performance of the control schemes in terms of the mean MOS of the five "Controlled UDP" flows in the network topology #1 with a CLR of ∼ 11%

13.6.4.1 Increasing the Percentage of "Controlled UDP" Flow Traffic in the UDP Traffic Mix while Keeping the Overall Percent of UDP in the Total Traffic Constant

A set of experiments is carried out with the contribution of the UDP flows in the total network traffic remaining constant. The ratio of the traffic contributed by the "controlled UDP" with respect to the traffic contributed by the uncontrolled UDP flows is gradually increased. The effect of this increase in "controlled UDP" flow percentage is investigated by observing the goodput of the HTTP connections that exist between link R0-R1. The topology shown in Figure 13.6 is used for the purpose of this investigation. The percentage ratio of the controlled UDP traffic to the total UDP traffic varies from 0.67% to 62.77%. Total UDP contribution to the traffic does not undergo much change.

The scalability studies have been performed using the NCL control scheme. In the controller validation stage with networks of different packet loss percentages, NCL performed better than LCL. NCL reacts more aggressively to reduce bit-rate of the flow during congestion. Figures 13.16, 13.17, 13.18, and 13.19 show the results of the current set of scalability experiments. It can be observed from the figures that the goodput of the hundred HTTP flows between the HTTP nodes connected to

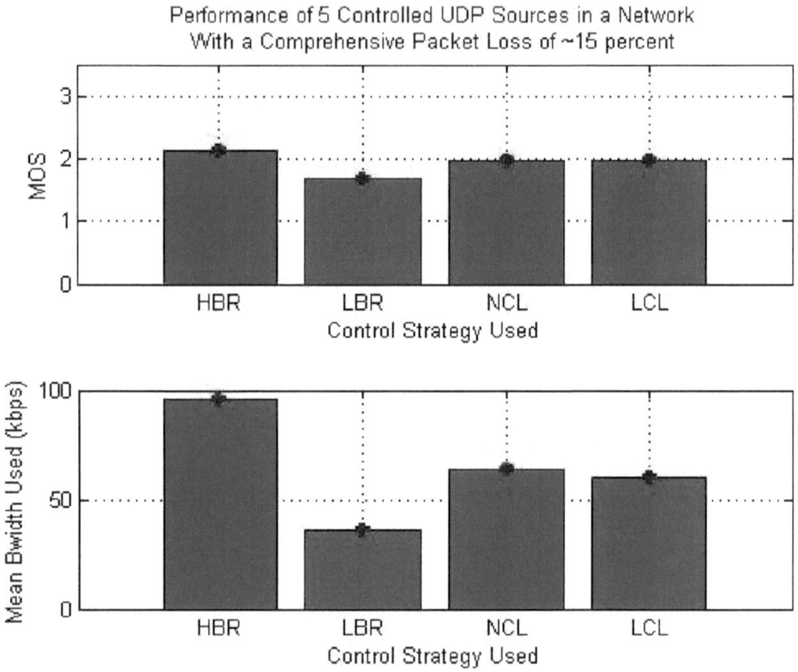

Fig. 13.13 Relative performance of the control schemes in terms of the mean MOS of the five "Controlled UDP" flows in the network topology #1 with a CLR of $\sim$ 15%

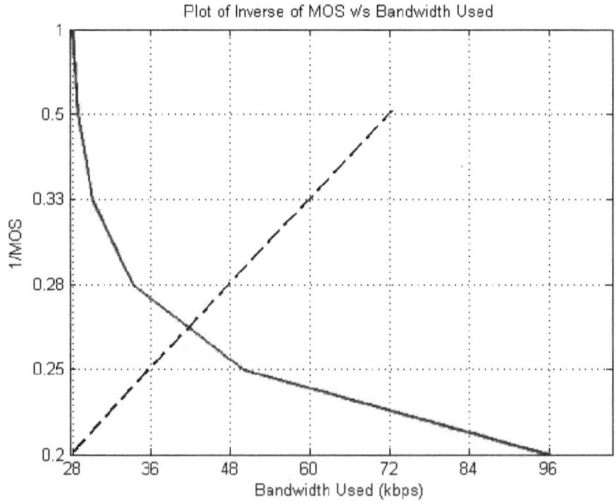

Fig. 13.14 Ideal plot of MOS^{-1} versus bandwidth utilized by the flows

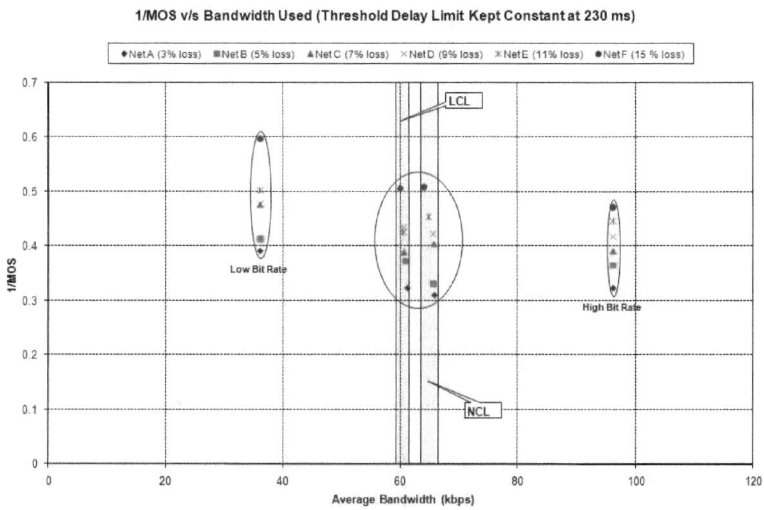

Fig. 13.15 Plot of inverse of mean MOS of 5 UDP flows employing different control schemes versus mean bandwidth utilized by them during each experiment

routers R0 and R1 has not been affected significantly by changing the percentage of the controlled UDP flows in the UDP flow mix. The control strategies do not affect the HTTP flows adversely.

Table 13.8 shows that as the number of controlled UDP flows increase in the network, the performance of the NCL flow control becomes better.

Table 13.8 Summary of the performance of the control laws for scalability tests when the contribution of UDP flows to the total traffic is kept constant

Cont. UDP Srcs.	MOS (HBR)	MOS (LBR)	MOS (NCL)
5	3.17	2.56	2.85
36	2.88	2.37	2.72
72	2.73	2.23	2.71
144	2.76	2.45	2.55

13.6.4.2 Increasing the Percentage of "Controlled UDP" Traffic in the Total Traffic Mix

The last part of the current work is to determine the scalability of the new control schemes when the percentage of UDP composition in the total network traffic goes up. Four sets of experiments are performed by changing the number of "controlled

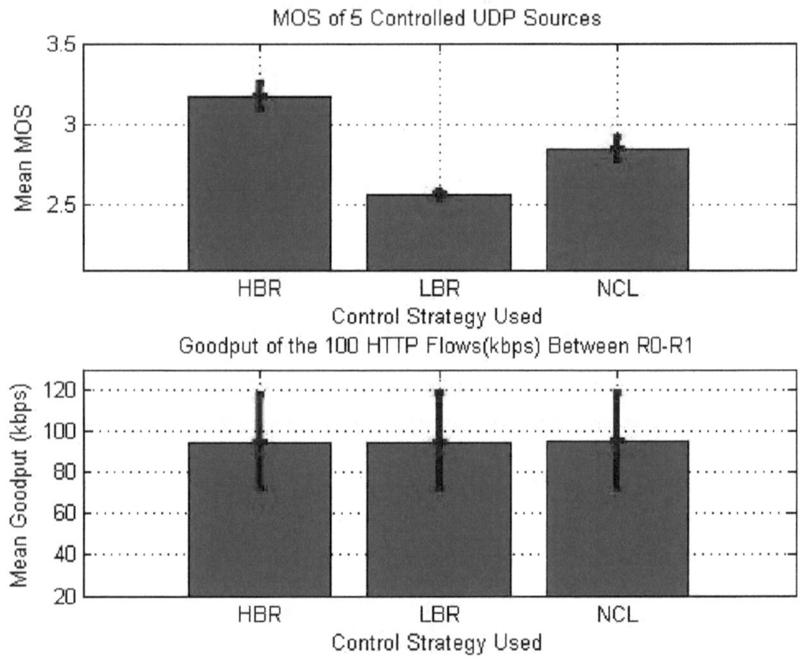

Fig. 13.16 Effect of 5 controlled UDP sources on the goodput of the HTTP flows between the nodes attached to routers, R0 and R1

UDP" flows to 5, 20, 50, and 80. As the number of "controlled UDP" sources are increased from 5 to 80, the traffic component comprising of the TCP flows drops from 81.31% to 73.46% percent when all the "controlled UDP" nodes transmit at the highest constant bit-rate of 96 kbps without responding to the network congestion.The percentage of UDP traffic increases from 18.69% to 26.54%, respectively. As the number of "controlled UDP" nodes in the simulation are increased, the amount of bandwidth saved by deploying the control schemes like NCL becomes more evident. The savings in the bandwidth is not visible when the number of "controlled UDP" flows is small. However, this saving increases as the number of flows deploying the end-to-end adaptive flow control schemes increases. This means that real-time multimedia flows deploying adaptive flow control schemes in the network will help the network managers adapt to the increase in the UDP usage in the future.

The quality of the real-time multimedia sessions during the different experiments need to be examined in order to see the efficacy of the flow control schemes when the percentage of the UDP flow contribution increases in the total traffic.

Average MOS of the five "controlled UDP" sources and their effect on the goodput of the 100 HTTP flows between routers, R0 and R1 has been shown in Figure 13.16. The figure shows that the quality achieved by the five flows when they transmit at the highest constant bit-rate of 96 kbps cannot be matched by the same flows

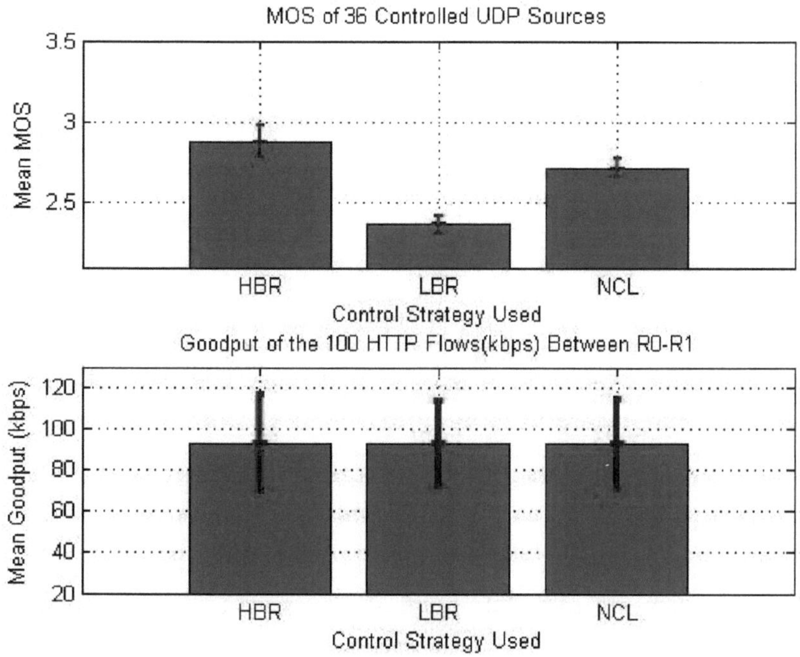

Fig. 13.17 Effect of 36 controlled UDP sources on the goodput of the HTTP flows between the nodes attached to routers, R0 and R1. Percentage of UDP traffic comprising the total traffic in the network has been kept approximately constant

when they are subjected to the end-to-end adaptive flow control schemes. Figure 13.20 shows the mean MOS of the 20 "controlled UDP" sources. The case with 20 "controlled UDP" flows using NCL control scheme does not perform as well as when they are uncontrolled. The 20 UDP flows with control inputs determined using the lower reference signal outperform the 20 UDP flows transmitting at the highest constant bit rate of 96 kbps. The congestion in the network has increased with the addition of 15 more flows in the forward direction. This means that a higher reference accumulation signal is not able to distinguish between a normal system and a congested system. A more conservative control scheme using a lower reference signal takes care of the congestion better while delivering better quality flows.

Figure 13.21 shows the mean MOS of the 50 "controlled UDP" sources. This shows the benefits of using adaptive end-to-end flow control in order to deliver better quality while saving bandwidth.

Figure 13.22 shows the mean MOS of the 80 "controlled UDP" sources. The bandwidth savings due to flow control schemes is substantial for 80 UDP sources. If each flow transmits at a constant bit rate of 96 kbps, then eighty flows consume 7.68 Mbps of available bandwidth. However, when these eighty UDP flows employ NCL control schemes they consume 4.34 Mbps of the available bandwidth. This results

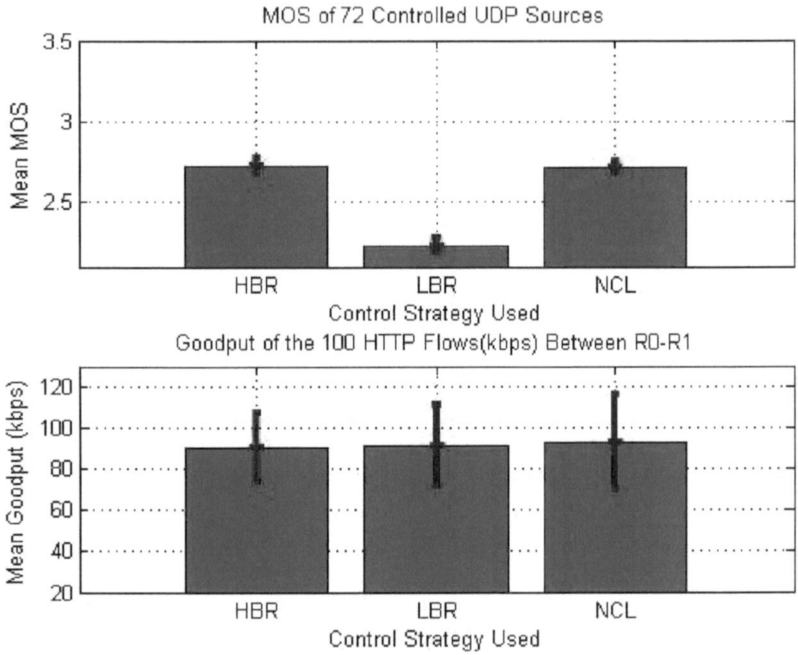

Fig. 13.18 Effect of 72 controlled UDP sources on the goodput of the HTTP flows between the nodes attached to routers, R0 and R1. Percentage of UDP traffic comprising the total traffic in the network has been kept approximately constant

into a saving of 3.34 Mbps of bandwidth that can be used by other flows. Table 13.9 summarizes the mean MOS achieved by the UDP sources as the contribution of the UDP traffic to the total traffic is increased. Flows using MPC9 and NCL controllers scale up well as the number of flows deploying them increase.

Table 13.9 Summary of the performance of the control laws for scalability tests when the contribution of UDP flows to the total traffic increases

Cont. UDP Srcs.	MOS (HBR)	MOS (LBR)	MOS (NCL)
5	3.17	2.56	2.85
20	2.83	2.30	2.61
50	2.70	2.21	2.50
80	2.78	2.37	2.51

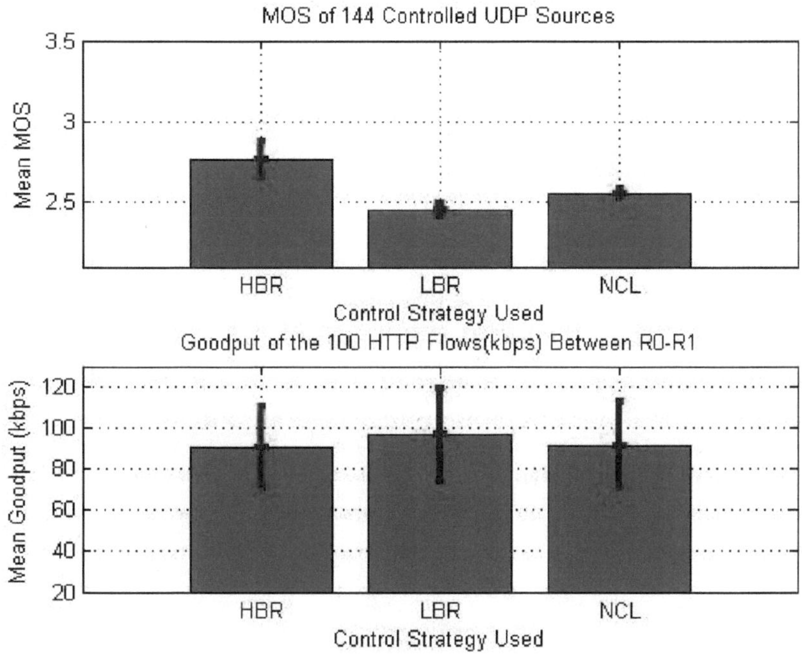

Fig. 13.19 Effect of 144 controlled UDP sources on the goodput of the HTTP flows between the nodes attached to routers, R0 and R1. Percentage of UDP traffic comprising the total traffic in the network has been kept approximately constant

13.7 Summary and Conclusions

There are many ways to provide better QoS for real-time multimedia flows. These can be categorized broadly into scheduling, shaping, policing, control, and synchronization of the flows. The current work focuses on flow control. The problems related to QoS cannot be solved by allocating more resources. At some point, efficiency of the utilization of the resources need to be considered. This work proposes end-to-end flow control algorithms for unicast real-time multimedia applications transmitting over best-effort networks.

Two simple flow control schemes, LCL and NCL, are proposed. These simple schemes perform well. Their relative performance is competitive as the proportion of "controlled UDP" flows in the network is increased. The performance of the proposed flow control schemes is determined by changing the network topology to increase the percentage of comprehensive packet losses of the observed flows from 3%to 15%. NCL shows the best performance.

The NCL control scheme is put through a series of tests that aim to study its behavior as the characteristics of traffic change in the network. The first set of experiments test the performance of the NCL control scheme when the percentage of

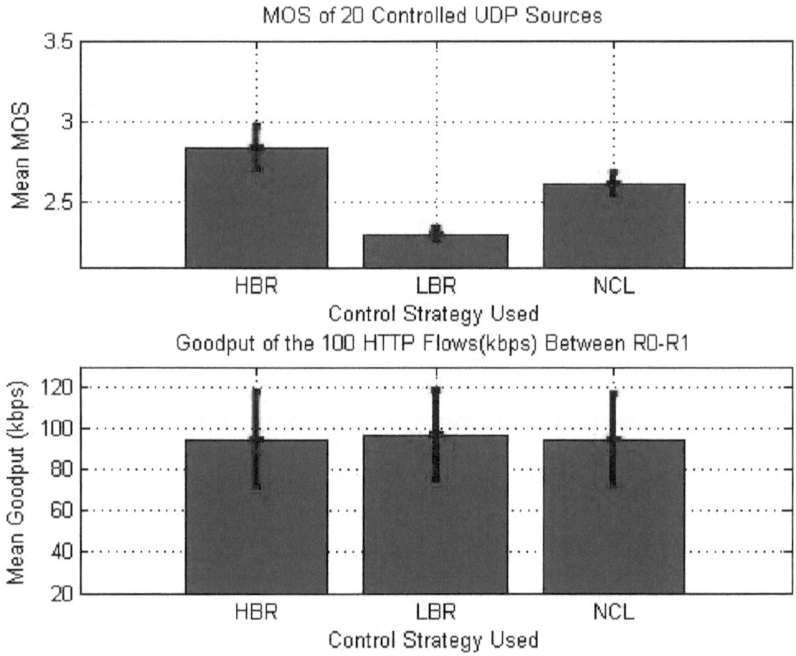

Fig. 13.20 MOS of 20 "controlled UDP" flows and their effect on the goodput of the 100 HTTP flows between routers, R0 and R1

the "controlled UDP" traffic in the network as compared to the uncontrolled UDP traffic increases steadily. The proportion of UDP to TCP traffic is kept constant. At a higher percentage of "controlled UDP" traffic, the bandwidth saved by the deployment of the NCL control scheme is substantial and the performance is at par with the flows that deliver information at the highest constant bit-rate of 96 kbps.

The next set of experiments investigate the performance of the NCL control scheme when the composition of traffic in the network topology is changed by increasing the ratio of UDP traffic with respect to the TCP traffic. The NCL controller is friendly to the TCP flows as it does not lead to any abnormal and unexplained decline of the goodput.

Active control of hard real-time flows deliver the same or somewhat better QoS as HBR (no control), but with a lower average bit rate. Consequently, the proposed active flow control helps reduce bandwidth use of controlled real-time flows by anywhere between 31.43% and 43.96%.

The following conclusions can be derived from this work:

1. Actively controlled hard real-time flows deliver the same or somewhat better QoS as HBR, or flows with no control.

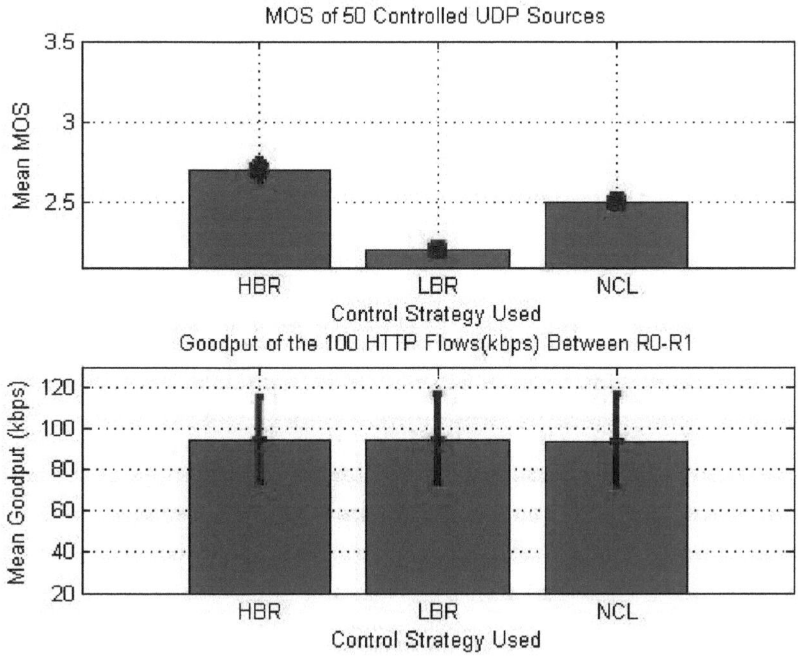

Fig. 13.21 MOS of 50 "controlled UDP" flows and their effect on the goodput of the 100 HTTP flows between routers, R0 and R1

2. For the same level of QoS delivered, actively controlled flows utilize a lower average bit rate. Consequently, active flow control helps reduce bandwidth use by controlled real-time flows.

The presented research is preliminary in nature and has some limitations as follows:

1. Lack of real-world data: all experiments have been done using the ns-2 simulation framework. Significant time has been spent in developing network topologies that reflect the conditions of the Internet. In spite of the wide acceptability of ns-2 in the networking community, a simulator cannot replicate real world traffic conditions entirely.
2. Lack of real-world implementation: multimode variable bit-rate codecs that can change their bit-rate dynamically during a VoIP session must be developed. This is a very difficult task. Without the availability of such codecs, it is not feasible to implement the proposed flow control algorithms.

Acknowledgements The authors would like to acknowledge the financial support provided by the State of Texas Advanced Technology Program, Grants No. $999903-083$, $999903-084$, and $512-0225-2001$, the US Department of Energy, Grant No. $DE-FG07-98ID13641$, the National

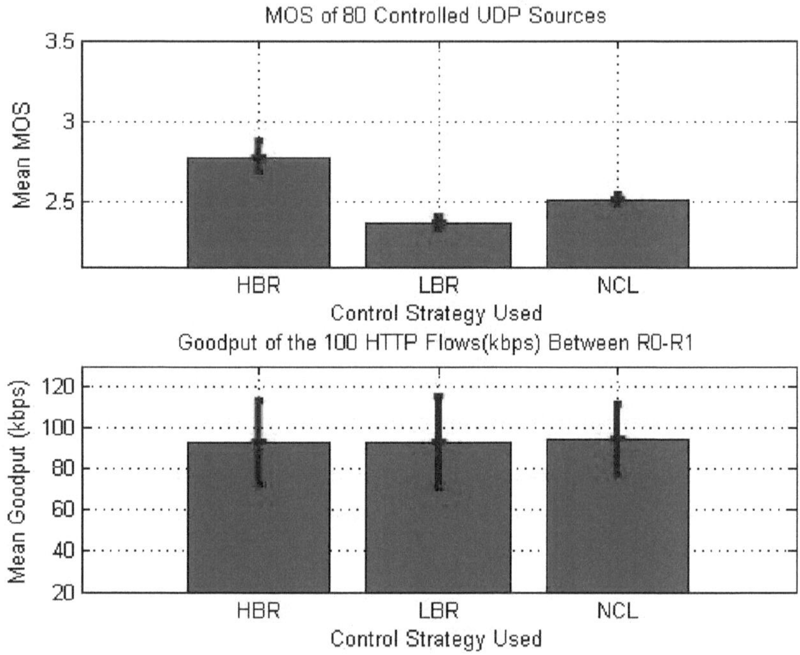

Fig. 13.22 MOS of 80 "controlled UDP" flows and their effect on the goodput of the 100 HTTP flows between routers, R0 and R1

Science Foundation Grants No. $CMS - 0100238$ and $CMS - 0097719$, and the US Department of Defense, Contract No. $C04 - 00182$.

References

1. D. Minoli and E. Minoli, *Delivering voice over* IP *networks*, 2nd ed. New York, USA: Wiley, 2002.
2. T. Yensen, R. Goubran and I. Lambadaris, "Synthetic stereo acoustic echo cancellation structure for multiple participant VoIP conferences, *IEEE Transactions on Speech Audio Processing*, vol. 9, pp. 168–174, February, 2001.
3. S. Poretsky, J. Perser, S. Erramilli, and S. Khurana, Terminology for benchmarking network-layer traffic control mechanisms, IETF RFC 4689 (Informational), October 2006.
4. R. Ramjee, J. F. Kurose, D. F. Towsley, and H. Schulzrinne, Adaptive playout mechanisms for packetized audio applications in wide-area networks, in *INFOCOM (2)*, pp. 680–688, 1994.
5. J. Pinto and K. J. Christensen, An algorithm for playout of packet voice based on adaptive adjustment of talkspurt silence periods, in *24th Conference on Local Computer Networks*, pp. 224–231, October 1999.
6. H. Sanneck, A. Stenger, K. Ben Younes, and B. Girod, A new technique for audio packet loss concealment, *IEEE GLOBECOM*, pp. 48–52, November 1996.

7. Y. J. Liang, N. Färber, and B. Girod, Adaptive playout scheduling using time-scale modification in packet voice communications, *ICASSP 2001: IEEE International Conference on Acoustics, Speech, and Signal Processing*, vol. 3, pp. 1445–148, 2001.
8. M. K. Ranganathan and L. Kilmartin, Neural and fuzzy computation techniques for playout delay adaptation in VoIP networks, *IEEE Transactions on Neural Networks*, vol. 16, no. 5, pp. 1174–1194, September 2005.
9. D. Sisalem and H. Schulzrinne, The loss-delay based adjustment algorithm: A TCP-friendly adaptation scheme, in *Proceedings of NOSSDAV*, Cambridge, UK. 1998.
10. P. DeLeon and C. J. Sreenan, An adaptive predictor for media playout buffering, in *International Conference on Acoustics, Speech, and Signal Processing*, vol. 6, pp. 3097–3100, March 1999.
11. C. Casetti, J. C. De Martin, and M. Meo, A Framework for the analysis of adaptive Voice over IP, in *International Conference on Communications*, New Orleans, USA, June 2000.
12. C. H. Wang, J. M. Ho, R. I. Chang, and S. C. Hsu, A control-theoretic mechanism for rate-based flow control of89z real-time multimedia communication, in *Multimedia, Internet, Video Technologies WSES/IEEE International Multi-conference*, September 2001.
13. F. Beritelli, G. Ruggeri, and G. Schembra, TCP-friendly transmission of voice over IP, in *ICC 2002: IEEE International Conference on Communications*, pp. 1204–1208, April–May 2008.
14. K. Homayounfar, Rate adaptive speech coding for universal multimedia access, *IEEE Signal Processing Magazine*, vol. 20, no. 2, pp. 30–39, March 2003.
15. V. Abreu-Sernandez and C. Garcia-Mateo, Adaptive multi-rate speech coder for VoIP transmission, *IEE Electronic Letters*, vol. 36, no. 23, pp. 1978–1980, November 2000.
16. Y. Bai and M. R. Ito,QoS control for video and audio communication in conventional and active networks: approaches and comparison, *IEEE Communication Surveys*, vol. 6, no. 1, pp. 42–49, 2004
17. L. Roychoudhury and E. S. Al-Shaer, Adaptive rate control for real-time packet audio based on loss prediction, in *GLOBECOM 2004: IEEE Global Telecommunications Conference*, vol. 2, pp. 634–638, November–December, 2004.
18. W. C. Chu, Speech Coding Algorithms: Foundation and evolution of standardized coders, Wiley Interscience, Hoboken, New Jersey, 2003.
19. ITU-T Rec. G.107, The E-model, a computational model for use in transmission planning, March 2003.
20. R. G. Cole and J. H. Rosenbluth, Voice over IP performance monitoring, Computer Communication Review, vol. 31, no. 2, pp. 9–24, April 2001.
21. L. Ding and R. Goubran, Speech quality prediction in VoIP using the extended E-model, in *IEEE GLOBECOM*, vol. 7, pp. 3974–3978, December 2003.
22. S. Tao, K. Xu, A. Estepa, T. Fei, L. Gao, R. Guerin, J. Kurose, D. Towsley, Z. Zhang, Improving VoIP quality through path switching, in *IEEE INFOCOM 2005: 24th Annual Joint Conference of the IEEE Computer and Communications Societies*, vol. 4, pp. 2268-2278, March 2005.
23. L. Sun and E. C. Ifeachor, Voice quality prediction models and their application in VoIP networks, IEEE Transactions on Multimedia, vol. 8, no. 4, pp. 809–820, August 2006.
24. D. Ye, Control of real-time multimedia applications in best-effort networks, Texas A&M University, College Station, TX, 2006.
25. J. M. Valin, The Speex codec manual, www.Xiph.org, ed. 1.2-beta1, August 2006.
26. ANSI T1.521-1999, Packet Loss Concealment for Use with ITU-T Recommendation G.711, December 1999.
27. PLANETLAB: An open platform for developing, deploying, and accessing planetary-scale services, www.planet-lab.org.
28. K. Park and W. Willinger, *Self-Similar Network Traffic and Performance Evaluation*, New York, USA: John Wiley and Sons, 2000.
29. H. Ohsaki, M. Morita, and M. Murata, On modeling Round-Trip Time dynamics of the Internet using system identification, in *ICOIN '02: Revised Papers from the International Conference on Information Networking, Wireless Communications Technologies and Network Applications-Part I*, pp. 359–371, London, UK: Springer-Verlag, 2002.

30. E. F. Camacho and C. Bordons, *Model Predictive Control*, Advanced Textbooks in Control and Signal Processing, Springer, 1998.

Chapter 14
Online Synchronous Policy Iteration Method for Optimal Control

Kyriakos G. Vamvoudakis and Frank L. Lewis

Abstract In this chapter, we discuss an online algorithm based on policy iteration (PI) for learning the continuous-time (CT) optimal control solution for nonlinear systems with infinite horizon costs. We present an online adaptive algorithm implemented as an actor/critic structure which involves simultaneous continuous-time adaptation of both actor and critic neural networks. We call this "synchronous" PI. A persistence of excitation condition is shown to guarantee convergence of the critic to the actual optimal value function. Novel tuning algorithms are given for both critic and actor networks, with extra terms in the actor tuning law being required to guarantee closed-loop dynamical stability. The convergence to the optimal controller is proven, and stability of the system is also guaranteed. Simulation examples show the effectiveness of the new algorithm.

14.1 Introduction

The allocation of human and physical resources over time is a fundamental problem that is central to engineering science. Cost-effective solutions are regularly required in the control loops which are placed at the high levels of the hierarchy of a complex integrated control applications; the so called money-making loops. Such a controller is given by the solution of an optimal control problem. Optimal control policies satisfy the specified system performances while minimizing a structured cost index which describes the balance between desired performances and available control resources.

Kyriakos G. Vamvoudakis
Automation and Robotics Research Institute, University of Texas at Arlington, Arlington, TX, USA, e-mail: kyriakos@arri.uta.edu

Frank L. Lewis
Automation and Robotics Research Institute, University of Texas at Arlington, Arlington, TX, USA, e-mail: lewis@uta.edu

From a mathematical point of view the solution of the optimal control problem is based on the solution of the underlying Hamilton-Jacobi-Bellman (HJB) equation. Until recently, due to the intractability of this nonlinear differential equation for CT systems, which form the object of interest in this chapter, only particular solutions were available (e.g. for the linear time-invariant case). For this reason considerable effort has been devoted to developing algorithms which approximately solve this equation (e.g. [1, 3, 13]). Far more results are available for the solution of the discrete-time HJB equation. Good overviews are given in [4, 15].

Some of the methods involve a computational intelligence technique known as PI [7, 9, 10, 16]. PI refers to a class of algorithms built as a two-step iteration: policy evaluation and policy improvement. Instead of trying a direct approach to solving the HJB equation, the PI algorithm starts by evaluating the cost of a given initial admissible (in the sense defined herein) control policy and then uses this information to obtain a new improved control policy (i.e., which will have a lower associated cost). These two steps of policy evaluation and policy improvement are repeated until the policy improvement step no longer changes the actual policy, thus convergence to the optimal controller is achieved. One must note that the cost can be evaluated only in the case of admissible control policies, admissibility being a condition for the control policy which is used to initialize the algorithm.

Actor/critic structures based on value iteration have been introduced and further developed by Werbos [19, 20, 21] with the purpose of solving the optimal control problem online in real-time. Werbos defined four types of actor-critic algorithms based on value iteration, subsumed under the concept of approximate or adaptive dynamic programming (ADP) algorithms. Adaptive critics have been described in [14] for discrete-time systems and [2, 8, 17, 18] for continuous-time systems.

In the linear CT system case, when quadratic indices are considered for the optimal stabilization problem, the HJB equation becomes the well known Riccati equation and the PI method is in fact Newton's method [9] which requires iterative solutions of Lyapunov equations. In the nonlinear systems case, successful application of the PI method was limited until [3], where Galerkin spectral approximation methods were used to solve the nonlinear Lyapunov equations describing the policy evaluation step in the PI algorithm. Such methods are known to be computationally intensive and cannot handle control constraints.

The key to solving practically the CT nonlinear Lyapunov equations was in the use of neural networks [1] which can be trained to become approximate solutions of these equations. In fact the PI algorithm for CT systems can be built on an actor/critic structure which involves two neural networks: one, the critic neural networks, is trained to become an approximation of the Lyapunov equation solution at the policy evaluation step, while the second one is trained to approximate an improving policy at the policy improving step.

In [17, 18] an online PI algorithm was developed for CT systems which converges to the optimal control solution without making explicit use of any knowledge on the internal dynamics of the system. The algorithm was based on sequential updates of the critic (policy evaluation) and actor (policy improvement) neural networks (i.e., while one is tuned the other one remains constant).

This chapter is concerned with developing approximate online solutions, based on PI, for the infinite horizon optimal control problem for CT nonlinear systems. We present an online adaptive algorithm which involves simultaneous tuning of both actor and critic neural networks (i.e. both neural networks are tuned at the same time). We term this "synchronous" PI. Knowledge of the system dynamics is required for this algorithm. This approach is a version of the generalized policy iteration (GPI), as introduced in [16]. An almost synchronous version of PI has been described in [8], without providing explicit training laws for neither actor or critic networks, nor convergence analysis.

There are two main contributions in this chapter. The first involves introduction of a nonstandard "normalized" critic neural network tuning algorithm, along with guarantees for its convergence based on a certain persistence of excitation condition. The second involves adding nonstandard extra terms to the actor neural network tuning algorithm that are required to guarantee close loop stability, along with stability and convergence proofs.

14.2 The Optimal Control Problem and the Policy Iteration Problem

14.2.1 Optimal Control and the HJB Equation

Consider the nonlinear time-invariant affine in the input dynamical system given by

$$\begin{aligned} \dot{x}(t) &= f(x(t)) + g(x(t))u(x(t)) \\ x(0) &= x_0 \end{aligned} \tag{14.1}$$

with $x(t) \in \Re^n$, $f(x(t)) \in \Re^n$, $g(x(t)) \in \Re^{n\times m}$ and control input $u(t) \in \bigcup \subset R^m$.

We assume that $f(0) = 0$, $g(0) = 0$, $f(x) + g(x)u$ is Lipschitz continuous on a set $\Omega \subseteq \Re^n$ that contains the origin, and that the system is stabilizable on Ω i.e. there exists a continuous control function $u(t) \in \bigcup$ such that the system is asymptotically stable on Ω.

Define the infinite horizon integral cost

$$V(x_0) = \int_0^\infty r(x(\tau), u(\tau))d\tau \tag{14.2}$$

where $r(x,u) = Q(x) + u^T Ru$ with Q(x) positive definite, i.e. $\forall x \neq 0, Q(x) > 0$ and $x = 0 \Rightarrow Q(x) = 0$, and $R \in \Re$.

Definition 14.1. [1] (Admissible Policy) A control policy $\mu(x)$ is defined as admissible with respect to (14.2) on Ω, denoted by $\mu \in \Psi(\Omega)$ if $\mu(x)$ is continuous on $\Omega, \mu(0) = 0, \mu(x)$ stabilizes (14.1) on Ω and $V(x_0)$ is finite $\forall x_0 \in \Omega$.

For any admissible control policy $\mu \in \Psi(\Omega)$, if the associated cost function

$$V(x_0) = \int_0^\infty r(x(\tau), \mu(x(\tau)))d\tau \tag{14.3}$$

is C^1, then an infinitesimal version of (14.3) is

$$0 = r(x, \mu(x)) + V_x^{\mu T}(f(x) + g(x)\mu(x)), V^\mu(0) = 0 \tag{14.4}$$

where V_x^μ denotes the partial derivative of the value function V^μ with respect to x. (Note that the value function does not depend explicitly on time.) (14.4) is a Lyapunov equation for nonlinear systems which, given a controller $\mu \in \Psi(\Omega)$, can be solved for the value function $V^\mu(x)$ associated with it. Given that $\mu(x)$ is an admissible control policy, if $V^\mu(x)$ satisfies (14.4) with $r(x, \mu(x)) \geq 0$ then $V^\mu(x)$ is a Lyapunov function for the system (14.1) with control policy $\mu(x)$.

The optimal control problem can now be formulated: Given the continuous-time system (1), the set $\mu \in \Psi(\Omega)$ of admissible control policies and the infinite horizon cost functional (14.2), find an admissible control policy such that the cost index (14.2) associated with the system (14.1) is minimized.

Defining the Hamiltonian of the problem

$$H(x, u, V_x) = r(x(t), u(t)) + V_x^T(f(x(t)) + g(x(t))u(t)), \tag{14.5}$$

the optimal cost function $V^*(x)$ satisfies the HJB equation

$$0 = min_{u \in \Psi(\Omega)}[H(x, u, V_x^*)]. \tag{14.6}$$

Assuming that the minimum on the right hand side of (14.6) exists and is unique then the optimal control function for the given problem is

$$u^*(x) = -R^{-1}g^T(x)V_x^*(x). \tag{14.7}$$

Inserting this optimal control policy in the Hamiltonian we obtain the formulation of the HJB equation in terms of the optimal cost

$$\begin{gathered} 0 = Q(x) + V_x^{*T}(x)f(x) - \tfrac{1}{4}V_x^{*T}(x)g(x)R^{-1}g^T(x)V_x^*(x) \\ V^*(0) = 0. \end{gathered} \tag{14.8}$$

This is a necessary and sufficient condition for the optimal value function [12]. For the linear system case, considering a quadratic cost functional, the equivalent of this HJB equation is the well known Riccati equation.

In order to find the optimal control solution for the problem one only needs to solve the HJB equation (14.8) for the value function and then substitute the solution in (14.7) to obtain the optimal control. However, due to the nonlinear nature of the HJB equation finding its solution is generally difficult or impossible. In subsequent sections, we present an online method for approximately solving the HJB equation online to obtain the optimal cost (value) and control.

14.2.2 Neural Network Approximation of the Value Function

PI is an iterative method of reinforcement learning [16] for solving (14.8), and consists of policy improvement based on (14.7) and policy evaluation based on (14.4). See [1, 2, 3, 8, 9, 13, 17].

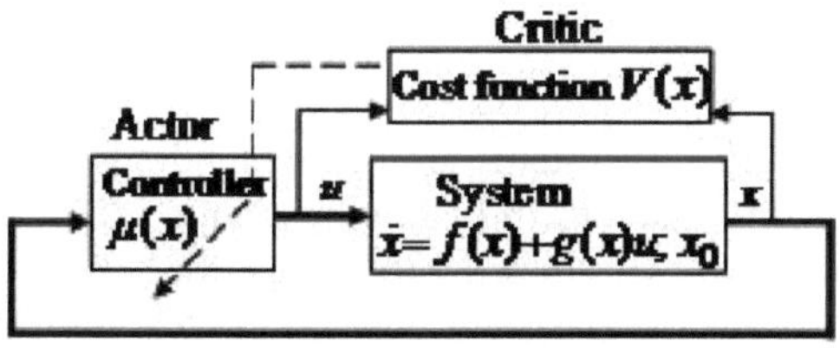

Fig. 14.1 Actor/critic structure

In the actor/critic structure the critic and the actor functions are approximated by neural networks, and the PI algorithm consists in tuning alternatively each of the two neural networks. The critic neural network is tuned to evaluate the performance of the current control policy. The critic neural networks is based on value function approximation (VFA). Thus, assume there exist weights W_1 such that the value $V(x)$ is approximated by a neural network as

$$V(x) = {W_1}^T \phi_1(x) + \varepsilon(x) \tag{14.9}$$

where $\phi_1(x) : \Re^n \to \Re^N$ is the activation functions vector, N the number of neurons in the hidden layer, and $\varepsilon(x)$ the neural networks approximation error. It is well known that $\varepsilon(x)$ is bounded by a constant on a compact set. Select the activation functions to provide a complete basis set such that $V(x)$ and its derivative

$$\frac{\partial V}{\partial x} = \nabla \phi_1^T W_1 + \frac{\partial \varepsilon}{\partial x}$$

are uniformly approximated. According to the Weierstrass higher-order approximation theorem [1], such a basis exists if $V(x)$ is sufficiently smooth. This means that, as the number of hidden-layer neurons $N \to \infty$, the approximation error $\varepsilon \to 0$ uniformly.

Using the neural networks value function approximation, considering a fixed control policy, the Hamiltonian (14.5) becomes

$$H(x,u,W_1) = W_1^T \nabla \phi_1 (f + gu) + Q(x) + u^T Ru = \varepsilon_H \tag{14.10}$$

where the residual error due to the function approximation error is

$$\varepsilon_H = (\nabla \varepsilon)^T (f + gu)$$

Under the Lipschitz assumption on the dynamics, this residual error is bounded on a compact set. Moreover, [1] has shown that, under certain reasonable assumptions, as the number of hidden layer neurons $N \to \infty$ one has $\varepsilon_H \to 0$.

14.3 Online Generalized PI Algorithm with Synchronous Tuning of Actor and Critic Neural Networks

Standard PI algorithms for CT systems are offline methods that require complete knowledge on the system dynamics to obtain the solution (i.e., the functions $f(x)$, $g(x)$ in (14.1) need to be known). In order to change the offline character of PI for CT systems, and thus make it consistent with online learning mechanisms in the mammal brain, we present an adaptive learning algorithm that uses simultaneous continuous-time tuning for the actor and critic neural networks. We term this "synchronous" online PI for CT systems. Our algorithm also needs knowledge of the system dynamics, yet can approximately solve the optimal control problem online using real-time measurements of closed-loop signals.

14.3.1 Critic Neural Networks

The weights of the critic neural networks, W_1, which solve (14.10) are unknown. Then the output of the critic neural network is

$$\hat{V}(x) = \hat{W}_1{}^T \phi_1(x) \tag{14.11}$$

where $\hat{W}_1$ are the current known values of the critic neural networks weights. Recall that $\phi_1(x) : \Re^n \to \Re^N$ is the activation functions vector, with N the number of neurons in the hidden layer. The approximate Hamiltonian is then

$$H(x, \hat{W}_1, u) = \hat{W}_1{}^T \nabla\phi_1(f + gu) + Q(x) + u^T Ru = e_1.$$

It is desired to select $\hat{W}_1$ to minimize the squared residual error $E_1 = \frac{1}{2} e_1^T e_1$. Then $\hat{W}_1(t) \to W_1$. We select the tuning law for the critic weights as the normalized gradient descent algorithm

$$\dot{\hat{W}}_1 = -a_1 \frac{\partial E_1}{\partial \hat{W}_1} = -a_1 \frac{\sigma_1}{(\sigma_1{}^T \sigma_1 + 1)^2} [\sigma_1{}^T \hat{W}_1 + Q(x) + u^T Ru] \tag{14.12}$$

where $\sigma_1 = \nabla\phi_1(f + gu)$. This is a modified Levenberg-Marquardt algorithm where $(\sigma_1{}^T \sigma_1 + 1)^2$ is used for normalization instead of $(\sigma_1{}^T \sigma_1 + 1)$. This is required in the proofs, where one needs both appearances of $\frac{\sigma_1}{(\sigma_1{}^T \sigma_1 + 1)}$ in (14.12) to be bounded. Define the critic weight estimation error $\tilde{W}_1 = W_1 - \hat{W}_1$ and note that, from (14.10),

$$Q(x) + u^T R u = -W_1{}^T \nabla\phi_1(f + gu) + \varepsilon_H. \tag{14.13}$$

Substitute (14.13) in (14.12) and, with the notation $\bar{\sigma}_1 = \frac{\sigma_1}{(\sigma_1{}^T \sigma_1 + 1)}$ and $m_s = \sigma_1{}^T \sigma_1 + 1$ we obtain the dynamics of the critic weight estimation error as

$$\dot{\tilde{W}}_1 = -a_1 \bar{\sigma}_1 \bar{\sigma}_1^T \tilde{W}_1 + a_1 \bar{\sigma}_1 \frac{\varepsilon_H}{m_s}.$$

Though it is traditional to use critic tuning algorithms of the form (14.12), it is not generally understood when convergence of the critic weights can be guaranteed. In this chapter, we address this issue in a formal manner. To guarantee convergence of $\hat{W}_1$ to W_1, the next persistence of excitation (PE) assumption and associated technical lemmas are required.

Let the signal $\bar{\sigma}_1$ be persistently exciting over the interval $[t, t+T]$, i.e., there exist constants $\beta_1 > 0, \beta_2 > 0, T > 0$ such that, for all t,

$$\beta_1 I \le S_0 \equiv \int_t^{t+T} \bar{\sigma}_1(\tau) \bar{\sigma}_1{}^T(\tau) d\tau \le \beta_2 I. \tag{14.14}$$

It is PE assumption.

Lemma 14.1. *Consider the error dynamics system with output defined as*

$$\begin{aligned} \dot{\tilde{W}}_1 &= -a_1 \bar{\sigma}_1 \bar{\sigma}_1^T \tilde{W}_1 + a_1 \bar{\sigma}_1 \tfrac{\varepsilon_H}{m_s} \\ y &= \bar{\sigma}_1^T \tilde{W}_1. \end{aligned} \tag{14.15}$$

The PE condition (14.14) is equivalent to the uniform complete observability (UCO) of this system, that is there exist constants $\beta_3 > 0$, $\beta_4 > 0$, $T > 0$ such that, for all t,

$$\beta_3 I \le S_1 \equiv \int_t^{t+T} \Phi^T(\tau, t) \bar{\sigma}_1(\tau) \bar{\sigma}_1^T(\tau) \Phi(\tau, t) d\tau \le \beta_4 I$$

with $\Phi(t_1, t_0)$, $t_0 \le t_1$ the state transition matrix of (14.15).

Proof. System (14.15) and the system defined by $\dot{\tilde{W}}_1 = a_1 \bar{\sigma}_1 u$, $y = \bar{\sigma}_1^T \tilde{W}_1$ are equivalent under the output feedback $u = -y + \varepsilon_H / m_s$. Note that (14.14) is the observability gramian of this last system. □

The importance of UCO is that bounded input and bounded output implies that the state $\tilde{W}_1(t)$ is bounded. In Theorem 14.1 we shall see that the critic tuning law (14.12) indeed guarantees boundedness of the output in (14.15).

Lemma 14.2. *Consider the error dynamics system (14.15). Let the signal $\bar{\sigma}_1$ be persistently exciting. Then:*

1. *The system (14.15) is exponentially stable. In fact if $\varepsilon_H = 0$ then $\|\tilde{W}(kT)\| \le e^{-\alpha kT} \|\tilde{W}(0)\|$ with*

$$\alpha = -\frac{1}{T} ln(\sqrt{1 - 2a_1\beta_3}).$$

2. *Let* $\|\varepsilon_H\| \leq \varepsilon_{max}$ *and* $\|y\| \leq y_{max}$ *then* $\|\tilde{W}_1\|$ *converges exponentially to the residual set*

$$\tilde{W}_1(t) \leq \frac{\sqrt{\beta_2 T}}{\beta_1}\left\{[y_{max} + \delta\beta_2 a_1(\varepsilon_{max} + y_{max})]\right\}.$$

where δ *is a positive constant of the order of* 1.

The next result shows that the tuning algorithm (14.12) is effective, in that the weights $\hat{W}_1$ converge to the actual unknown weights W_1 which solve the Hamiltonian equation (14.10) for the given control policy $u(t)$. That is, (14.11) converges close to the actual value function of the current control policy.

Theorem 14.1. *Let* $u(t)$ *be any admissible bounded control input. Let tuning for the critic neural networks be provided by (14.12) and assume that* $\bar{\sigma}_1$ *is persistently exciting. Let the residual error in (14.10) be bounded* $\|\varepsilon_H\| \leq \varepsilon_{max}$. *Then the critic parameter error is practically bounded by*

$$\tilde{W}_1(t) \leq \frac{\sqrt{\beta_2 T}}{\beta_1}\left\{[1 + 2\delta\beta_2 a_1]\,\varepsilon_{max}\right\}. \tag{14.16}$$

Proof. Consider the following Lyapunov function candidate

$$L(t) = \frac{1}{2}tr\left\{\tilde{W}_1^T a_1{}^{-1}\tilde{W}_1\right\}.$$

The derivative of is given by

$$\begin{aligned}
\dot{L} &= -tr\left\{\tilde{W}_1^T \tfrac{\sigma_1}{m_s{}^2}[\sigma_1{}^T\tilde{W}_1 - \varepsilon_H]\right\}\\
\dot{L} &= -tr\left\{\tilde{W}_1^T \tfrac{\sigma_1\sigma_1{}^T}{m_s{}^2}\tilde{W}_1\right\} + tr\left\{\tilde{W}_1^T \tfrac{\sigma_1}{m_s}\tfrac{\varepsilon_H}{m_s}\right\}\\
\dot{L} &\leq -\left\|\tfrac{\sigma_1{}^T}{m_s}\tilde{W}_1\right\|^2 + \left\|\tfrac{\sigma_1{}^T}{m_s}\tilde{W}_1\right\|\left\|\tfrac{\varepsilon_H}{m_s}\right\|\\
\dot{L} &= -\left\|\tfrac{\sigma_1{}^T}{m_s}\tilde{W}_1\right\|\left[\left\|\tfrac{\sigma_1{}^T}{m_s}\tilde{W}_1\right\| - \left\|\tfrac{\varepsilon_H}{m_s}\right\|\right].
\end{aligned}$$

Therefore $\dot{L} \leq 0$ if

$$\left\|\frac{\sigma_1{}^T}{m_s}\tilde{W}_1\right\| > \varepsilon_{max} > \left\|\frac{\varepsilon_H}{m_s}\right\|. \tag{14.17}$$

Note that $\|m_s\| \geq 1$. This provides an effective practical bound for $\|\bar{\sigma}_1{}^T\tilde{W}_1\|$ since $L(t)$ decreases if (14.17) holds.

Consider the estimation error dynamics (14.15) with the output bounded effectively by $\|y\| < \varepsilon_{max}$, as just shown. Now Lemma 14.2 shows that

$$\tilde{W}_1(t) \leq \frac{\sqrt{\beta_2 T}}{\beta_1}\left\{[1 + 2\delta\beta_2 a_1]\,\varepsilon_{max}\right\}. \tag{14.18}$$

This completes the proof. □

Remark 14.1. Note that, as $N \to \infty$, $\varepsilon_H \to 0$ uniformly [1]. This means that ε_{max} decreases as the number of hidden layer neurons in (14.11) increases.

Remark 14.2. This theorem requires the assumption that the control input $u(t)$ is bounded, since $u(t)$ appears in ε_H. In the upcoming Theorem 14.2 this restriction is removed.

14.3.2 Action Neural Networks

The policy improvement step in PI is given by substituting (14.9) into (14.7) as

$$u(x) = -\frac{1}{2}R^{-1}g^T(x)\nabla\phi_1{}^T W_1$$

with critic weights W_1 unknown. Therefore, define the control policy in the form of an action neural network which computes the control input in the structured form

$$u_2(x) = -\frac{1}{2}R^{-1}g^T(x)\nabla\phi_1{}^T \hat{W}_2$$

where $\hat{W}_2$ denotes the current known values of the actor neural networks weights.

Based on (14.8), define the approximate HJB equation

$$\begin{aligned} Q(x) + W_1{}^T\nabla\phi_1(x)f(x) - \tfrac{1}{4}W_1{}^T\bar{D}_1(x)W_1 &= \varepsilon_{HJB}(x) \\ W_1{}^T\phi_1(0) + \varepsilon(0) &= 0 \end{aligned}$$

with the notation $\bar{D}_1(x) = \nabla\phi_1(x)g(x)R^{-1}g^T(x)\nabla\phi_1{}^T(x)$, where W_1 denotes the ideal unknown weights of the critic and actor neural networks which solve the HJB.

We now present the main theorem, which provides the tuning laws for the actor and critic neural networks that guarantee convergence to the optimal controller along with closed-loop stability. The next notion of practical stability is needed.

Definition 14.2. The equilibrium point $x_e = 0$ is said to be uniformly ultimately bounded (UUB) if there exists a compact set $S \subset \Re^n$ so that for all $x_0 \in S$ there exists a bound B and a time $T(B, x_0)$ such that $\|x(t) - x_e\| \leq B$ for all $t \geq t_0 + T$.

Theorem 14.2. *Let tuning for the critic neural networks be provided by*

$$\dot{\hat{W}}_1 = -a_1 \frac{\sigma_2}{(\sigma_2{}^T\sigma_2 + 1)^2}\left[\sigma_2{}^T\hat{W}_1 + Q(x) + u_2{}^T R u_2\right]$$

where $\sigma_2 = \nabla\phi_1(f + gu_2)$, *and assume that* $\bar{\sigma}_2 = \sigma_2/(\sigma_2{}^T\sigma_2 + 1)$ *is persistently exciting. Let the actor neural networks be tuned as*

$$\dot{\hat{W}}_2 = -a_2\left[\frac{1}{2}\bar{D}_1(x)\left(\hat{W}_2 - \hat{W}_1\right) - \frac{1}{4}\bar{D}_1(x)\hat{W}_2\frac{\bar{\sigma}_2^T}{m_{s2}}\hat{W}_1\right]$$

where $m_{s2} = 1 + \sigma_2{}^T \sigma_2$. Then the closed-loop system state is UUB, the critic parameter error $\tilde{W}_1 = W_1 - \hat{W}_1$ and the actor parameter error $\tilde{W}_2 = W_1 - \hat{W}_2$ are UUB, and (14.16) holds asymptotically so that convergence of $\hat{W}_1$ to the approximate optimal critic value W_1 is obtained.

Proof. The convergence proof is based on Lyapunov analysis. We consider the Lyapunov function

$$L(t) = V(x) + \frac{1}{2} tr\left(\tilde{W}_1^T a_1{}^{-1} \tilde{W}_1\right) + \frac{1}{2} tr\left(\tilde{W}_2^T a_2{}^{-1} \tilde{W}_2\right).$$

With the chosen tuning laws one can then show that the errors $\tilde{W}_1$ and $\tilde{W}_2$ are UUB and convergence is obtained. □

14.4 Simulation results

To support the new synchronous online PI algorithm for CT systems, we offer two simulation examples, one linear and one nonlinear. In both cases we observe convergence to the actual optimal value function and control.

14.4.1 Linear System Example

Consider the linear system with quadratic cost function used in [8]

$$\dot{x} = \begin{bmatrix} -1 & -2 \\ 1 & -4 \end{bmatrix} x + \begin{bmatrix} 1 \\ -3 \end{bmatrix} u$$

where Q and R in the cost function are identity matrices of appropriate dimensions. In this linear case the solution of the HJB equation is given by the solution of the algebraic Riccati equation. Since the value is quadratic in the LQR case, the critic neural networks basis set $\phi_1(x)$ was selected as the quadratic vector in the state components. That is $\phi_1(x) = [x_1^2 \;\; x_1x_2 \;\; x_2^2]^T$. The parameters of the optimal critic are then $W_1{}^* = [0.3199 \;\; -0.1162 \;\; 0.1292]^T = [p_{11} \;\; 2p_{12} \;\; p_{22}]^T$, where the Riccati equation solution is $P = \begin{bmatrix} p_{11} & p_{12} \\ p_{21} & p_{22} \end{bmatrix}$.

The synchronous PI algorithm is implemented as in Theorem 14.2. PE was ensured by adding a small probing noise to the control input. Figure 14.2 shows the critic parameters, denoted by $\hat{W}_1 = [W_{11} \;\; W_{12} \;\; W_{13}]^T$, converging to the optimal values. In fact after 200s the critic parameters converged to

$$\hat{W}_1(t_f) = [0.3192 \;\; -0.1174 \;\; 0.128]^T.$$

The evolution of the system states is presented in Figure 14.3. One can see that after 200s convergence of the neural networks weights in both critic and actor has occurred. Then, the PE condition of the control signal is no longer needed, and the probing signal was turned off. After that, the states remain very close to zero, as required.

14.4.2 Nonlinear System Example

Consider the following affine in control input nonlinear system, with a quadratic cost used in [5]

$$\dot{x} = f(x) + g(x)u, x \in \Re^2$$

where

$$f(x) = \begin{bmatrix} x_2 \\ -x_1(\frac{\pi}{2} + arctan(5x_1)) - \frac{5x_1{}^2}{2(1+25x_1{}^2)} + 4x_2 \end{bmatrix}, g(x) = \begin{bmatrix} 0 \\ 3 \end{bmatrix}$$

One selects

$$Q = \begin{bmatrix} 0 & 0 \\ 0 & 1 \end{bmatrix}, R = 1.$$

As detailed in [5] the optimal value function is

$$V^*(x) = x_1{}^2(\frac{\pi}{2} + arctan(5x_1)) + x_2{}^2$$

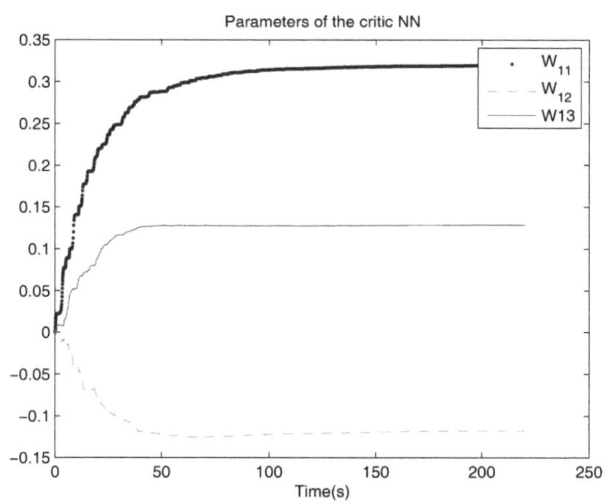

Fig. 14.2 Convergence of the critic parameters to the parameters of the optimal critic

and the optimal control signal is

$$u^*(x) = -3x_2.$$

One selects the critic neural networks vector activation function as

$$\phi(x) = \begin{bmatrix} x_1 & x_2 & x_1x_2 & {x_1}^2 & {x_2}^2 & {x_1}^2 arctan(5x_1) & {x_1}^3 \end{bmatrix}.$$

In order to have the desired signals to approximate the optimal value function and the optimal control signal. The rationale for selecting this basis set follows from inverse optimality theory [7]. There, it is known that the optimal value generally contains terms like the nonlinearities in $f(x)$.

Figure 14.4 shows the critic parameters, denoted by

$$\hat{W}_1 = \begin{bmatrix} W_{c1} & W_{c2} & W_{c3} & W_{c4} & W_{c5} & W_{c6} & W_{c7} \end{bmatrix}^T.$$

These converge to the values of

$$\hat{W}_1(t_f) = [0.5674 \quad -0.00239 \; 0.0371 \; 1.7329 \; 0.9741 \; 0.3106 \; 0.4189]^T.$$

The evolution of the system states is presented in Figure 14.5. One can see that after 3300s convergence of the neural networks weights in both critic and actor has occurred. Then, the PE condition of the control signal is no longer needed, and the probing signal was turned off. After that, the states remain very close to zero, as required.

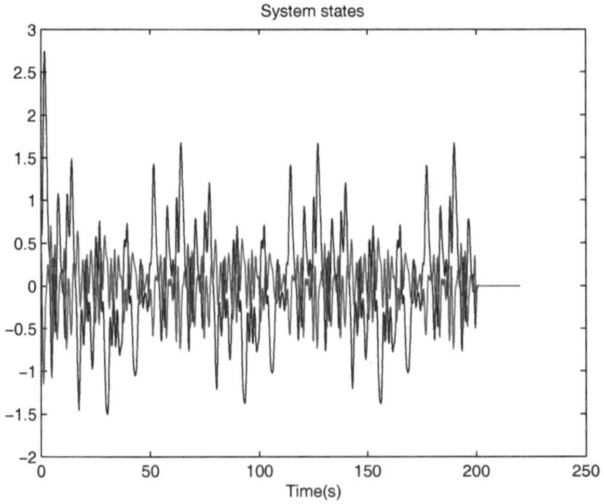

Fig. 14.3 Evolution of the system states for the duration of the experiment

Figure 14.6 show the optimal value function. The identified value function given by $\hat{V}_1(x) = \hat{W}_1(x)^T \phi_1(x)$ is virtually indistinguishable. In fact, Figure 14.7 shows the 3D plot of the difference between the approximated value function, by using the online algorithm, and the optimal one. This error is close to zero. Good approximation of the actual value function is being evolved. The actor neural networks also converged to the optimal control $u^*(x) = -3x_2$.

14.5 Conclusions

In this chapter we have proposed a new adaptive algorithm which solves the continuous-time optimal control problem for affine in the inputs nonlinear systems. We call this algorithm synchronous online PI for CT systems. The algorithm is built on an actor/critic structure and involves simultaneous online training of both actor and critic networks. This means that initiation of the tuning process for the parameters of either actor or critic does not require complete convergence of the adaptation process of critic or respectively actor, which is regularly required in PI algorithms where only one neural network, critic or actor, is tuned until convergence, while the other neural network is held constant. This is a form of generalized CT policy iteration as defined in [16]. The algorithm requires complete knowledge of the system model. For this reason our research efforts will now be directed towards integrating a third neural network with the actor/critic structure with the purpose of approximating in an online fashion the system dynamics, as suggested by [19, 20, 21].

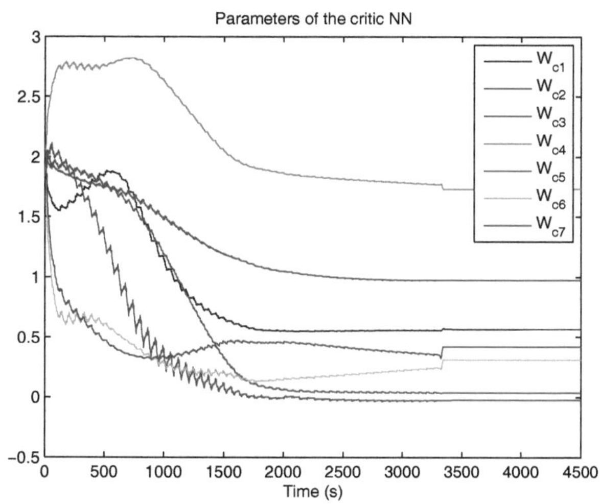

Fig. 14.4 Convergence of the critic parameters

Acknowledgements This work was supported by the National Science Foundation ECS-0501451 and the Army Research Office W91NF-05-1-0314. The authors would like to acknowledge the insight and guidance received from Draguna Vrabie, currently a PhD student at UTA's Automation and Robotics Research Institute.

Appendix 1

Proof for Lemma 14.2 part 1.

Set $\varepsilon_H = 0$ in (19). Take the Lyapunov function

$$L = \frac{1}{2}\tilde{W}_1^T {a_1}^{-1} \tilde{W}_1. \tag{14.19}$$

The derivative is

$$\dot{L} = -\tilde{W}_1^T \bar{\sigma}_1 \bar{\sigma}_1^T \tilde{W}_1.$$

Integrating both sides

$$\begin{aligned} L(t+T) - L(t) &= -\int_t^{t+T} \tilde{W}_1^T \sigma_1(\tau) {\sigma_1}^T(\tau) \tilde{W}_1 d\tau \\ L(t+T) = L(t) - \tilde{W}_1(t)^T &\int_t^{t+T} \Phi^T(\tau,t) \sigma_1(\tau) {\sigma_1}^T(\tau) \Phi(\tau,t) d\tau \tilde{W}_1(t) \\ &= L(t) - \tilde{W}_1^T(t) S_1 \tilde{W}_1 \le (1 - 2a_1\beta_3) L(t). \end{aligned}$$

So

$$L(t+T) \le (1 - 2a_1\beta_3) L(t). \tag{14.20}$$

Define $\gamma = (1 - 2a_1\beta_3)$. By using norms we write (14.20) in terms of $\tilde{W}_1$ as

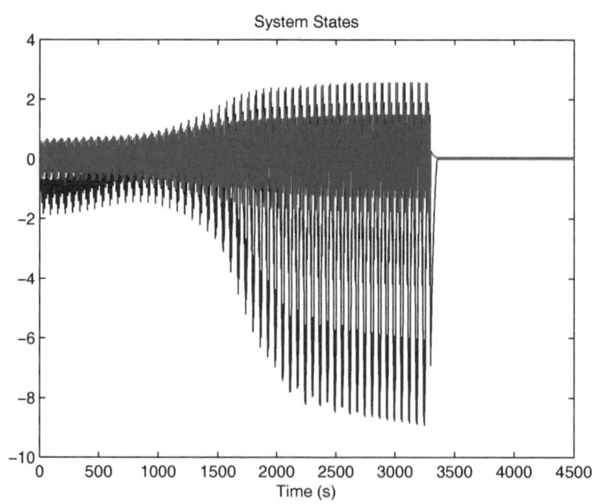

Fig. 14.5 Evolution of the system states for the duration of the experiment

$$\begin{aligned}\frac{1}{2a_1}\left\|\tilde{W}(t+T)\right\|^2 &\leq \sqrt{(1-2a_1\beta_3)}\frac{1}{2a_1}\left\|\tilde{W}(t)\right\|^2\\ \left\|\tilde{W}(t+T)\right\| &\leq \sqrt{(1-2a_1\beta_3)}\left\|\tilde{W}(t)\right\|\\ \left\|\tilde{W}(t+T)\right\| &\leq \gamma\left\|\tilde{W}.(t)\right\|.\end{aligned}$$

Therefore

$$\left\|\tilde{W}(kT)\right\| \leq \gamma^k\left\|\tilde{W}(0)\right\|$$

i.e., $\tilde{W}(t)$ decays exponentially. To determine the decay time constant in continuous-time, note that

$$\left\|\tilde{W}(kT)\right\| \leq e^{-\alpha kT}\left\|\tilde{W}(0)\right\|$$

where $e^{-\alpha kT} = \gamma^k$. Therefore the decay constant is

$$\alpha = -\frac{1}{T}ln(\gamma) \Leftrightarrow \alpha = -\frac{1}{T}ln(\sqrt{(1-2a_1\beta_3)}).$$

This completes the proof. □

Appendix 2

Proof of Lemma 14.2 part 2.

Consider the system

$$\begin{cases}\dot{x} = B(t)u(t)\\ y = C^T(t)x(t).\end{cases} \tag{14.21}$$

The state and the output are

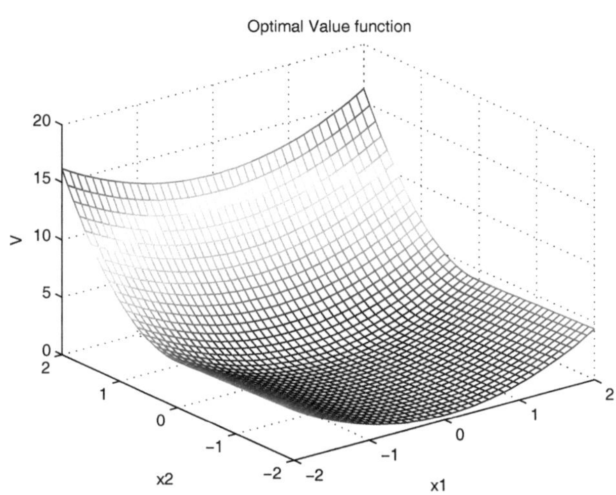

Fig. 14.6 Optimal value function

$$\begin{cases} x(t+T) = x(t) + \int_t^{t+T} B(\tau)u(\tau)d\tau \\ y(t+T) = C^T(t+T)x(t+T). \end{cases}$$

Let $C(t)$ be PE, so that

$$\beta_1 I \le S_C \equiv \int_t^{t+T} C_1(\lambda)C^T(\lambda)d\lambda \le \beta_2 I. \tag{14.22}$$

Then,

$$\begin{aligned} y(t+T) &= C^T(t+T)x(t) + \textstyle\int_t^{t+T} C^T(t+T)B(\tau)u(\tau)d\tau \\ \textstyle\int_t^{t+T} C(\lambda)\left(y(\lambda) - \int_t^{\lambda} C^T(\lambda)B(\tau)u(\tau)d\tau\right)d\lambda &= \textstyle\int_t^{t+T} C(\lambda)C^T(\lambda)x(t)d\lambda \\ \textstyle\int_t^{t+T} C(\lambda)\left(y(\lambda) - \int_t^{\lambda} C^T(\lambda)B(\tau)u(\tau)d\tau\right)d\lambda &= S_C x(t) \\ x(t) &= S_C{}^{-1}\left\{\textstyle\int_t^{t+T} C(\lambda)\left(y(\lambda) - \int_t^{\lambda} C^T(\lambda)B(\tau)u(\tau)d\tau\right)d\lambda\right\}. \end{aligned} \tag{14.23}$$

Now consider

$$\dot{\tilde{W}}_1(t) = a_1\bar{\sigma}_1 u. \tag{14.24}$$

Note that setting $u = -y + \frac{\varepsilon_H}{m_s}$ with output given $y = \bar{\sigma}_1^T\tilde{W}_1$ turns (14.24) into (14.15). Set $B = a_1\bar{\sigma}_1$, $C = \bar{\sigma}_1$, $x(t) = \tilde{W}_1$ so that (14.21) yields (14.24). Then,

$$\|u\| \le \|y\| + \left\|\frac{\varepsilon_H}{m_s}\right\| \le y_{max} + \varepsilon_{max}.$$

Since $\|m_s\| \ge 1$,

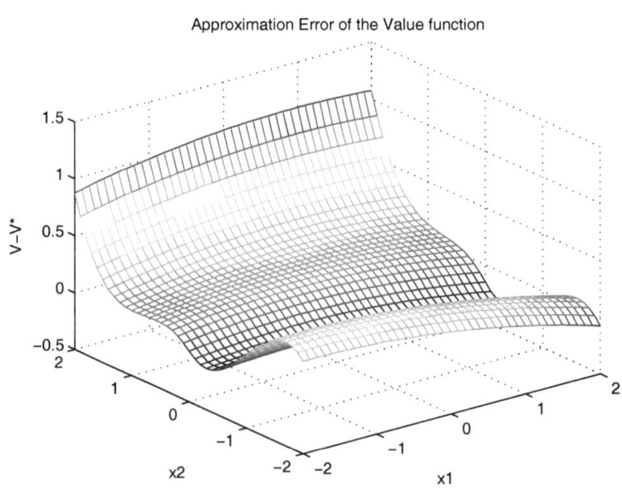

Fig. 14.7 Approximation error of the value function

$$\begin{aligned} N \equiv \int_t^{t+T} \|B(\tau)\| \, \|u(\tau)\| \, d\tau &= \int_t^{t+T} \|a_1 \bar{\sigma}_1(\tau)\| \, \|u(\tau)\| \, d\tau \\ &\leq a_1 (y_{max} + \varepsilon_{max}) \int_t^{t+T} \|\bar{\sigma}_1(\tau)\| \, d\tau \\ &\leq a_1 (y_{max} + \varepsilon_{max}) \left[\int_t^{t+T} \|\bar{\sigma}_1(\tau)\|^2 d\tau \right]^{\frac{1}{2}} \left[\int_t^{t+T} 1 d\tau \right]^{\frac{1}{2}}. \end{aligned}$$

By using (14.22),

$$N \leq a_1 (y_{max} + \varepsilon_{max}) \sqrt{\beta_2 T}. \tag{14.25}$$

Finally (14.23) and (14.25) yield,

$$\tilde{W}_1(t) \leq \frac{\sqrt{\beta_2 T}}{\beta_1} \{ [y_{max} + \delta \beta_2 a_1 (\varepsilon_{max} + y_{max})] \}.$$

This completes the proof. □

References

1. M. Abu-Khalaf, F. L. Lewis, Nearly Optimal Control Laws for Nonlinear Systems with Saturating Actuators Using a Neural Network HJB Approach, *Automatica*, vol. 41, no. 5, pp. 779-791, 2005.
2. L. C. Baird III, Reinforcement Learning in Continuous Time: Advantage Updating, *Proc. Of ICNN*, Orlando FL, June 1994.
3. R. Beard, G. Saridis, J. Wen,, Galerkin approximations of the generalized Hamilton-Jacobi-Bellman equation, *Automatica*, vol. 33, no. 12, pp. 2159-2177, 1997.
4. D. P. Bertsekas and J. N. Tsitsiklis, *Neuro-Dynamic Programming*, Athena Scientific, MA, 1996.
5. J. W. Curtis, R. W. Beard, Successive Collocation: An Approximation to Optimal Nonlinear Control, *IEEE Proc. ACC01*, IEEE, 2001.
6. K. Doya, Reinforcement Learning In Continuous Time and Space, *Neural Computation*, 12(1), 219-245, 2000.
7. W. M. Haddad and V. Chellaboina, *Nonlinear Dynamical Systems and Control: A Lyapunov-Based Approach.*, Princeton, NJ: Princeton University Press, 2008.
8. T. Hanselmann, L. Noakes, and A. Zaknich, Continuous-Time Adaptive Critics, *IEEE Transactions on Neural Networks*, 18(3), 631-647, 2007.
9. R. A. Howard, *Dynamic Programming and Markov Processes*, MIT Press, Cambridge, Massachusetts, 1960.
10. D. Kleinman, On an Iterative Technique for Riccati Equation Computations, *IEEE Trans. on Automatic Control*, vol. 13, pp. 114- 115, February, 1968.
11. F. L. Lewis, K. Liu, and A.Yesildirek, Neural Net Controller with Guaranteed Tracking Performance, *IEEE Transactions on Neural Networks*, vol. 6, no. 3, pp. 703-715, 1995.
12. F. L. Lewis, V. L. Syrmos, *Optimal Control*, John Wiley, 1995.
13. J. J. Murray, C. J. Cox, G. G. Lendaris, and R. Saeks,Adaptive Dynamic Programming, *IEEE Trans. on Systems, Man and Cybernetics*, vol. 32, no. 2, pp 140-153, 2002.
14. D. Prokhorov, D. Wunsch, Adaptive critic designs, *IEEE Trans. on Neural Networks*, vol. 8, no 5, pp. 997-1007, 1997.
15. J. Si, A. Barto, W. Powel, D. Wunch, *Handbook of Learning and Approximate Dynamic Programming*, John Wiley, New Jersey, 2004.
16. R. S. Sutton, A. G. Barto, *Reinforcement Learning - An Introduction*, MIT Press, Cambridge, Massachusetts, 1998.

17. D. Vrabie, F. Lewis,Adaptive Optimal Control Algorithm for Continuous-Time Nonlinear Systems Based on Policy Iteration, *IEEE Proc. CDC08*, IEEE, 2008 (Accepted).
18. D. Vrabie, O. Pastravanu, F. Lewis, M. Abu-Khalaf, Adaptive Optimal Control for Continuous-Time Linear Systems Based on Policy Iteration, Automatica (to appear).
19. P. J. Werbos, *Beyond Regression: New Tools for Prediction and Analysis in the Behavior Sciences*, Ph.D. Thesis, 1974.
20. P. J. Werbos, Approximate dynamic programming for real-time control and neural modeling, *Handbook of Intelligent Control*, ed. D.A. White and D.A. Sofge, New York: Van Nostrand Reinhold, 1992.
21. P. J. Werbos, Neural networks for control and system identification, *IEEE Proc. CDC89*, IEEE, 1989.

Index

MIX
Papier aus verantwortungsvollen Quellen
Paper from responsible sources
FSC® C105338

If you have any concerns about our products,
you can contact us on
ProductSafety@springernature.com

In case Publisher is established outside the EU,
the EU authorized representative is:
Springer Nature Customer Service Center GmbH
Europaplatz 3, 69115 Heidelberg, Germany

Printed by Libri Plureos GmbH
in Hamburg, Germany